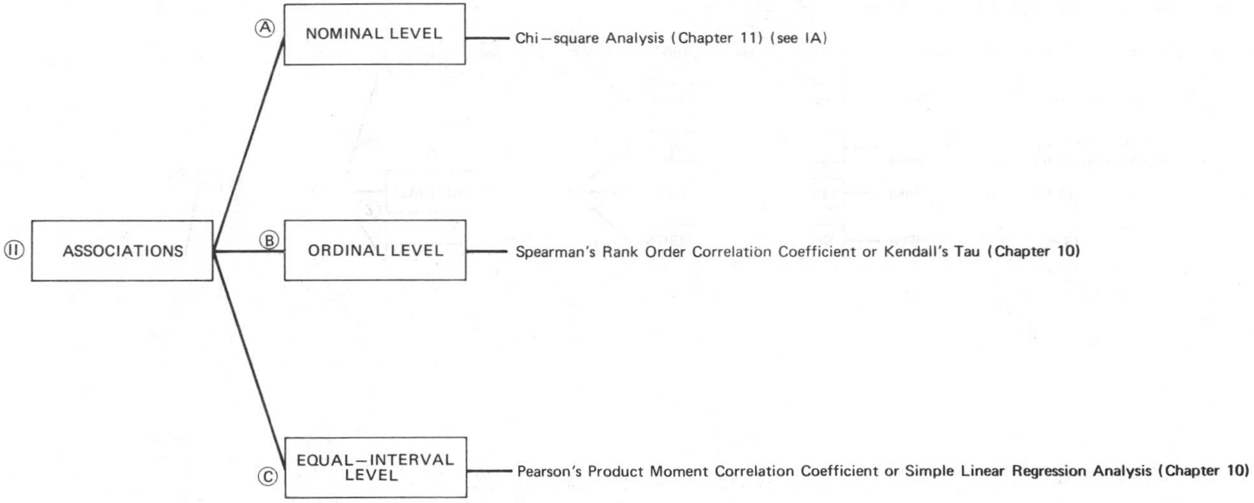

DECISION TREE ALGORITHM FOR SELECTING AN APPROPRIATE STATISTICAL ANALYSIS

BASIC STATISTICS FOR NURSES

BASIC STATISTICS FOR NURSES

SECOND EDITION

REBECCA GRANT KNAPP
Department of Biometry
Medical University of South Carolina
Charleston, South Carolina

A WILEY MEDICAL PUBLICATION
JOHN WILEY & SONS
New York • Chichester • Brisbane • Toronto • Singapore

Library of Congress Cataloging in Publication Data

Knapp, Rebecca Grant.
Basic statistics for nurses.

Includes index.
1. Nursing—Statistical methods. 2. Statistics.
I. Title.

RT68.K53 1985 519.5′024613 84-7274
ISBN 0-471-87563-5

Printed in the United States of America

10 9 8 7 6 5 4 3 2

PREFACE TO THE SECOND EDITION

The second edition of *Basic Statistics For Nurses,* like the first edition, is intended for the undergraduate nursing student who must become a critical and knowledgeable consumer of nursing research. For the ever-increasing number of students who plan to pursue master's or doctoral degrees in nursing, this text should provide a foundation for advanced statistics courses required for such degrees. By including the sections that cover more difficult and theoretical topics, indicated in the text by a dagger (†) preceding the paragraph or heading, the second edition may also be suitable for those enrolled in graduate programs in nursing. Finally, because of the emphasis on worked examples and exercises, this text may be used by those in nursing who wish to review statistical concepts without enrolling in a formal statistics course.

The format of the first edition has been retained. Each chapter contains an overview, a detailed list of objectives, worked examples with solutions, student exercises, and a summary of critical concepts presented in the chapter. However, several major additions and improvements have been made in the second edition. These are (1) an expanded discussion of level of measurement and its role in selecting a statistical analysis; (2) a discussion of the method for calculating standardized scores and their associated probabilities using the normal curve; (3) a new chapter on estimating population parameters, including estimation of a population proportion and differences in population proportions; (4) an expanded discussion of statistical decisions and their outcomes, including a discussion of power of a test and the determination of sample size; (5) the inclusion of a discussion of the Randomized Block Design and multiple comparison procedures in the chapter on Analysis of Variance; and (6) the addition of a chapter outlining a decision tree approach for selecting a statistical analysis.

This new chapter (Chapter 7) approaches the selection of an appropriate analysis by having students work through the series of questions:

16.50

(1) What type of research question is being asked? (2) What is the level of measurement of the response variable? (3) What is the design of the study? Depending on the response to each of these questions, students may trace a path through a "decision tree" to arrive at a suggested statistical procedure. Most introductory statistical texts concentrate on the "how to" aspect of a particular procedure with little or no attention given to what is often the most difficult task faced by the student in a statistical methodology course, the selection process itself. The author feels that this new chapter will be extremely helpful to the student because it formalizes in a step-by-step algorithm the procedure for selecting the appropriate statistical analysis. After the decision tree is described, the procedure is illustrated by use of many examples followed by student exercises.

Finally, as in the first edition, the emphasis in this text is on teaching nursing students to be consumers of research, not on transforming them into "practicing statisticians." The use of highly mathematical and theoretical arguments is avoided. Rather, students are guided through the analyses by means of worked examples and illustrations.

I wish to thank Ms. Deborah Gilliard for her assistance in the typing of the second edition; Dr. M. Clinton Miller III, my department chairman, for his suggestions and patience in allowing me the time to prepare the manuscript; and, finally, Dr. Daniel R. Knapp, without whose assistance this work could not have been completed.

Rebecca Grant Knapp

PREFACE TO THE FIRST EDITION

The ever-increasing demand from nursing professionals for more extensive and comprehensive nursing research has been accompanied by an increased emphasis on research training. Historically, this training has been reserved for students in master's or doctoral programs. Presently, however, there is a movement toward research training at the baccalaureate level as well. More and more schools of nursing are requiring an undergraduate research/statistics course either at the junior or senior level.

The statistics component of research/statistics training is usually offered either as a separate course taught by a department of statistics, psychology, or public health, or as part of a research methodology course taught in the school of nursing. Applications to the profession of nursing and to nursing research are apt to be sketchy when commonly used texts are selected by the outside departments. Few introductory statistics texts are sufficiently oriented specifically toward the profession of nursing. Many nursing research texts include chapters on statistical topics. However, due to the breadth of material that must be covered in such texts, only a broad overview of statistical topics can be presented. In contrast, this text will be suitable for an introductory statistics course offered by departments outside the nursing school with students primarily from nursing and the allied health professions, or may be used as a companion text for a research methodology course taught in the college of nursing.

Examples have been selected from actual nursing research studies, either ongoing or reported in the professional literature. These examples, especially in the first three chapters, often have been highly simplified for purposes of explanation and ease of computation. The simple structure of some of the examples serves merely as a means of introducing concepts necessary for more sophisticated applications to current research efforts in the nursing profession.

Each chapter contains an overview, a detailed list of objectives, worked examples followed by student exercises, and a summary of critical concepts presented in the chapter. A detailed solution set for all exercises is presented in the appendix. Some of the chapters contain specially marked paragraphs or sections that cover more difficult or theoretical topics.* These special sections may be omitted without loss of continuity in the presentation of remaining topics. Certain topics, such as the analysis of variance, which are not ordinarily covered in most introductory statistical texts, are briefly described in order to make the student aware of the existence of these more advanced techniques.

In covering each chapter, the student is urged to study each example carefully and then attempt to work the exercises that follow. In this way, the exercises function as both a teaching aid and a mechanism by which the students may assess their mastery of the written material before progressing to the next section.

This book has evolved over the past three years as a result of the author's experience with teaching a research/statistics methodology course for senior nursing students. At the end of each semester students were asked for criticism and the manuscript was revised accordingly. In addition, numerous members of the nursing faculty have made valuable contributions to the development of the text and the accompanying examples.

In summary, this text is suitable for undergraduate nursing students in an introductory statistics course or as a companion text for most research methodology courses taught in schools of nursing. In addition, it is felt that because of the emphasis on worked examples and exercises, this text will also be suitable for post-baccalaureate nurses who wish to review statistical concepts without enrolling in a formal statistics course.

I wish especially to thank Dr. Daniel R. Knapp and Dr. Robert C. Duncan for their careful reading of the manuscript and their many valuable suggestions. I wish also to express my appreciation to Ms. Barbara Wright, who spent many long hours typing the manuscript.

Rebecca Grant Knapp

* The special sections are marked with a dagger (†) preceding the paragraph or heading.

CONTENTS

BASIC STATISTICS FOR NURSES

CHAPTER 1

ORIENTATION TO STATISTICS

OVERVIEW

The most commonly heard complaint for students beginning a course in statistics is "I haven't had math since high school!" or "I just can't do mathematics, so I know I won't do well in statistics!"

First of all, very few mathematical skills are needed in an elementary applied statistics course. Most "high level" mathematics is reserved for advanced theory courses. With few exceptions, you will need only a knowledge of basic arithmetic, such as addition, subtraction, multiplication, and division.

Probably beginning students are intimidated most by the use of symbols and formulas. In reality every statistical formula is simply an English sentence translated into symbolic notation. The use of symbols to represent names is a shorthand way of saving time and work. Statistical formulas are to be neither feared nor avoided.

Two questions which are frequently encountered in an introductory statistics course are, "Just what is statistics?" and "Why do I have to take this course?"

The answer to the first question may be stated very simply. Statistics is a *tool* that aids the researcher in organizing and summarizing data, in making inferences about the data, and finally in communicating research findings clearly and meaningfully to others. The statistical techniques used for organizing, summarizing, inferring, and communicating do not necessarily require an extensive background in mathematics. To use statistics effectively as a research tool, one must acquire a working knowledge of certain basic statistical techniques, the rules, or assumptions for using these techniques, and finally how meaningfully to interpret the results obtained using these techniques.

Why is it important for you, the professional nurse, to acquire a knowledge of statistics? A popular answer to this question is that it will enable you to do "research." It is well known that research is essential in building a solid theoretical framework for the nursing profession. In this sense, the term "research" implies the formal application of experimental principles usually in an academic institution. The term "research" may, however, be interpreted in a much broader sense. A pediatric nurse practitioner who reads a pediatric journal in order to learn about new practices and procedures in the field is engaging in research in the broad sense. Critical evaluation of relevant studies is necessary in order to determine whether the incorporation of a new procedure into the nursing care plan is warranted. Unfortunately, not all studies which are reported, even by the most prestigous journals, were carried out correctly. The lack of statistical expertise on the part of some investigators is perpetuated by the statistical inexperience of many of the readers. All too often, a reader will skip over the statistical methodology section of a paper or presentation, assuming the methods were applied correctly. As the statistical sophistication of the "research consumer" increases so also, of necessity, will the statistical expertise of the active researcher increase.

The purpose of Chapter 1 is to present definitions and examples of basic statistical terminology. These definitions are intended only to provide the reader with a general overview of the topics. This chapter thus serves as an orientation to the statistical analyses that will be presented in subsequent chapters.

OBJECTIVES

Upon completion of this chapter, the student will be able to:

1. Recognize the differences in and give examples of the nominal, ordinal, interval, and the ratio scales of measurement.
2. Recognize the differences in and give examples of continuous and discrete data.
3. Specify the variable of interest in a research study.
4. Describe the difference between descriptive and inferential statistics.
5. Perform simple manipulations using Σ notation.
6. Translate a statistical formula into an English sentence.

DISCUSSION

A Typology of Research

In a general sense, research may be considered as an activity whose purpose is to find a valid answer to some question that has been raised.*

* Faye G. Abdellah and Eugene Levine, *Better Patient Care Through Nursing Research* (London: The MacMillan Co., Collier-MacMillan Ltd, 1965), p. 30.

In a stricter sense, research is the empirical investigation of the relationship between or among several variables.* To further explore the meaning of the term "research," we may classify it according to purposes of research, methods of conducting research, and types of research settings. The following outline presents a modified version of the typology of research presented by Abdellah and Levine:

I. Purposes of research
   A. Descriptive research
   B. Explanatory research
II. Methods of research
   A. Experimental method
   B. Nonexperimental method
   C. Partially experimental method
III. Settings for research
   A. Highly controlled setting
   B. Uncontrolled setting
   C. Partially controlled setting

PURPOSES OF RESEARCH

The aim of *descriptive* research is to obtain accurate and meaningful descriptions of a phenomenon by means of the collection of reliable, factual information. A descriptive study is conducted primarily for the purpose of describing a situation or event and attempts to make no generalizations beyond the observations at hand. An *explanatory* study, on the other hand, attempts to discover why a phenomenon occurred. As an illustration of the distinction between descriptive and explanatory research, consider the following situations:

A: Nursing staff patterns at a large university hospital, i.e., types of nurses (baccalaureate, diploma, etc.) on various shifts
B: Patient satisfaction with care received in above hospital

Descriptive studies carried out in A simply attempt to describe the educational level of nurses on the various shifts and in B the patient satisfaction in the particular hospital. Each of the events A and B are studied independently. In an explanatory study, the events A and B would be studied in terms of their *relationship* to each other. An explanatory study, going beyond a simple description of each situation, may attempt to answer questions such as: Are patients whose care is

* Marigold Litton and P. S. Gallo, Jr., *The Practical Statistician: Simplified Handbook of Statistics* (Monterey, Calif.; Brooks/Cole Publishing Company, 1975), p. 10.

given by a professional nurse more satisfied than patients whose care is given by a licensed practical nurse? Are patients more satisfied with day shift care than with night shift care? Explanatory research involves the comparison of one phenomenon with another in terms of their effects on the subjects being studied.* In the clinical setting, explanatory studies may attempt to determine the effects of different modes of treatment on patients.

RESEARCH METHODS OR DESIGNS

Research methods may be broadly classified according to two types: the *experimental* and the *nonexperimental* design. The fundamental distinction between these two designs is the degree to which the conditions of the study are "controlled" or manipulated. In an experimental research design, the investigator manipulates or controls the conditions of the study, whereas in a nonexperimental design, the investigator has no such control over the research conditions. The purest application of the experimental method is a study carried out in a laboratory setting where all factors affecting subject response may be manipulated by the investigator. For example, in a study of the effect of smoking on blood pressure carried out on mice in a laboratory setting, the investigator may control such factors as amount of tar and nicotine given the animals, their diet, heredity, etc. In this situation, "nature" is consciously interfered with. A purely nonexperimental approach to studying smoking and blood pressure would be carried out in a "natural" setting. That is, the investigator would select a group of smokers and nonsmokers from the population and determine their blood pressures. In this situation, the researcher is unable to control the other conditions the subjects experience.

The purely experimental method carried out in a highly controlled laboratory setting and the purely nonexperimental method carried out in a "natural" setting represent the two extremes of research methodology. Intermediate to these two extremes is the partially experimental design in which the setting may be natural but the investigator is able to exert some degree of influence on conditions affecting subject response.

SETTINGS FOR RESEARCH

As stated in the previous section, one of the major distinctions between experimental and nonexperimental designs is the research setting or research environment. At one end of the spectrum is the highly controlled, laboratory setting. This category, which may include

* Abdellah and Levine, *Better Patient Care*, p. 40.

test units, research centers, or other experimental centers in a clinical setting, is an artificially constructed environment established for the purposes of research. Because this type of artificial setting is outside an actual life situation, it has the advantage of providing increased purity of research observation; that is, the investigator may control extraneous factors that may affect the outcome of the study. At the same time, the artificial nature of the highly controlled setting, especially in studies involving humans, may produce unpredictable and unnatural responses in the study subjects. People involved in a research study in a highly controlled setting may react to the study conditions in an entirely different way than they would if the conditions were imposed in a natural setting. This reaction to the study setting may result in entirely erroneous conclusions being drawn from the study.

At the other extreme in the spectrum of research settings is the purely "natural" setting. The natural setting is an uncontrolled setting that has the advantage of a real-life situation. The investigator does not manipulate the study conditions as he does in the controlled setting; the study setting is observed without modification.

The highly controlled and the natural or uncontrolled setting have both advantages and disadvantages. The freedom of the investigator to manipulate the environment in a highly controlled setting, thus obtaining increased purity of observation, must be weighed against the chance of obtaining an unnatural subject response due to the artificial nature of the setting.

SUMMARY

The term "research" has been presented in terms of three components: purpose, design, and setting. The nature of the presentation should not be taken to indicate that these three components are separate, unrelated entities. In fact, they are highly related and interwoven concepts. A study that is *experimental* in method is carried out in a *highly controlled setting* and is usually *explanatory* in purpose. Descriptive research as an end in itself may also be carried out in a highly controlled setting but more frequently occurs in this setting as simply preliminary to an explanatory study. *Nonexperimental* designs that are carried out in *uncontrolled* settings tend to be *descriptive* in purpose. Both descriptive and explanatory studies may be carried out in an uncontrolled setting, although in this setting the purpose is much more likely to be descriptive.

The purpose of the above typology of research has been to give the reader a perspective of research as it relates to the statistical analyses presented in this text. The interrelationship between purpose, design, and setting play a major role in the determination of the type of statistical analysis that is appropriate to a given study. Descriptive studies rely almost exclusively on the techniques of descriptive statistics presented in Chapters 2 and 3. In addition, descriptive research studies may employ some of the correlational techniques presented in Chapter 8. Studies that employ nonexperimental research designs and studies that are carried out in an uncontrolled setting also rely heavily on descriptive statistical techniques. Studies that are explanatory in purpose, experimental in design, and carried out in a controlled setting rely most heavily on the techniques of inferential statistics presented in Chapters 4 through 12. These studies may also employ descriptive statistics as the initial step in the inferential analysis.

Independent and Dependent Variables

Research involves the description or explanation of relationships among phenomena being studied. These phenomena are called *variables*. A variable can be defined as a quality, property, or characteristic of the persons or things being studied that can be quantitatively measured or enumerated.* The value of the variable may be different for each individual in the study, hence the name "variable." In general, variables in a research study may be classified as *independent* (stimulus) variables or as *dependent* (response) variables. In a strict sense, the independent (stimulus) variable is the variable that is manipulated or applied by the investigator and the dependent (response) variable is the resulting response or behavior that is observed. The above definition applies in a strict sense only to experimental research. For the purposes of most research, however, a broader definition has been adopted. Under this broad definition, the term independent variable is any variable that is assumed to produce an effect on, or be related to, a behavior of interest (dependent variable).** The following are examples of independent and dependent variables:

Dependent Variable	*Independent Variable*
1. Patient response to exercise treadmill	a. Overall physical condition of patient
	b. Amount of time for practice
	c. Amount of exertion prior to exercise
	d. Physical coordination of patient

* Abdellah and Levine, *Better Patient Care*, p. 122.
** Litton and Gallo, *The Practical Statistician*, p. 10.

2. Blood pressure	a. Type of antihypertensive drug b. Dose level of drug c. Stress level of patient d. Diet
3. Effectiveness of patient education program	a. Educational level of nurse involved in program b. IQ of patient c. Length of program

In the above examples, the investigator may consider one or many independent variables as they relate to the behavior of the dependent variable. The value of each independent variable is varied and the resulting changes in the dependent variable are observed.

Levels of Measurement

A variable has been defined as a characteristic of a person or thing being studied that can be quantitatively measured or enumerated. An important question in any research study is how the variable(s) is to be measured. The answer to this question influences the decision of which statistical analysis is appropriate in a given situation. Therefore, before beginning a discussion of statistical techniques, we must first look at the ways in which numbers can be used to measure.

Basically, there are four types of measurements. They are the nominal, the ordinal, the interval, and the ratio scales. These scales, as listed above, represent an increasing refinement of the measurement process.

NOMINAL SCALE

Measurement in which a "name" (hence, "nominal" scale) is assigned to each observation in the study belongs to the nominal scale of measurement. This scale simply involves the classification of subjects according to specified categories of a given variable. There is no necessary relationship among the categories. For example, if patients are classified according to the variable blood type, the measurement assigned to each patient consists of the words "Type A," "Type B," "Type AB," or "Type O." Other examples of variables that would use the nominal scale of measurement are type of illness, marital status, sex, medical diagnosis, and cause of death.

ORDINAL SCALE

The ordinal scale differs from the nominal scale in that it ranks the different categories specified in the scale in terms of a graded order (greater than, less than, equal to). For example, a variable that may employ the ordinal scale of measurement is "patient satisfaction." The categories for the variable may be "very satisfied," "moderately satisfied," "very unsatisfied." If Patient A falls in the first category and Patient B falls in the second, we know that A is "more satisfied" than B; however we do not know how much more satisfied he is.

Ordinal data are obtained when the subjects in a research study are asked to respond using a "Likert"-type scale, that is, when responses consist of the extent of agreement or disagreement with a series of statements expressing opinions of a topic. Extent of agreement with a given statement is usually on a five-point scale ranging from strongly agree to strongly disagree. Often the data consist of the numerals 1 to 5 corresponding to the degree-of-agreement categories. When numerals are used to represent categories, they do not possess true arithmetic properties as do numbers obtained in a true measuring process. This will be discussed more fully in the next section.

Finally, when the data we have collected are in the form of *rankings,* such as rankings of students' performances of a task, the level of measurement is ordinal.

INTERVAL SCALE AND RATIO SCALE

The ordinal scale, which is used when values of a variable are categories that stand in ordered relationships to each other or are ranks, does not specify the "distance" between the given categories. That is, we may say that Patient A is "more satisfied" than Patient B but we cannot say that he is twice as satisfied, or three times as satisfied, etc. Measurement scales in which the distances between any two numbers on the scale are of known and equal size are the *interval* scale and the *ratio* scale. For example, suppose the variable being measured is temperature. If Patient A's temperature is 96.4° F and Patient B's temperature is 103.4° F, we know that Patient A's temperature is 7 units higher than Patient B's. If the variable being measured is weight, we know that a patient who weighs 200 pounds is twice as heavy as a patient who weighs 100 pounds. Thus, the interval and ratio scales are capable of producing more refined distinctions in the measurement process than those offered by the nominal and ordinal scales.

The primary distinction between the interval and ratio scales is the presence of an absolute zero point on the scale, a point at which the variable being measured is totally absent. Measurement scales without a true zero point are *interval* scales, while those with a true zero point are ratio scales. For example, the variable temperature, measured in Farenheit or Centigrade, has an interval scale, since both of these scales begin with an arbitrary zero point (the zero point of the Centigrade scale is placed at the freezing temperature of water and does not represent an absolute absence of heat). The variable weight, on the other hand, is measured on a ratio scale, since a weight of 0 represents the true absence of weight. This distinction between the interval and ratio scales will not be of importance in this text. For our purposes it is sufficient to know that the interval and ratio scales are *numerical* scales in which the distances between any two points are of known and equal size.

SCORE DATA

A common type of data encountered in nursing is "score" data. Included in this category are psychological test scores, course grades, scores on standardized exams, and performance scores on audit forms. In a strict sense, such score data possess an underlying ordinal scale of measurement, since we cannot say that a difference in scores from 70 to 80 reflects a difference in performance exactly equal to a difference from 90 to 100 (i.e., the ordinal scale does not require *equal* intervals as do the interval and ratio scales). However, in many practical applications, analyses that require at least an interval scale (such as averaging) are also applied to score data, based upon the assumption that no serious errors will be incurred. (Applications of such techniques to ordinal data are in error to the extent that the intervals of measurement differ from true equality.) Since in most instances this assumption is probably a safe one, score data will be treated in this text as if it possessed equal interval scaling.

As stated in the beginning of this section, a very important question in any research study is how the variable of interest is to be measured (i.e., what measurement scale). The type of statistical analysis is influenced by the answer to this question. With the nominal scale we are generally restricted to only those analyses that deal with frequencies such as the X^2 analyses in Chapter 11 and a few of the nonparametric tests in Chapter 12. The ordinal scale, since observations can be ranked, allows greater flexibility in the number of statistical tests that can be employed (many nonparametric tests require only an ordinal scale of measurement). With both the interval and ratio scales almost all statistical analyses are possible.

Continuous and Discrete Measurements

Another characteristic of measurement that influences statistical analyses is the type of variable being studied. Variables, in a broad sense, may be classified as either *qualitative* or *quantitative*. Qualitative variables are those on which measurement is made at the nominal level. For example, blood type, marital status, and cause of death are qualitative variables. Quantitative variables are those for which the level of measurement is ordinal or equal-interval. Quantitative variables measured on an equal-interval scale may be further categorized as *discrete* or *continuous*. Data collected on such variables are labeled as *numerical discrete* data or *numerical continuous* data, respectively. Numerical *discrete* data occur when the measurements are integers that correspond to a count of some sort. Examples are the number of children per family, number of heart beats within a given time interval, number of pregnancies (gravidity), number of live births (parity), and number of episodes of illness for a patient within a specified time period. Measurements on a variable are discrete if only a countable number of distinct values are possible.

In contrast, measurements on a variable that result from the process of *measuring* rather than counting are *continuous* rather than discrete. Theoretically, each measurement on a continuous variable falls somewhere along a continuum. That is, theoretically, each measurement is capable of being subdivided into smaller and smaller units (there is no indivisible unit). Unlike a discrete variable, a continuous variable is not limited to particular values such as the integers. For example, the variable weight may be measured to the nearest gram, but it could also be measured to the nearest tenth of a gram, or nearest hundredth of a gram, etc. Theoretically, the unit gram may be subdivided into smaller and smaller units. One is restricted only by the degree of accuracy of the measuring instrument. Other examples of continuous variables are height, blood pressure, temperature, age, and serum cholesterol level.

In Chapter 7 we will describe the intertwining of level of measurement and type of variable and the role of each of these in the selection of a statistical test.

The Process of Measurement

In the preceding sections, we have seen that types of measurement scales (nominal, ordinal, interval, ratio) and the nature of the units of measurement (continuous or discrete) will affect the type of statistical analysis to be employed in a study. Before proceeding to a discussion of these statistical techniques, let us first briefly identify sources of error within the measurement process. All measurements are subject to error, and it is because of this that statistical analyses are particularly important. Statistical techniques allow the researcher to take measurement error into account in reaching conclusions about the information collected in an investigation.

One source of error inherent in the measurement process is the accuracy with which the measurements may be obtained. This potential source of error is common to all measurement scales. For example, using a nominal scale to classify patients according to type of illness, an investigator may commit a "classification" error; that is, a patient may be classified as having one disease when in fact he has another. The use of an ordinal scale of measurement requires the ordering or ranking of measurements with respect to certain criteria such as degree of nursing competency. This type of ranking usually requires subjective judgment on the part of an observer and again, as for the nominal scale, is subject to "classification" or "observer" error. The interval and ratio scales by their nature are more amenable to mechanization and thus are more objective and less prone to observer error than are the nominal and ordinal scales. Potential sources of measurement error for these scales are equipment malfunction and observer error in "taking a reading."

The final source of measurement error that will be discussed involves the selection of "representative" units on which the measurements are made. For example, if it is of interest to determine the "average" weights of patients suffering from a particular disease by weighing a selected group of patients with the disease, the choice of which patients to actually weigh can drastically affect the information obtained. This potential source of measurement error plays a vital role in the area of *inferential* statistics, which will be discussed in the following section.

Descriptive and Inferential Statistics

Statistical methodology is usually divided into two main branches: descriptive statistics and inferential statistics. Descriptive statistics are concerned with the organization, presentation, and summarization of data. Techniques of descriptive statistics include the presentation of data in tables and graphs as well as the summarization of a set of data by means of one or two meaningful numerical values such as average, median, percentages, and ranges. Inferential statistics is a set of procedures used to draw conclusions about a large body of data, called a *population*, based on a smaller set of data, a *sample*, taken from the population. The calculation of descriptive statistics may precede the application of inferential techniques. Common examples of statistical inferences include:

EXAMPLE 1.1 The test of a new nursing procedure for geriatric patients. Based on results obtained from a sample of geriatric patients, a conclusion may be drawn regarding the procedure's effectiveness among *all* geriatric patients under similar conditions.

EXAMPLE 1.2 A public health survey. From a sample of households within a given community, inferences about the existence of chronic and acute illness for the entire community may be drawn.

EXAMPLE 1.3 A blood sample analysis. A hemoglobin determined from a small volume of a patient's blood is taken to be representative of his entire blood volume.

Descriptive statistics may be employed in the above examples if one is interested only in describing the characteristics of the sample at hand and not in making statements about a larger group from which the sample was selected.

The assumption that must be satisfied in order to infer properties of a population from the properties of a sample is that the sample is *representative* of the population of interest. The procedures for obtaining a representative sample are described in Chapter 13.

Chapters 2 and 3 present techniques of descriptive statistics. Chapters 4 and 5 lay the theoretical groundwork for statistical inference, and Chapter 6 discusses estimation of population parameters. Chapter 7 is particularly noteworthy in that it presents a decision tree algorithm for selecting a statistical test based on criteria described in this chapter, specifically type of research and level of measurement of the response variable. Chapters 8–13 discuss common inferential statistical tests and methods of sample selection. The final chapter presents an overview of applications of computers to nursing. The remainder of this chapter is devoted to the introduction of statistical notation.

Using Σ(Sigma) the Summation Operator

One of the most frequent arithmetic manipulations encountered in statistics is that of summing a set of values. The mathematical symbol used to denote the summing operation is Σ (sigma). The presence of Σ in a formula indicates that a group of numbers is to be added or summed. Before proceeding with examples of the use of Σ, we must first introduce some necessary terminology.

As we have seen, data that are collected in a research study usually consist of measurements or counts of particular characteristics of individuals or items. This characteristic of interest in a research study, called a *variable*, is usually designated by a letter or symbol (this text will use the letter Y). If two or more characteristics are measured, then different letters are used for each variable.

A distinction between different measurements on a given variable may be made by adding a subscript to the corresponding symbol. This will be illustrated by means of hypothetical data for weight gain in ounces for premature infants following 10 days on a supplemental diet.

12, 9, 4, 7, 8

The first measurement is designated Y_1, the second by Y_2, and so on. In general, weight gain would be represented by Y_i where Y_i refers to the ith value in the set. The i may be replaced by any number from 1 to 5 in order to specify any particular value within the set. Thus the weight gain for the third individual in the sample is given by $Y_3 = 4$ and the weight gain for the last individual is given by $Y_5 = 8$.

If we wish to say "take the sum of all the weight gains (Y values)," we would write:

$$\Sigma Y_i$$

To be technically correct, the summation expression should be written:

$$\sum_{i=1}^{n} Y_i$$

which reads "take the sum of the Y values beginning with the subscript $i = 1$ and ending with the subscript $i = n$ where n is the number of observations in the set." The sum of all the weight gains is written:

$$\sum_{i=1}^{5} Y_i = Y_1 + Y_2 + Y_3 + Y_4 + Y_5$$

$$= 12 + 9 + 4 + 7 + 8$$

$$= 40$$

For our purposes, it will be assumed that the summation will always be carried out over all cases. Therefore it will be necessary to write only:

$$\Sigma Y_i$$

Using the weight gain data, the following examples illustrate the most common uses of Σ notation:

EXAMPLE 1.4 $\Sigma Y_i^2 = Y_1^2 + Y_2^2 + Y_3^2 + Y_4^2 + Y_5^2$

$= 12^2 + 9^2 + 4^2 + 7^2 + 8^2$

$= 354$

EXAMPLE 1.5 $\Sigma Y_i - 8 = (Y_1 + Y_2 + Y_3 + Y_4 + Y_5) - 8$

$= (12 + 9 + 4 + 7 + 8) - 8$

$= 40 - 8 = 32$

The above expression is read "sum all the Y values and then subtract the value 8 from the results." The summation operation applies only to the variable or expression immediately following the Σ symbol. Note the distinction between Example 1.5 and 1.6.

EXAMPLE 1.6 $\Sigma(Y_i - 8) = (Y_1 - 8) + (Y_2 - 8) + (Y_3 - 8) + (Y_4 - 8) + (Y_5 - 8)$

$= (12 - 8) + (9 - 8) + (4 - 8) + (7 - 8) + (8 - 8)$

$= (4) + (1) + (-4) + (-1) + 0$

$= 0$

In this example, the parentheses imply that the entire expression within the parentheses is to be summed. Therefore, in this case, the value 8 is subtracted from each Y value before the summation is carried out.

EXAMPLE 1.7 $\Sigma(Y_i - 8)^2 = (Y_1 - 8)^2 + (Y_2 - 8)^2 + (Y_3 - 8)^2 + (Y_4 - 8)^2 + (Y_5 - 8)^2$

$= (12 - 8)^2 + (9 - 8)^2 + (4 - 8)^2 + (7 - 8)^2 + (8 - 8)^2$

$= 4^2 + 1^2 + (-4)^2 + (-1)^2 + 0^2$

$= 16 + 1 + 16 + 1 + 0$

$= 34$

EXAMPLE 1.8 $(\Sigma Y_i)^2 = (Y_1 + Y_2 + Y_3 + Y_4 + Y_5)^2$

$= (12 + 9 + 4 + 7 + 8)^2$

$= (40)^2$

$= 1600$

In this example, the parentheses imply sum all the values of Y and then square the results. Note that ΣY_i^2 (Example 1.4) is not the same as $(\Sigma Y_i)^2$.

With mastery of the use of the Σ symbol, you will now be able to translate statistical formulas into English sentences that are easily understandable. For example, a common statistical procedure is to find the "average" of a set of numbers. To find the average of five numbers, we all know that we add the five numbers and divide the result by 5. In summation notation this would be written:

$$\frac{\Sigma Y_i}{5}$$

In statistical terminology the above quantity is called the *mean* of the five numbers. In our weight gain example, the mean weight gain for the five infants on the supplemental diet is given by:

$$\frac{\Sigma Y_i}{5} = \frac{Y_1 + Y_2 + Y_3 + Y_4 + Y_5}{5}$$

$$= \frac{(12 + 9 + 4 + 7 + 8)}{5} = \frac{40}{5} = 8$$

Thus the mean weight gain is 8 ounces.

EXERCISE 1.1

Five patients on a certain unit were asked to rate the quality of nursing care on a scale from 0 to 5. The following data were recorded:

2, 1, 1, 2, 4

Perform the following manipulations using Σ notation where Y represents the ratings:

a. $\Sigma Y_i =$

b. $\Sigma Y_i^2 =$

c. $\dfrac{(\Sigma Y_i)^2}{5} =$

d. $\Sigma(Y_i - 2) =$

EXERCISE 1.2

Using the above problem, translate the following into Σ notation and determine the results:

a. Sum the ratings and divide the results by 5.
b. Subtract 2 from each rating, square this value, and sum the results.
c. Sum all the ratings, square the results, and divide by the total number of observations.
d. Sum the squares of all the ratings, from this subtract the square of the sum of all the ratings divided by five (the square of the sum, not each rating, is divided by 5). Divide all the above by one less than the number of ratings.

EXERCISE 1.3

For the following situations, specify the variable being measured and the measurement scale (nominal, ordinal, interval, or ratio):

a. A nurse wished to determine the effect of a particular formula on the weights of newborns.
b. In a study of the effect of deprivation on children, a nurse investigator determined the number of children in a particular section of Appalachia who had various health defects such as in hearing, sight, etc.
c. The Progress Committee of a College of Nursing wished to determine the clinical expertise of senior nursing students. Students were classified according to the categories—highly competent, competent, not competent.
d. In a study of the effect of fondling on body temperature of infants, an investigator determined the body temperature of infants after they had been held for an extended period of time.

EXERCISE 1.4

Classify the following variables as to the nature of their units of measurement (i.e., continuous or discrete):

a. See Exercise 1.3a.
b. See Exercise 1.3d.
c. A public health nurse conducted a study in which she determined the number of children per family in a particular county.

d. In a study to determine the reliability of a particular method for recording blood pressure, 20 patients each had their blood pressures recorded by different nurses.
e. A nurse wished to determine the effect of increasing visiting time on the anxiety level of cardiac patients. The data consisted of anxiety scores of 25 cardiac patients on a standardized anxiety test before and after being visited.
f. In a cardiac care unit an investigator monitored the heart rate of 10 patients following the administration of an antiarrythmic drug.
g. In a study to determine the effect of economic status on tooth decay in children, a nurse classified a sample of 100 children according to economic status and number of cavities.

SUMMARY OF CRITICAL CONCEPTS IN CHAPTER 1

The following concepts were presented in Chapter 1:

1. A variable is a quality, property, or characteristic of the persons or things being studied that can be quantitatively measured or enumerated.
2. The value of a variable may be different for each individual in the study.
3. An important question in any research is how the variable(s) is to be measured.
4. There are four types of measurement scales: the nominal, the ordinal, the interval, and the ratio scales.
   a. Measurement on the *nominal* scale results when a "name" (category) is assigned to each observation in the study (example: blood type).
   b. Measurement on the *ordinal* scale results when the different categories in the scale are specified in terms of a graded order (example: patient satisfaction).
   c. Measurement on the *interval* scale results when distances between any two numbers on the scale are of known and equal size. The interval scale does not have a true zero point (total absence of the quantity being measured) (example: temperature in °F or °C).
   d. Measurement on the *ratio* scale has all the characteristics

specified by the interval scale. In addition, the ratio scale has a true zero point (example: weight).

5. The nature of the unit of measurement (discrete or continuous) is a characteristic of measurement that also influences the type of statistical analysis.
   a. Numerical *discrete* data occur when the measurements are integers that result from a count of some sort (example: number of children per family).
   b. *Continuous* data result when each measurement theoretically is capable of being subdivided into smaller and smaller units (example: weight).
6. Statistics is a tool used for organizing and summarizing data, making inferences about the data, and communicating research findings clearly and meaningfully to others.
7. Statistical methodology is divided into *descriptive* and *inferential* statistics.
8. Descriptive statistics is concerned with the organization, presentation, and summarization of data.
9. Inferential statistics is a set of procedures used to make inferences about a population based on information obtained from a sample of measurements drawn from the population.
10. The presence of the symbol Σ (sigma) in a mathematical formula indicates that the expression immediately following the symbol is to be summed or added.

CHAPTER **2**

DESCRIPTIVE STATISTICS: TABULAR AND GRAPHICAL PRESENTATIONS

OVERVIEW

The raw data that result from a sample survey, a census, or an experiment are usually in the form of large sets of unorganized numerical values. Anxiety scores for a group of newly hospitalized patients, blood pressure readings using two different procedures, and attitude scores of nursing students toward alcoholics are examples of data that may be collected in a research project. Before any meaningful interpretation or presentation of these data can be made, it is first necessary to organize and summarize them so that important features may be grasped at a glance.

Just as a person or object may be described by using a photograph or a few numerical values such as height and weight, so also may a set of data be described. Pictorial descriptions of data include frequency tables, histograms, bar charts, and circle graphs. These techniques will be discussed in Chapter 2. Numerical summaries for sets of data usually consist of measures of some sort of "average" data value (called measures of central tendency) and a measure of how spread out the values are from the average (measures of dispersion). Measures of central tendency and measures of dispersion will be discussed in Chapter 3.

OBJECTIVES

Upon completion of this chapter, the student will be able to:

1. Define the term frequency distribution.

2. Construct an absolute frequency table and a relative frequency table.
3. Present a frequency table graphically by means of a histogram.
4. Present data by means of bar charts, circle graphs, and frequency polygons.
5. Identify symmetric and skewed distributions of data.

DISCUSSION

Frequency Tables and Frequency Distributions

To demonstrate the principles and techniques of tabular and graphical presentations we will use data representing attitude scores of 16 senior nursing students toward alcoholic patients. The students were asked how they feel about working with alcoholic patients (1 = very negative, 5 = very positive). The attitude scores are as follows:

1, 3, 5, 2, 4, 3, 1, 4, 3, 2, 3, 3, 4, 2, 2, 3

At a glance, it would be difficult to answer such questions as: What is the "range" of data values? The most frequently occurring value? The least frequently occurring? The average score?

By tabulating the number of times each score occurs in the sample, or the *frequency* of each score, and displaying these frequencies, as in Table 2.1, the above questions may be answered more readily. Table 2.1 is called a *frequency table* since it displays the *frequency distribution* of

TABLE 2.1 FREQUENCY DISTRIBUTION OF ATTITUDE SCORES OF SENIOR NURSING STUDENTS WORKING WITH ALCOHOLIC PATIENTS

Score	*Frequency*
1	2
2	4
3	6
4	3
5	1
Total	16

the 16 attitude scores. In its simplest form, a frequency distribution consists of a series of predetermined values (the scores in the lefthand column) and the frequency with which these values occur in the sample (the righthand column). Later, we will see that this concept may be expanded to include the use of intervals of data values rather than individual scores.

Returning to Table 2.1, we see at a glance that the scores range from 1 to 5, the most frequently occurring score is 3, the least frequently occurring score is 5, and the "average" or mean score is probably close to 3.

The frequency distribution of attitude scores may be represented pictorially by means of a *frequency diagram*, shown in Figure 2.1. In a frequency diagram, frequency is on the vertical axis and the measurement of interest (attitude score) is placed on the horizontal axis. Thus, the height of each bar represents the frequency of occurrence of each score.

The frequency diagram in Figure 2.1 contains the same information as the frequency table, Table 2.1. In addition, the frequency diagram conveys a sense of the "shape" of the distribution of the frequencies among the possible attitude scores.

The "shape" of a frequency diagram, or frequency distribution is an important characteristic of a set of data values. Consider the examples shown in Figure 2.2. In (*a*) and (*d*) in the figure the frequencies are equal (or nearly equal) on either side of a center point. A distribution of values with this shape is called a *symmetric distribution*. A *skewed* distribution of values is one whose shape is nonsymmetric—(*b*) and (*c*) in the figure.

Before proceeding to a discussion of the class interval approach to

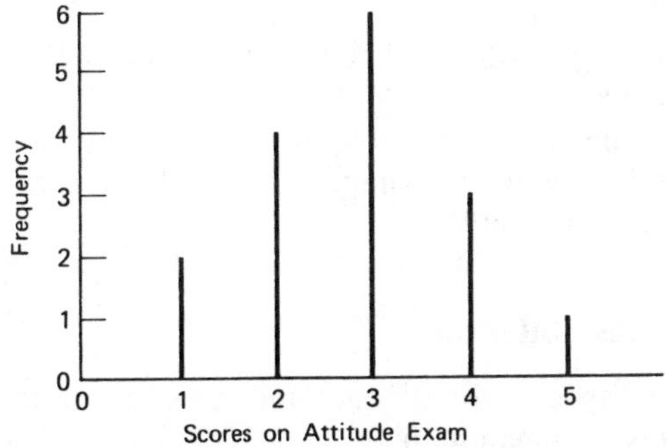

FIGURE 2.1 FREQUENCY DIAGRAM OF ATTITUDE SCORES OF 16 SENIOR NURSING STUDENTS TOWARD WORKING WITH ALCOHOLIC PATIENTS

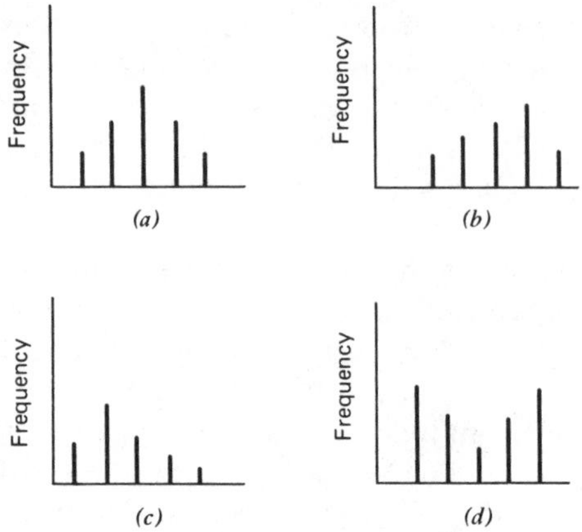

FIGURE 2.2 SYMMETRIC AND NONSYMMETRIC FREQUENCY DIAGRAMS

frequency distributions, you may wish to check your understanding of frequency distributions and frequency diagrams by completing the following exercise.

EXERCISE 2.1

In a study of the efficiency of the nursing staff on a particular unit, 10 patients were asked to ring for a nurse and to record the waiting time before her arrival. The waiting times to the nearest minute recorded by the 10 patients were:

8, 6, 7, 4, 8, 8, 7, 8, 7, 6

a. Make a frequency table for the data. Be sure to include a title for the table.
b. Draw the frequency diagram for the waiting times. Be sure to label both axes.
c. Describe the shape of the distribution, i.e., is it symmetric or skewed? What does the shape tell us about the efficiency of the nursing staff?

Frequency Tables Using Class Intervals

In Table 2.1 the frequencies associated with each individual score in the sample were recorded. This is meaningful only when the number of possible data values is small. When a large sample of data has been taken, the number of distinct sample values may be too large to be meaningful. When this is the case, the possible data values may be divided into groups called classes or intervals and the number of data values falling into each of the predetermined intervals may be tabulated. The data below, representing birth weight in pounds of 35 infants, will be used to illustrate.

5.6	6.6	11.7	6.3	4.4
7.5	6.8	2.7	4.9	6.7
6.0	4.9	4.0	6.0	7.3
4.2	7.6	8.4	5.9	6.4
6.6	7.4	5.2	5.1	5.2
7.6	7.2	10.2	3.6	7.5
10.3	5.9	7.3	4.8	3.7

In the data above you will notice that most of the values (birth weights) occur only once or twice in the sample. Thus, a tabulation of the frequency of occurrence of each possible birth weight would provide little more information than is contained in the raw data alone. A more meaningful way to summarize the data is to group the possible birth weights into intervals, or classes, and determine the frequency with which the actual sample values fall into each of these intervals.

A frequency table for the birth weight data using class intervals is shown in Table 2.2. Let us now see how the intervals used in Table 2.2 are determined. The first step in the construction of frequency tables is to order the raw birth weights from smallest to largest, as shown in Table 2.3.

After the data have been arranged in order, the intervals of birth weight values must be determined. Unfortunately, there is no ironclad

TABLE 2.2 FREQUENCY DISTRIBUTION OF BIRTH WEIGHTS OF 35 INFANTS

Birth Weight (Pounds)	*Frequency*
1.0–2.9	1
3.0–4.9	8
5.0–6.9	14
7.0–8.9	9
9.0–10.9	2
11.0–12.9	1
Total	35

TABLE 2.3 BIRTH WEIGHTS (IN POUNDS) ARRANGED IN ORDER FROM SMALLEST TO LARGEST VALUE

2.7	4.9	5.9	6.7	7.5
3.6	4.9	6.0	6.8	7.6
3.7	5.1	6.0	7.2	7.6
4.0	5.2	6.3	7.3	8.4
4.2	5.2	6.4	7.3	10.2
4.4	5.6	6.6	7.4	10.3
4.8	5.9	6.6	7.5	11.7

"formula" for determining either the size of each interval or the number of intervals to be used. As a general rule of thumb, usually 6 to 20 intervals will be sufficient to convey the distribution and "shape" of data values. Too many intervals result in the same problem encountered when tabulating the frequency of occurrence of each data value; too few intervals may result in loss of valuable information about the shape and distribution of the data. Often, the process of arriving at a satisfactory number of intervals is by trial and error. Usually, it is better to begin with a larger number of intervals having small "width" since the smaller intervals may be combined into larger ones without a retabulation of the raw data. (The *width* of an interval is the distance between the two endpoints of the interval.)

Referring to Table 2.2, we see that the intervals 1.0 to 2.9, 3.0 to 4.9, etc., have been selected. These values, called *class intervals* or *class limits*, are the limits actually used in tabulation of the frequency table. The level of precision for these values reflects the degree of precision of the raw data. However, in actuality, the measurements were made to two decimal places and *rounded* to *the nearest tenth of a pound*. The measurement 4.56 pounds was recorded as 4.6 pounds while the measurement 4.54 was recorded as 4.5 pounds. Therefore, to reflect the true level of the measurement and the underlying continuous nature of the measuring scale, the *true class limits* of Table 2.2 would be as shown in Table 2.4. From this table, we can more readily see that true interval width is $2.95 - .95 = 2.0$ units.

Suppose we had elected to use 11 intervals of width 1 unit for the birth weight values. The frequency table would be as shown in Table 2.5. If we had selected interval widths of 0.5 units rather than widths of 1 unit, we would have had twice as many intervals and many more intervals with 0 frequency.

TABLE 2.4 FREQUENCY DISTRIBUTION OF BIRTH WEIGHTS OF 35 INFANTS SHOWING TRUE CLASS LIMITS

Birth Weight (Pounds)	*Frequency*
.95–2.95	1
2.95–4.95	8
4.95–6.95	14
6.95–8.95	9
8.95–10.95	2
10.95–12.95	1
Total	35

Suppose we decide to use intervals of width 3. The result would be shown by Table 2.6. Clearly, this does not give an adequate description of the frequency of occurrence of the birth weights in the sample.

Probably, the most information is given by Table 2.2, where intervals of width 2 have been used. From the previous examples, it may be seen that the construction of frequency tables requires the manipula-

TABLE 2.5 FREQUENCY TABLE OF BIRTH WEIGHTS USING INTERVAL WIDTHS OF 1 UNIT

Birth Weight (Pounds)	*Frequency*
1.0–1.9	0
2.0–2.9	1
3.0–3.9	2
4.0–4.9	6
5.0–5.9	6
6.0–6.9	8
7.0–7.9	8
8.0–8.9	1
9.0–9.9	0
10.0–10.9	2
11.0–11.9	1
Total	35

TABLE 2.6 FREQUENCY TABLE OF BIRTH WEIGHTS USING INTERVAL WIDTHS OF 3 UNITS

Birth Weight (Pounds)	*Frequency*
1.0–3.9	3
4.0–6.9	20
7.0–9.9	9
10.0–12.9	3
Total	35

tion of interval width and number until the researcher is satisfied that the maximum information is being presented. For the remainder of this text frequency tables will be presented in terms of true class limits.

Some general guidelines for the construction of frequency tables are:

1. Usually, the number of intervals should be from 6 to 20. Convenience and clear presentation of the data dictate the optimum number of class intervals.
2. Intervals of equal width, although not essential, are preferred because they facilitate graphical presentation and other mathematical manipulations. Occasionally, the use of unequal intervals cannot be avoided.
3. The end points of the intervals must not overlap. For example, for data that have been recorded to the nearest whole number, the intervals 1 to 3, 3 to 5, 5 to 7, etc., would be inappropriate since it is not clear whether the value 3 would fall in the first or second interval.
4. True class limits may be used to reflect the true measurement precision of the data. Note: Using true class limits would appear to violate guideline 3 above (see, for example, Table 2.4). Notice, however, that the true class limits are expressed to one more decimal place than the recorded data. Therefore, no piece of recorded data will fall on one of the boundaries and there will be no question about which interval any data point belongs to.
5. Open-ended intervals should be avoided whenever possible. An entry in Table 2.2 that reads "greater than 10.9" would give no indication as to where the last measurement falls. It could be

11, or 111. Further, the use of open-ended intervals complicates graphical presentation of the frequency table.

6. All tables should be clearly *labeled*. The reader should know exactly what was tabulated without having to read the accompanying text.
7. The total number of observations should be included in the table. This gives the reader a frame of reference when looking at the frequencies recorded in the table.
8. Unit of measurement (e.g., pounds, inches, centimeters, etc.) should be indicated at the beginning of the table.

EXERCISE 2.2

In a study of the age distribution of patients in a coronary care facility, the following ages were recorded:

66	25	63
59	46	35
39	37	50
26	40	41
54	38	42

Arrange the data above in a frequency table. Be sure to observe all the guidelines listed above.

Relative Frequency

Often, it is of interest to compare the frequency distributions of two different sets of data. For example, suppose we have a frequency table showing the distribution of ages of 100 males who have suffered a myocardial infarction (MI), and we wish to compare this with a frequency table of ages for 500 normal males selected at random from the population of normal males. The age interval 30.0 to 39.9 might contain 30 individuals from the MI group and 30 from the normal group. However, for the MI group the 30 males are from a total of 100 males and the 30 males for the normal group are from a total of 500. You can easily see that the *proportion* of men with MI's in the age range 30.0 to 39.9 is much greater than that for the normals. For ease of interpretation and comparison, frequency tables usually contain a column called *relative frequency*, which is the proportion or percentage of the total number of observations falling into each interval. By definition,

$$\text{Relative frequency (\%)} = \frac{\text{Absolute frequency}}{\text{Total no. of observations}} \times 100\%$$

Table 2.7 presents both frequency and relative frequency for the birth weight data on page 22. From Table 2.7 we see that 22.9% of the birth weights were greater than or equal to 2.95 pounds but less than or equal to 4.95 pounds. To determine the percent of birth weights less than 6.95 pounds, we add all the percentages up to and including 6.95. Thus 65.8% of the birth weights in the sample were less than or equal to 6.95, i.e. (2.9 + 22.9 + 40.0).

EXERCISE 2.3

For the table in Exercise 2.2, complete the relative frequency column.

Histograms

Pictorial or graphical representations of either frequency or relative frequency tables are called *histograms*. Histograms are simply extensions of the frequency diagram previously discussed. The true class intervals are placed on the horizontal scale and the frequencies or relative frequencies are placed on the vertical scale. The height of each bar is equal to the frequency (or relative frequency) of each interval. The histogram for the birth weight example is shown in Figure 2.3.

Two important points to remember in the construction of a histogram are:

1. A histogram must be self-explanatory. It should have a concise *title* containing sufficient information to allow the reader to identify relevant aspects of the data.

TABLE 2.7 FREQUENCY DISTRIBUTION OF BIRTH WEIGHTS FOR 35 INFANTS

Birth Weight (Pounds)	*Frequency*	*Relative Frequency (%)*
.95– 2.95	1	2.9
2.95– 4.95	8	22.9
4.95– 6.95	14	40.0
6.95– 8.95	9	25.7
8.95–10.95	2	5.7
10.95–12.95	1	2.9
Total	35	100.1*

* In theory, the total of the column should be 100%, but often due to rounding error the total is slightly different from this value.

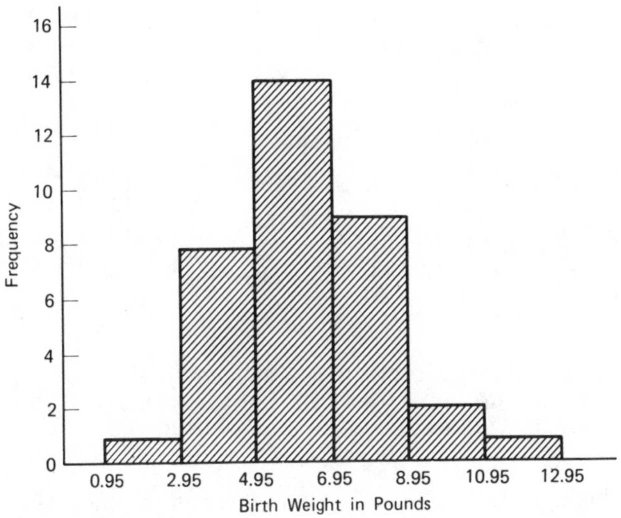

FIGURE 2.3 HISTOGRAM OF BIRTH WEIGHTS (IN POUNDS) OF 35 INFANTS

2. Each axis should be *labeled* and the scale of each axis should be marked.

When *comparing* distributions of sets of data, relative rather than absolute frequencies should be plotted if the total number of values in the two sets is not the same. Two histograms that appear quite dissimilar when plotted with absolute frequencies may in reality be very similar with respect to relative frequencies. (Note: This is not a problem when the total for both groups is the same.)

EXERCISE 2.4

Draw and completely label the histogram for the frequency table in Exercise 2.2.

EXERCISE 2.5

The two histograms below show a comparison of attitudes toward nursing school (on a 5-point scale) of 50 junior and 20 senior nursing students. The students were asked to rate their experience in nursing school on a scale from 1 to 5 with 1 = very negative to 5 = very positive.

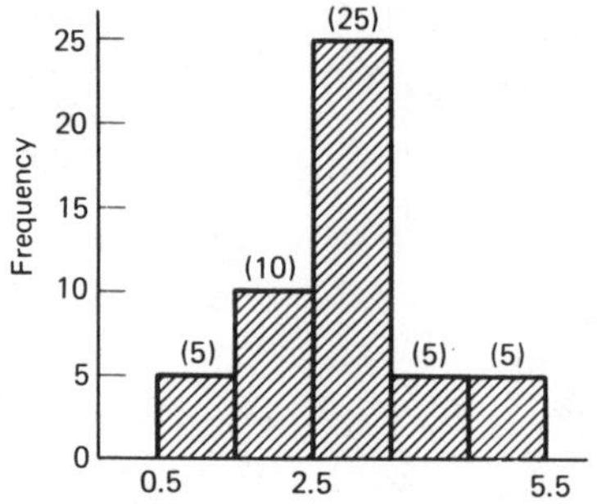

(a)

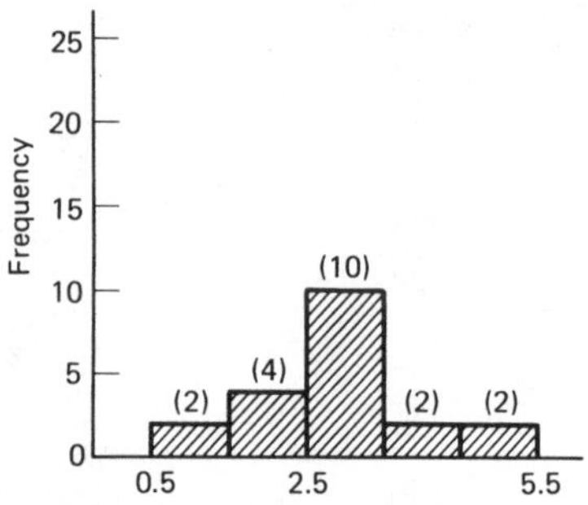

(b)

(*A*) ATTITUDES OF 50 JUNIOR NURSING STUDENTS TOWARD NURSING SCHOOL (*B*) ATTITUDES OF 20 SENIOR NURSING STUDENTS TOWARD NURSING SCHOOL

Plot both histograms in terms of relative rather than absolute frequency. (The absolute frequencies are given in parentheses above each bar.) Which gives a more meaningful comparison of the scores of the two groups of students, the absolute frequency or the relative frequency?

EXERCISE 2.6

A nurse was interested in determining the effect of prolonged periods of crying on blood pressures of infants. For a group of 10 normal infants, systolic blood pressures were recorded at the end of three minutes of crying and again after seven minutes of crying. The blood pressure data (in mmHg) were as follows:

After Three Minutes		*After Seven Minutes*	
104	118	90	122
110	120	108	134
90	108	140	124
124	128	112	128
102	140	120	130

Summarize in both tabular and graphical form the results of the study.

† A Histogram with Unequal Class Intervals

Consider Table 2.8. The histogram, if drawn for this frequency table, may give the misleading impression that deaths due to Disorder X rise

† Sections marked by a dagger (†) cover more difficult or theoretical topics.

TABLE 2.8 DEATHS DUE TO DISORDER X

Age (Months)	*Number of Deaths*
0– .9	12
1.0– 1.9	10
2.0– 2.9	8
3.0– 3.9	6
4.0– 4.9	4
5.0– 5.9	2
6.0– 6.9	2
7.0–12.9	6
Total	50

sharply after seven months of age. Upon closer examination of the table, we see that the age span covered in the interval 7.0 to 12.9 months is almost as large as that covered in the other seven intervals combined. To correct for this in presenting the data by means of a histogram, we plot average number of deaths per month for this interval. The interval 7.0 to 12.9 spans a six-month period and contains six deaths. Thus, there was an average of one death per month for this interval. The histogram would be as shown in Figure 2.4 (true class limits are plotted on the horizontal axis).

Other Forms of Graphical Representations

FREQUENCY POLYGON

A frequency polygon, like a histogram, is a graphical representation of a frequency table. For a frequency polygon, class intervals are laid out along the horizontal axis and frequency (or relative frequency) along the vertical axis. The frequency associated with each interval is indicated by a dot placed above the midpoint of the interval. The dots are then connected by straight lines. The frequency polygon for the data in Table 2.7 is shown in Figure 2.5. Frequency polygons or relative frequency polygons are useful when one is interested in comparing the frequency distributions of two or more sets of measurements. Relative frequency polygons may be plotted on the same graph whereas several histograms superimposed on the same graph are often misleading and difficult to interpret.

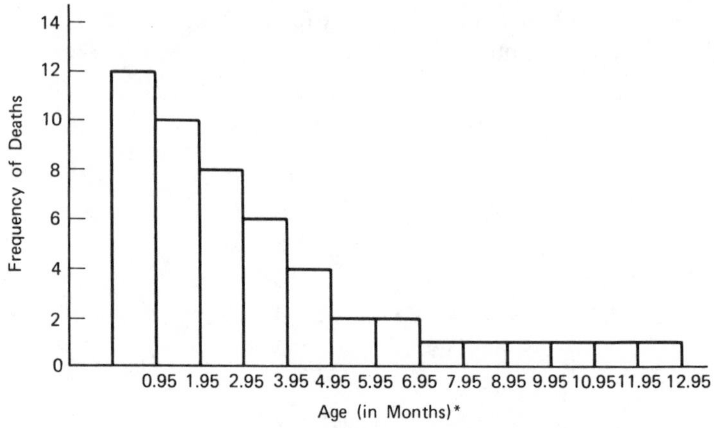

FIGURE 2.4 A HISTOGRAM SHOWING FREQUENCY OF DEATHS ACCORDING TO AGE (IN MONTHS) FOR DISORDER X. NOTE THAT THE FREQUENCIES FOR THE AGE INTERVALS 7.0 MONTHS TO 13 MONTHS REPRESENT THE AVERAGE NUMBER OF DEATHS PER MONTH FOR THE 6-MONTH PERIOD SO THAT THE TOTAL OF SIX DEATHS ARE SPREAD EQUALLY OVER EACH MONTH IN THIS RANGE.

EXERCISE 2.7

Draw the frequency polygons for the data in Exercise 2.6.

CIRCLE CHARTS, OR PIE CHARTS

One of the most common graphical presentations is the *pie chart*, or *circle chart*. A pie chart is simply a circle which has been partitioned into percentages of the total corresponding to percentages of some measurement of interest. For example, suppose the distribution of grade point averages (GPA) for senior nursing students at a particular college of nursing is as shown below.

GPA	*Relative Frequency* (%)
2.01–2.50	18
2.51–3.00	9
3.01–3.50	43
3.51–4.00	30

The pie chart representing the above proportions is shown in Figure 2.6.

In constructing pie charts, the *percentage* (relative frequencies) corresponding to each category rather than the absolute frequency of each category is used. The sum of the percentages for the slices on the

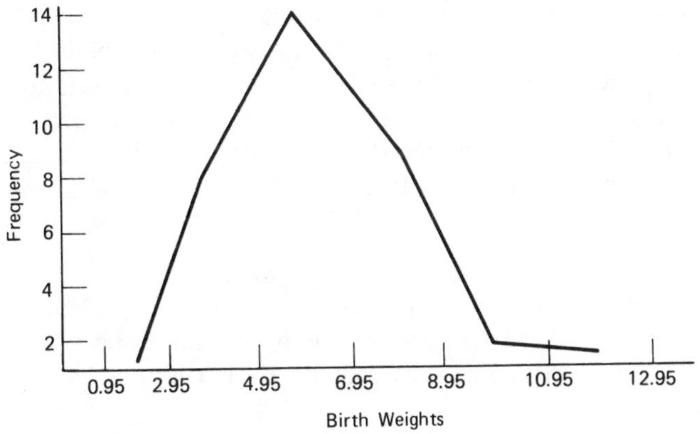

FIGURE 2.5 FREQUENCY POLYGON OF BIRTH WEIGHTS (IN POUNDS) OF 35 INFANTS

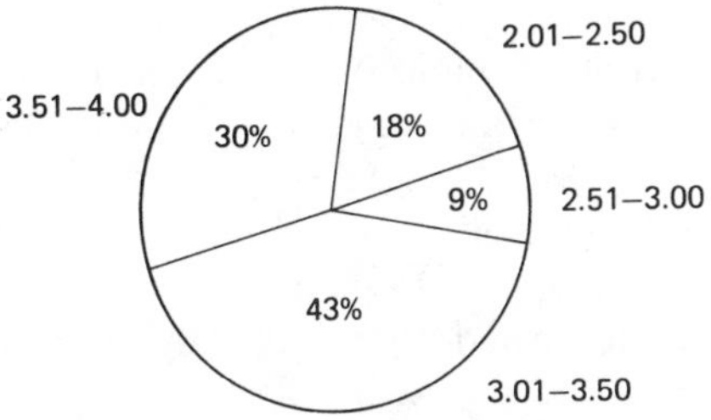

FIGURE 2.6 PIE CHART SHOWING GRADE POINT AVERAGE DISTRIBUTION FOR SENIOR NURSING STUDENTS

circle should add up to 100%. Also, as an aid to interpretation, the categories, when possible, should be presented on the circle in either ascending or descending order.

EXERCISE 2.8

Of 200 nursing students at a particular institution, 76 planned to go into a medical-surgical specialty, 60 onto ob-pediatrics, 40 into psychiatric nursing, and 24 into public health. Display this information using a pie chart.

BAR CHARTS

Histograms are graphical representations of frequency tables with intervals plotted on the horizontal axis and frequencies (or relative frequencies) on the vertical scale. A bar chart is very similar to a histogram except that the horizontal scale is a group of distinct categories or groups rather than continuous numerical intervals. Examples of categories are sex, economic status, and year in school (freshmen, sophomore, etc.). Figure 2.7 shows the frequency distribution of 1977 nursing graduates of a particular college of nursing according to marital status.

EXERCISE 2.9

Of 50 patients in a particular hospital who were asked which type of room they preferred, 25 said private room, 20 said semiprivate, and 5 said other. Present this data by means of a bar graph.

SUMMARY OF CRITICAL CONCEPTS IN CHAPTER 2

The following concepts were presented in Chapter 2:

1. The raw data that result from a sample survey, a census, or an experiment are usually in the form of large sets of unorganized numerical values.
2. A set of data must be organized and summarized so that important features may be grasped at a glance.
3. A set of data may be described pictorially by means of

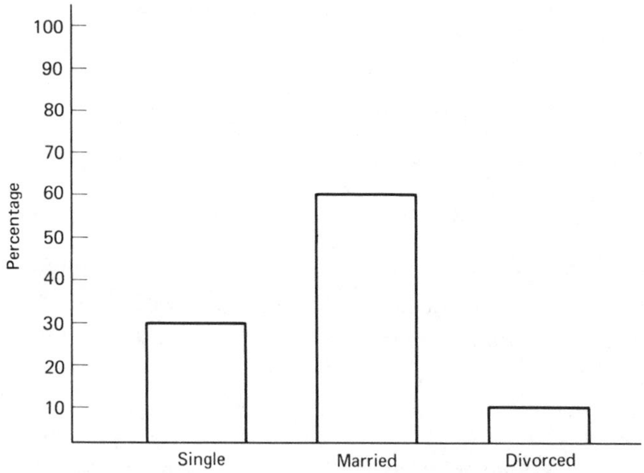

FIGURE 2.7 MARITAL STATUS OF 1977 NURSING GRADUATES

frequency tables, histograms, frequency polygons, circle charts, and bar charts.

4. Data may be described by numerical values, called numerical summaries. These include measures of central tendency (mean, median, mode) and measures of dispersion (range, variance, and standard deviation).
5. A frequency distribution consists of a series of predetermined values or intervals and the frequency with which these values occur.
6. A frequency table is a tabular presentation of the frequency distribution of a set of data.
7. A symmetric distribution is one in which the frequencies are equal (or nearly equal) on either side of a center point.
8. A skewed distribution is one that is not symmetric.
9. Relative frequency is the proportion or percent of the total number of observations falling into each interval in a frequency table.
10. A histogram is a graphical representation of either a frequency or relative frequency table.
11. When comparing two sets of data, relative rather than absolute frequencies should be used if the total number of values in the set is not the same.
12. Frequency polygons are graphical representations of frequency tables. The frequency associated with each interval is indicated by a dot placed above the midpoint of the interval. The dots are connected by straight lines.
13. A circle chart, or pie chart, is a circle which has been partitioned into percentages of the total corresponding to percent of some measurement of interest.
14. A bar chart is a graphical presentation whose horizontal scale is a group of distinct categories rather than continuous numerical intervals. The vertical axis is frequency or relative frequency.

CHAPTER **3**

DESCRIPTIVE STATISTICS: NUMERICAL SUMMARIES

OVERVIEW

In the previous chapter, it was learned that a set of measurements may be described by means of frequency distributions, frequency diagrams, and histograms. Although these techniques are valuable for displaying the salient features of the data, often it is desirable to describe a set of measurements using one, two, or more numbers called numerical summaries. Just as the values height and weight create a mental image of an individual, numerical summaries may be used to create a mental image of the frequency distribution of a set of measurements. In addition to providing a concise description of a set of data, numerical descriptive measures play a vital role in making inferences about a population. Numerical summaries calculated from measurements in the sample are used as "estimates" of corresponding values for the entire population.

The two most important types of numerical summaries are (1) those that locate the "center" of a distribution of data, called *measures of central tendency*; and (2) those that describe the "spread" of points about the center location, called *measures of dispersion* or *variability*. Measures of central tendency that will be discussed are mean, median, and mode and measures of dispersion that will be presented are range, variance, and standard deviation.

OBJECTIVES

Upon completion of this chapter, the student will be able to:

1. Define and compute measures of central tendency:
   a. Mean
   b. Median
   c. Mode

2. Define and compute measures of variability:
   a. Range
   b. Variance
   c. Standard deviation
3. Choose the appropriate descriptive statistic for summarizing a set of data.
4. Summarize data by means of percents, percentiles, rates, and ratios.

DISCUSSION

Measures of Central Tendency

An important descriptive characteristic associated with a set of numerical data is the central value or center point about which the other values tend to group themselves. These central values, which are used to locate the "middle" of the frequency distribution, are referred to as *measures of location*, or *measures of central tendency*.

The three most common measures of central tendency are the *mean*, the *median*, and the *mode*. The measure used for any given situation depends on the particular situation and how we wish to describe the "middle" of the distribution. The definitions, computations, and advantages of each will be discussed.

MEAN

The most widely used measure of location or central tendency is the arithmetic mean, or simply, the *mean* of a set of data. The mean of a set of measurements is the arithmetic average of the measurements. By definition, *the mean of a set of measurements is the sum of the measurements divided by the number of measurements in the set.* Symbolically, this may be represented using Σ notation as:

$$\overline{Y} = \frac{\Sigma Y_i}{n}$$

where $\overline{Y}$, read "Y bar," denotes the mean; ΣY_i represents the sum of all the Y values (measurements of interest); and n is the number of measurements in the set.

EXAMPLE 3.1 Length of hospital stay (in days) was recorded for five patients in a clinical research unit. The data are given below.

3, 5, 2, 3, 2

Find the mean length of hospital stay for the five patients.

SOLUTION

$$\text{Mean} = \bar{Y} = \frac{\Sigma Y_i}{n}$$

$$= \frac{3 + 5 + 2 + 3 + 2}{5}$$

$$= 3.0$$

The mean length of stay for the five patients is three days.

EXAMPLE 3.2 For the data in Example 3.1, calculate $\Sigma(Y_i - \bar{Y})$.

SOLUTION

$$\Sigma(Y_i - \bar{Y}) = (3 - 3) + (5 - 3) + (2 - 3) + (3 - 3) + (2 - 3)$$

$$= 0 + 2 + (-1) + 0 + (-1)$$

$$= 0$$

Example 3.2 illustrates an important characteristic of the mean: If the mean is subtracted from all the sample values, the sum of these differences is zero. The difference between a sample value and the mean is called a *deviation*. The importance of this property of the mean will become more apparent when measures of dispersion are discussed.

EXAMPLE 3.3 For Example 3.1 suppose Patient No. 1 stayed thirteen days rather than three days. How does this affect the mean stay for the five patients?

SOLUTION

$$\bar{Y} = \frac{\Sigma Y_i}{n}$$

$$= \frac{13 + 5 + 2 + 3 + 2}{5}$$

$$= 5.0$$

Because of the one very large data value, the mean value has been increased considerably. Four of the five data values are less than the mean for the set. Thus, the one "extreme" value gives a somewhat misleading picture of the average hospital stay for patients in the research unit.

The above example illustrates one disadvantage of the mean as a measure of the "center" of a set of data. The mean is affected by extreme values, especially when the number of observations in the sample is small.

MEDIAN

A second measure of central tendency is the *median*. By definition, *the median of a set of measurements is the middlemost measurement when the values are arranged in order of size.* For an odd number of observations, the median is the middle measurement in the ordered set of measurements; for an even number of observations, the median is the mean of the two middle observations.

EXAMPLE 3.4 Find the median hospital stay for the five patients in the clinical research unit (Example 3.1).

SOLUTION The first step in finding the median is to arrange the data in order from smallest to largest.

2, 2, 3, 3, 5

Since there are five observations, the median is the third value in the ordered list. The median hospital stay is three days.

EXAMPLE 3.5 Given the values:

6, 3, 2, 6, 1, 8

find the median.

SOLUTION The data arranged in order from smallest to largest are:

1, 2, 3, 6, 6, 8

Since there are an even number of measurements in the set ($n = 6$), the median is the average of the third and fourth values. Thus,

$$\text{Median} = \frac{3 + 6}{2} = 4.5$$

The median may also be defined as that observation above and below which 50% of the observations fall. It divides a set of data into two equal parts *by count*.

EXAMPLE 3.6 Compute the median for the data described in Example 3.3. How do the mean and median compare for this data?

SOLUTION Arranged in order, the data are:

2, 2, 3, 5, 13

The median is 3; the mean for this set of data is 5. The median in this case is more representative of the "center" of the measurements.

Note that neither the mean nor the median is necessarily equal to one of the sample values.

MODE

The third, and least common, measure of central tendency is the mode. By definition, *the mode of a set of measurements is the most commonly occurring value or the value within the set that occurs with greatest frequency.* A set of data may have one mode, two modes (bimodal), or many modes (multimodal). If all the observations are different in a set of data, then there is no mode.

With grouped data, the modal value is the midpoint of the group in the histogram that has the highest frequency. For example in Figure 2.3, the modal value is the midpoint of the interval 4.95–6.95, that is, 5.95 pounds.

EXAMPLE 3.7 Find the mode for the data in Example 3.3 and Example 3.1.

SOLUTION In Example 3.3, the value 2 occurs most frequently; the mode is therefore equal to 2. In Example 3.1, the values 2 and 3 both occur twice. In this case, the distribution of data is bimodal, i.e., has two modes.

EXAMPLE 3.8 Find the mode for the data in Figure 2.4.

SOLUTION The modal value is .45, the midpoint of the first interval.

CHOOSING A MEASURE OF CENTRAL TENDENCY

A primary factor in selecting which of the measures—mean, median, or mode—is to be used in describing the center of a distribution of data is what we intend to use it for once it has been computed. As a general rule of thumb, the median is preferred when the data have the possibility of extreme values. For example, in describing the distribution of ages of senior nursing students, the large majority of seniors are in their early twenties; however, some seniors are very much older. In this case the median would serve as a better indicator of age of senior nursing students than the mean because the mean would possibly be inflated by the older students. As the size of the sample becomes larger, this disadvantage of the mean as a descriptive measure becomes less important. For purposes of statistical analysis and inference, the mean is more likely to be used since it is more amenable to mathematical manipulations.

The mode is probably most useful when describing qualitative data. For example, types of requests made to a nurse by patients on a cardiac care unit may be determined. The request(s) that occurs most frequently would be the modal request. The mode is rarely used as a single measure for describing the central tendency for a set of data.

When a given measure is to be used solely for the purpose of describing the sample at hand, often no one measure is clearly better than the others. In this case, it may be of benefit to report two, or even all, of these measures.

An important factor that influences the choice of a measure of central tendency is the underlying scale of measurement possessed by the variable of interest.

Table 3.1 summarizes measures of central tendency as related to the underlying scale of measurement of the variable being measured. The mode is the only measure of central tendency that can be used with a nominal measurement scale. The median is the most appropriate statistic for describing the central tendency of ordinal data. For example, the median may be used to describe the "middle" category if patients in a geriatric nursing care center are asked to rate overall quality of nursing care using the scale "very poor," "poor," "fair," "good," "very good." For the interval and ratio scales, all three measures of central tendency may be employed.

For Exercises 3.1–3.4, which measure(s) of central tendency is most appropriate for describing the center of the distribution of measurements?

EXERCISE 3.1

In a hospital where the CASH (Commission on Administrative Services in Hospitals) nursing audit is used, the following total audit scores were obtained from five patients on a maternity ward on a given day:

86, 70, 75, 82, 68

TABLE 3.1 MEASURES OF CENTRAL TENDENCY AS RELATED TO SCALE OF MEASUREMENT

Measure of Central Tendency	*Type of Data* Nominal	Ordinal	Interval and Ratio	
Mean			X	When distribution reasonably symmetric
Median		X	X	When distribution very skewed
Mode	X	X	X	Never used as sole measure

EXERCISE 3.2

A study was done on survival time for 10 patients following a new treatment for cancer. The time in months was:

24, 8, 12, 3, 20, 18, 24, 19, 27, 25

EXERCISE 3.3

In a study to ascertain which size hospital gown should be purchased, a hospital director recorded all the different sizes worn by patients currently in the hospital.

EXERCISE 3.4

In a study on patient satisfaction with a particular nursing procedure, 25 patients were selected at random and asked to rate the procedure on the scale "very negative," "negative," "neutral," "positive," "very positive."

Measures of Dispersion

Consider the two histograms in Figure 3.1. Both (*a*) and (*b*) are centered around the same mean value, but the two sets of measurements obviously do not have the same frequency distribution. Since the purpose of reporting numerical summaries is to provide a mental image of the frequency distribution of a set of measurements, we must calculate in addition to location of center (central tendency), a measure that reflects the scatter, or dispersion, among the observed values about the center point. The most common measures of dispersion, or variability, are the range, the variance, and the standard deviation. Each will be discussed in the following sections.

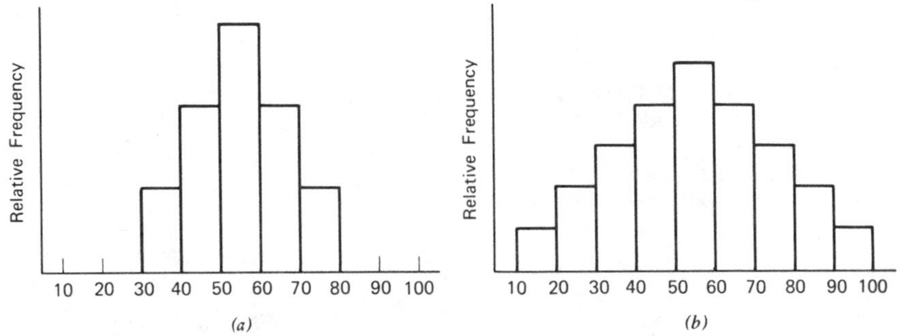

FIGURE 3.1 TWO HISTOGRAMS WITH EQUAL MEAN VALUES BUT DIFFERENT SPREAD OF VALUES ABOUT THE MEAN

RANGE

The range of a set of measurements is the difference between the largest and smallest values in the sample. In Figure 3.1*a* and *b*, the range would be approximately 80 to 30 = 50 and 100 to 10 = 90. (These are approximations since we do not have the actual data values.)

The range is simple to define and calculate. For this reason, it is considered a useful "rough and ready" measure of variability. However, the use of the range as the sole indicator of the variability of a set of measurements often gives a misleading impression of the true variability of the data. To illustrate, consider Figure 3.2. The range for both (*a*) and (*b*) is 100 to 10 = 90, yet the spread of the points about the center (the variability) of the two frequency distributions is markedly different. In (*a*) the majority of data values are clumped between 50 and 80; in (*b*) the data values are spread more evenly throughout the whole range. Thus, while the ranges for (*a*) and (*b*) are equal, the values in (*b*) have greater variability.

In addition, the range is very sensitive to the presence of extreme values (called outliers) in a set of measurements. The presence of one or two extreme values may result in a very large value for the range and consequently a misleading idea of the true variability of the data.

EXAMPLE 3.9 For the length of hospital stay in Example 3.1, the data in days were:

3, 5, 2, 3, 2

Find the range of data values for the above set.

SOLUTION The largest value is 5; the smallest is 2. Hence

$$\text{Range} = 5 - 2 = 3$$

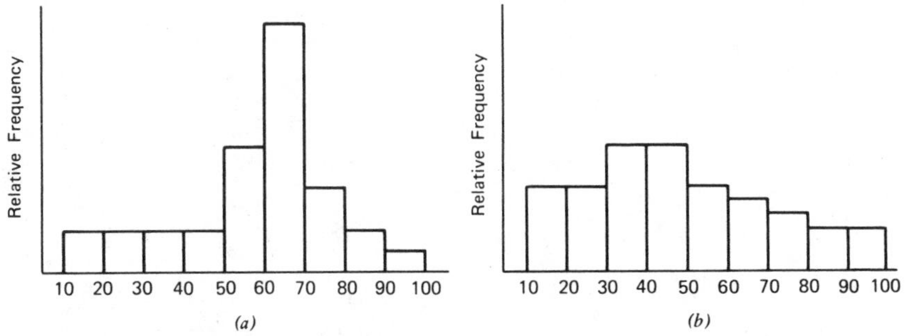

FIGURE 3.2 TWO HISTOGRAMS WITH EQUAL RANGES OF DATA VALUES

EXERCISE 3.5

Find the range of data values for the attitude score data reported on page 19.

VARIANCE, STANDARD DEVIATION

Since we use only two numbers in calculating the range, we have seen that the range often provides a misleading picture of the true variability, or spread, of a set of data. By making use of *all* the measurements in a set and their individual *deviations* from the mean, or center point, of the distribution, we may obtain a more valid measure of the variability of the measurements. (A *deviation* is the distance between any measurement in the set and the mean value for the set.)

To illustrate, we will use the hospital stay data of Example 3.1.

3, 5, 2, 3, 2

Recall from Chapter 1 that we may denote each of the above measurements by the symbol Y, where $Y_1 = 3$, $Y_2 = 5$, $Y_3 = 2$, etc. The mean value for the five measurements is:

$$\bar{Y} = \frac{\Sigma Y_i}{n} = \frac{15}{5} = 3$$

Figure 3.3 shows the five measurements in relation to their mean value. Each measurement in the set is represented by a dot located above its corresponding value. The numbers above the arrows are the deviations of each data point from the mean. The diagram in Figure 3.3 is shown in table form in Table 3.2.

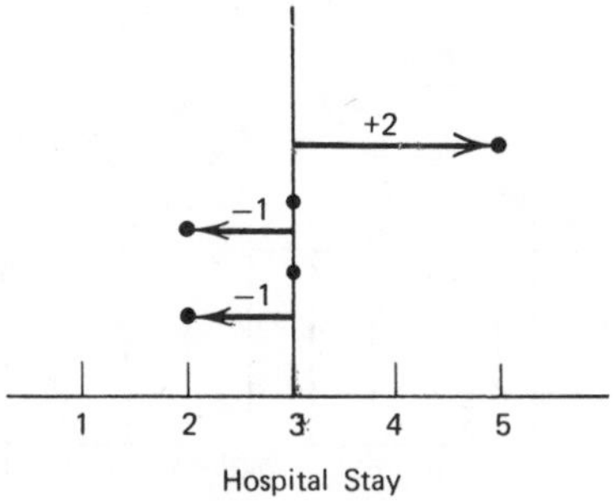

FIGURE 3.3 DIAGRAM OF DEVIATIONS OF HOSPITAL STAY DATA FROM THEIR MEAN

TABLE 3.2 LENGTH OF STAY IN HOSPITAL FOR FIVE PATIENTS: DEVIATIONS FROM THE MEAN

Patient	*Length of Stay in Days* (Y_i)	*Deviations* ($Y_i - \bar{Y}$)	($Y_i - \bar{Y})^2$
1	3	(3–3) = 0	0
2	5	(5–3) = 2	4
3	2	(2–3) = −1	1
4	3	(3–3) = 0	0
5	2	(2–3) = −1	1
	$\Sigma Y_i = 15$	$\Sigma(Y_i - \bar{Y}) = 0$	$\Sigma(Y_i - \bar{Y})^2 = 6$

The deviations of the individual measurements from the mean give an indication of how spread out the measurements are. From Figure 3.3, we see that the larger the deviations, the more dispersed the measurements are from their center, the mean.

We now wish to arrive at a measure of variability that takes into account the size of all the deviations. At first thought, we might wish to simply average the five deviations; a set of data having a large average deviation would be more variable than one having a small average deviation. However, what happens when we attempt to use a straightforward average of the deviations? As you recall, a property of the mean is that the sum of the deviations of values from the mean is *always zero*. Hence, the average deviation is always zero. As another possibility we might ignore all the positive and negative signs and simply average the actual numbers (absolute values.) This average, while relatively simple to compute, is difficult to manipulate mathematically. The most useful solution is first to square all the deviations (thus eliminating the minus signs), average the squared deviations, and then use the square root of the above to represent the "average" deviation. This final value is called the *standard deviation* of the measurements rather than the average deviation since it does not represent a true averaging of the deviations.

Symbolically, the average of the squared deviations is represented by:

$$s^2 = \frac{\Sigma(Y_i - \bar{Y})^2}{n - 1}$$

where s^2 is called the *variance* of a set of measurements. You will note that the denominator is $n - 1$ rather than n. In practice, when n is reasonably large, it makes little difference which is used. However, for theoretical reasons, $n - 1$ is preferred.

While the variance is a useful indicator of variability or dispersion, the most commonly used measure is the standard deviation (denoted by s).

$$s = \sqrt{s^2} = \sqrt{\frac{\Sigma(Y_i - \bar{Y})^2}{n - 1}}$$

The standard deviation, s, has the same units as the original data and is therefore easier to interpret than s^2, which is measured in square units.

EXAMPLE 3.10 The standard deviation of the hospital stay in Table 3.2 is:

$$s = \sqrt{\frac{\Sigma(Y_i - \bar{Y})^2}{n - 1}} = \sqrt{\frac{6}{4}} = \sqrt{1.5} = 1.22$$

Using the mean along with the standard deviation, we may now describe both the location of the center of a set of measurements and how the measurements are spread about the center.

EXAMPLE 3.11 In a study of the efficiency of two different emergency rooms, the waiting time (in minutes) before receiving attention was recorded for each patient and the mean and standard deviation reported.

	Emergency Room 1	*Emergency Room 2*
Mean ($\bar{Y}$)	60	50
Standard deviation (s)	10	100

Based on the above information, which of the emergency rooms would you say is the most efficient?

SOLUTION While Emergency Room 2 has the lowest mean waiting time, it also has a much larger spread of values about the mean. This means that for this set of measurements there are some very short waiting times and some very long ones. Emergency Room 1 has most of its values clumped more closely around the mean of 60, indicating that the waiting times do not vary much in either direction from

60 minutes. Based on the absence of very long waiting times, Emergency Room 1 would probably be judged the most efficient.

EXERCISE 3.6

Of the two histograms shown below which has the smallest standard deviation? Why?

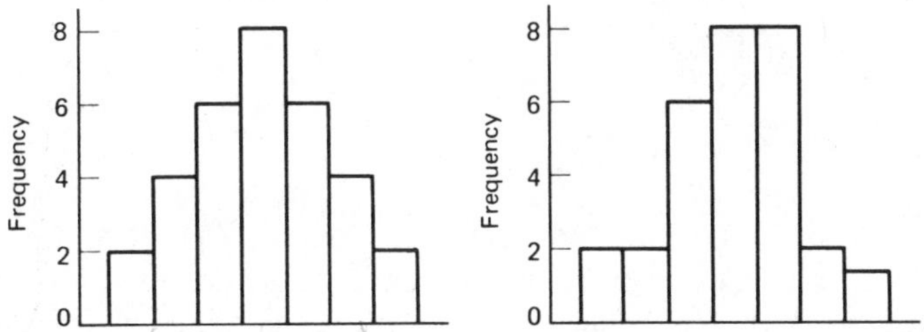

(a) (b)

EXERCISE 3.7

Find the mean and standard deviation for the data described in Exercise 3.2.

EXERCISE 3.8

Six mothers of premature infants were observed during their initial encounter with their infants. The observer rated the mothers' greeting behaviors on a scale from 1 to 10 with 1 = very anxious to 10 = not anxious. The scores were:

4, 3, 8, 5, 2, 2

Find:

a. Mean
b. Median
c. Mode
d. Range
e. Variance
f. Standard deviation

EXERCISE 3.9

A researcher was interested in comparing two techniques for measuring blood pressure. Ten patients received both Method A and Method B. The mean blood pressures for the groups were equal for the two methods, while the standard deviation for the measurements using Method B was twice as large as for those using Method A. Which method would you choose? Why?

COMPUTATIONAL FORMULA FOR STANDARD DEVIATION

For ease of computation, the following formula for standard deviation may be used:

$$s = \sqrt{\frac{\Sigma Y_i^2 - \frac{(\Sigma Y_i)^2}{n}}{n - 1}}$$

Returning to the hospital stay data of Example 3.1, let us find the standard deviation, s, using the computational formula given above. The calculations are shown in Table 3.3. You will notice that this is exactly the same answer we found by using the theoretical formula discussed earlier.

EXERCISE 3.10

For Exercise 3.8, calculate the standard deviation using the computational formula.

TABLE 3.3 LENGTH OF STAY IN HOSPITAL FOR FIVE PATIENTS

Patient	*Length of Stay in Days* (Y_i)	(Y_i^2)
1	3	9
2	5	25
3	2	4
4	3	9
5	2	4
	$\Sigma Y_i = 15$	$\Sigma Y_i^2 = 51$

$$s = \sqrt{\frac{\Sigma Y_i^2 - \frac{(\Sigma Y_i)^2}{n}}{n - 1}} = \sqrt{\frac{51 - \frac{(15)^2}{5}}{4}} = \sqrt{1.5} = 1.22$$

EXERCISES 3.11–3.12

For the data given in Exercises 3.1 and 3.2, find

a. Mean
b. Median
c. Mode
d. Range
e. Variance
f. Standard deviation (use both computational and theoretical formulas)

† Percentages, Percentiles, Rates, and Ratios

We have just seen how data may be summarized both tabularly and graphically by means of frequency tables, histograms, polygons, and bar charts, and numerically by such measures as mean and standard deviation. In addition, data may also be summarized by such measures as percentages, percentiles, rates, and ratios. These measures involve the expression of data in relative form, thus facilitating comparison between groups.

PERCENTAGES

A percent or proportion, as presented previously, is the *relative frequency* of occurrence of some event. (Relative frequency was defined as the absolute frequency of occurrence of some event divided by the total of the events.) Percentages as summary measures have the advantage of allowing comparison of data for different sized groups. For example, consider Table 3.4, which presents grade distribution of nursing students for three years at a particular college of nursing. From Table 3.4, we see that the percent of A's in 1975 is less than the percent of A's in 1974 even though the absolute frequencies for the two years are identical. This occurs because the total number of students for the two years are not equal. Absolute frequencies can meaningfully be compared only when the group totals are the same.

Often it is of interest to determine the "average" percent over a specified time period. Looking at Table 3.4, one may be tempted to average the values 10%, 6.7%, and 20%. This would be incorrect because the total numbers on which the percents are based are not equal. (In this example, the totals are 50, 75, and 100). The percents used to compute the "average" percent must be "weighted" according to the total on which each percent was based. This "weighting" is accomplished by summing the absolute frequencies of the three cate-

† Sections marked by a dagger (†) cover more difficult or theoretical topics.

TABLE 3.4 GRADE DISTRIBUTION BY YEAR FOR THREE YEARS FOR A COLLEGE OF NURSING

	1974		*1975*		*1976*	
Grade	*Absolute Frequency*	*Relative Frequency (%)*	*Absolute Frequency*	*Relative Frequency (%)*	*Absolute Frequency*	*Relative Frequency (%)*
A	5	10	5	6.7	20	20
B	10	20	20	26.7	20	20
C	10	20	20	26.7	30	30
D	20	40	20	26.7	30	30
F	5	10	10	13.3	0	0
	50		75		100	

gories and dividing by the sum of the three totals. In this example, the average percent of students making A's over the three-year period is:

$$\frac{5 + 5 + 20}{50 + 75 + 100} = \frac{30}{225} = 13.3\%$$

(Confirm that this value is not equal to the average of the three percents.)

In using average percent, one must realize that the average value is not necessarily indicative of the trends for the individual years. For example, the number of A's may sharply decrease in the second year and increase again sharply in the third year. A single average value for the three years will not reflect these changes.

PERCENT CHANGE

A useful measure that describes change over time is the *percent change* in some event over a specified time period. The percent change is calculated by subtracting the initial reading from the final reading and dividing by the initial reading. This value is multiplied by 100 in order to convert it to a percent. Hence,

$$\text{Percent change} = \frac{n_2 - n_1}{n_1} \times 100$$

where n_1 = initial number, n_2 = final number.

For the grade distribution given in Table 3.4, the percent change in number of A's from 1974 to 1976 is:

$$\text{Percent change} = \frac{20 - 5}{5} \times 100 = 300\%$$

Thus, there was a 300% *increase* in number of A's from 1974 to 1976.

PERCENTILES

Another useful means of comparison using percentages is *percentiles*. A percentile indicates the relative position of any individual measurement with respect to all measurements in the group. For example, suppose a group of patients are given a manual dexterity exam and it is of interest to compare Patient A's performance score with the performance scores of all other patients given the exam. To determine A's score in terms of percentile, we first determine the number of scores that are lower than Patient A's score, divide this number by the total number of scores in the group, and multiply by 100 to convert to percent.

$$\text{Percentile} = \frac{\text{Number of scores less than given score}}{\text{Total number of scores}} \times 100$$

Suppose 50 patients were given the dexterity exam and 10 patients scored lower than Patient A. His percentile score would be $\frac{10}{50} \times 100 =$ 20%. Thus, Patient A is in the lower fifth of the group (i.e., 20% of the scores were lower than his score). Suppose, instead, that 40 out of the 50 scores were lower than A's score. Then his percentile would be $\frac{40}{50} \times$ 100 = 80%. Since 80% of the scores are lower, a percentile score of 80 indicates that A is in the "top 20%" of the group.

Scores on various standardized tests are usually reported in terms of percentiles, thus facilitating comparison of an individual's scores on several different tests. For example, on a nursing proficiency exam a student nurse may score at the 90th percentile (upper 10%) on clinical expertise and at the 20th percentile on basic science knowledge, thus indicating a deficiency in basic science knowledge relative to clinical expertise.

RATES

Another useful percentage widely used in the fields of vital statistics and epidemiology is *rate*, which expresses the frequency of occurrence of a specific event (such as death from a disease) relative to the total group "at risk" of the event. A death rate, for example, is the frequency of deaths for the given disease in a given period divided by the total population (i.e., all people who are "at risk" of the disease).

A true rate is calculated according to the following basic formula:

$$\text{Rate} = \frac{A}{A + B}$$

where A = number of individuals in a given category of the variable and B = number in all other categories of the variable.

Thus, for the rate of death for a particular disease, the numerator is all who died from the disease (A) in the specified time period, and the denominator is all who died (A) plus all who did not die (B) from the disease in the specified time period. The value obtained by the above formula is usually multiplied by some preselected base number such as 1,000, 10,000, or 100,000 in order to avoid the very small decimal numbers that can occur when a small value of A is divided by a large value of ($A + B$). Thus, rates are usually reported in terms of "per 1,000 people," "per 10,000 people," etc. Table 3.5 gives some of the more important rates that may be encountered in nursing.

ADJUSTED RATES

Since basic characteristics of the population may differ substantially according to time periods or geographic locations, the comparison of crude rates (such as death rates) for different locations or different time periods may yield misleading information. For example, a comparison of the crude death rates for a particular disease among single and married women would obviously be complicated by the fact that married women in general are older than single women. A more meaningful comparison would be to remove the effect of the age differential by "adjusting" the rates for differences in ages in the two groups. Age-adjusted death (mortality) rates are determined by applying age-specific rates for both age groups to some specified standard population. The standard population is a standard of reference based on an actual census of a living population or on a theoretical population. For example, to determine the age-adjusted death rate from cancer for single and married women, the standard population may be chosen as the total female population in 1930. The resulting age-adjusted rate is an expression of deaths that would be expected for the specified group (e.g., single or married women) if the age distributions of the specified groups were the same as

TABLE 3.5 RATES THAT MAY BE ENCOUNTERED IN NURSING

$$\text{Crude death rate} = \frac{\text{All deaths in specified period}}{\text{Total population in middle of period}} \times 1{,}000$$

$$\text{Annual death rate (crude)} = \frac{\text{Deaths from all causes in calendar year}}{\text{Population of July 1}} \times 1{,}000$$

$$\text{Annual birth rate} = \frac{\text{Live births in calendar year}}{\text{Population of July 1}} \times 1{,}000$$

$$\text{Annual death rate from specific cause} = \frac{\text{Deaths in calendar year from specific cause}}{\text{Population of July 1}} \times 100{,}000$$

$$\text{Annual age specific death rate} = \frac{\text{Deaths in calendar year from all causes for given age group}}{\text{Population for given age group, July 1}} \times 1{,}000$$

$$\text{Infant mortality rate} = \frac{\text{Deaths under one year of age in year}}{\text{Live births in year}} \times 1{,}000$$

$$\text{Neonatal mortality rate} = \frac{\text{Deaths under one month of age in year}}{\text{Live births in year}} \times 1{,}000$$

$$\text{Prevalence rate of specific disease} = \frac{\text{All cases of specific disease at given time}}{\text{Population at given time}} \times 1{,}000$$

$$\text{Case fatality rate of specific disease} = \frac{\text{All deaths from specific disease in given period}}{\text{All cases of specific disease in given period}} \times 100$$

$$\text{Annual case incidence rate of specific disease} = \frac{\text{New cases of specific disease in year}}{\text{Population of July 1}} \times 1{,}000$$

the age distribution of the standard population. The actual procedure for computing adjusted rates will not be presented in this text.*

RATIOS

The final percentage, or relative frequency, to be discussed is the *ratio* of the occurrence of one event to the occurrence of some other event. For example, a rare disease may have a black-to-white ratio of 3.0 (3/1). This indicates that three times as many blacks are afflicted with the disease as whites. Other examples of commonly used ratios are number of births-to-number-of-deaths (called the vital index) and number-of-males-to-number-of-females for a given event.

SUMMARY OF CRITICAL CONCEPTS IN CHAPTER 3

The following concepts were presented in Chapter 3:

1. Numerical descriptive measures are used to describe a set of data (descriptive statistics) and are used as estimates of corresponding values in the population from which the sample data are selected.
2. Numerical summaries may locate the "center" of a distribution of data (measures of central tendency) or may describe the "spread" of values about the center location (measures of dispersion).
3. The most common measures of central tendency are the mean, the median, and the mode.
4. The *mean* is the arithmetic average of a set of data values; it is the sum of the measurements divided by the number of measurements in the set.
5. The existence of a few extremely large or extremely small data values relative to the rest of the data values can cause the mean to give a misleading indication of the center of the distribution of values, especially when the sample size is small.
6. The *median* of a set of measurements is the "middle most" measurement when the values are arranged in order of size.
7. The median is unaffected by extreme data values and is preferred over the mean when the possibility of extreme values exists.

* For a simplified discussion of both the direct and indirect methods of age-adjustment, the reader is referred to Theodore Colton, *Statistics in Medicine* (Boston: Little Brown and Co., 1974), pp. 44–51.

8. The *mode* of a set of measurements is the most commonly occurring value in the set.
9. The most common measures of dispersion are range, variance, and standard deviation.
10. The *range* of a set of measurements is the difference between the largest and smallest values in the sample.
11. A deviation is the distance between any measurement in a set of data and the mean of the set of measurements.
12. The larger the deviations of measurements from their mean, the more spread out or dispersed the values are.
13. The *standard deviation* is a kind of average of the magnitude of all deviations from the mean for a set of data.
14. The larger the deviations of measurements are from the mean, the larger is the value of the standard deviation.
15. The variance of a set of measurements is the square of the standard deviation of the measurements.

†16. A *percent* is the relative frequency of occurrence of some event.

†17. A *percentile* indicates the relative position of any individual measurement with respect to all measurements in the group.

†18. A *rate* expresses the frequency of occurrence of a specific event relative to the total group "at risk" of the event.

†19. A *ratio* is the relative frequency of occurrence of one event to the occurrence of some other event.

CHAPTER **4**

POPULATIONS, SAMPLES, AND THE NORMAL DISTRIBUTION

OVERVIEW

The measures described in Chapters 2 and 3 are techniques of *descriptive statistics*, since they may be used to describe a set of sample measurements. Chapters 5 through 11 present methods of *inferential statistics*, the branch of statistics concerned with making statements about a population based on information in a sample. In this chapter, the relationship between population and samples will be presented along with a discussion of a very important population frequency distribution—the normal distribution. Chapter 4 thus provides a bridge between descriptive and inferential statistics.

OBJECTIVES

Upon completion of this chapter, the student will be able to:

1. Define *population* and *sample* and state the relationship between them.
2. Describe the relationship between *parameters* and *statistics* and give examples of each.
3. State the difference between *theoretical* and *empirical* distribution.
4. Describe in words the meaning of *sampling error.*
5. List the properties of the normal distribution.
6. State the standard probabilities associated with the normal curve.

DISCUSSION

Populations and Samples

The relationship between population and samples is a cornerstone of statistics. The numerical quantities computed from a sample both describe the sample itself (descriptive statistics) and provide for inferences about the characteristics of the population from which the sample was collected (inferential statistics). Stated formally, the definitions of population and sample are: *a population is the set of all measurements of interest to an investigator; a sample is a subset of measurements selected from a given population.*

In the statistical sense, a population may be thought of not as a collection of people or objects but rather as a set of *measurements* or *counts* made on these people or objects. For example, a population might be the heights of all nurses in military service; the number of children per family in the Southeastern United States; the serum uric acid level of diabetics; the anxiety levels of parents with critically ill children.

The first step in the conduct of any research is first to specify the common characteristics that define the population of interest. Often a distinction is made between the population about which an investigator wishes to draw a conclusion, called the *target population*, and the population about which a conclusion *can* be made, called the *population sampled*.

EXAMPLE 4.1 In order to study the effects of different types of nursing interventions on the emotional state of burn patients, a nurse selected eight burn patients in a large university hospital burn facility. The target population is all burn patients, since this is the group about which the investigator would like to draw a conclusion. The population of burn patients from which the sample was taken is all burn patients in a university hospital. Statistically, this is the group about which inferences can be made (provided certain sampling requirements have been met).

EXAMPLE 4.2 A public health nurse interested in determining the nutritional habits of persons over 60 years old who live in inner city slumbs, interviewed 200 people over 60 years old who live in the inner city slum in a large United States city. The target population is all slum residents over 60 years old; the population sampled is all slum residents over 60 years old in the particular city.

The distinction between the two populations is a statistical one. The methods of statistical inference that will be described in Chapters 5 through 10 enable an investigator to generalize from the sample to the population sampled. Generalization from the population sampled to the target population is much more subjective and open to controversy. To make the final step in generalizing to the target population, the investigator must be assured that the characteristics of the population sampled and the target population are identical. The possibility for the existence of subtle or hidden differences between the two populations makes it mandatory that the investigator proceed with extreme caution when generalizing from the population sampled to the target population.

For the remainder of the statistical presentations, the term "population" will refer to the population sampled unless otherwise specified.

Once the population of interest has been specified, the next logical step is either to take a census (measure every element in the population) or select a sample of measurements from the population. A question that often arises is "why study a sample rather than the entire population?" There are several reasons why it is not usually feasible to measure every element in the population. Among these are:*

1. The size of the population often makes it impractical or impossible to study in its entirety.
2. The cost of making observations on all elements of the population may be prohibitive.
3. All the individual members of the population may not be observable.
4. The measurements may be destructive. For example, if one wishes to determine the functional life of a heart pacer, the instrument must be tested until failure. Hence, the testing procedure is destructive.

The objective in selecting a sample is to choose measurements that are representative of all the measurements in the population. The simplest and best known way to do this is to collect a *simple random sample*. By definition, a simple random sample is a sample selected in such a way that each observation or unit in the population has an *equal chance* of being selected in any sample. It is important to note that randomness in the statistical sense does not mean haphazardness. To prevent either conscious or unconscious bias on the part of the investi-

* From Robert C. Duncan, Rebecca G. Knapp, and M. Clinton Miller III, *Introductory Biostatistics for the Health Sciences*, 2nd ed. Copyright © 1983 John Wiley & Sons, Inc. Reprinted by permission of John Wiley & Sons, Inc.

gator, the selection of a truly random sample can only be achieved by mechanical means, for example, drawing names out of a hat. (Sample selection techniques will be discussed in more detail in Chapter 11.) By applying the methods of statistical inference to the set of sample measurements, the investigator may obtain information about the entire population of measurements from which the sample was drawn.

Parameters and Statistics

Statistical inference is a process by which one draws conclusions about a population based on information contained in a sample. Population values about which we commonly wish to obtain information are the mean or average value and the standard deviation. When measures are computed using the entire population they are called *parameters* and are usually denoted by Greek symbols. When measures are computed from a set of sample measurements, they are called *statistics*. Thus, the mean of a population is the parameter μ ("mu") and the mean of a sample is the statistic $\bar{Y}$. Likewise, the population standard deviation is a parameter denoted by σ, and the sample standard deviation is a statistic denoted by s. Table 4.1 shows these relationships.

In most research situations, we rarely know the value of the population parameters since we are rarely able to observe every element in the population. The only recourse is to take a sample of measurements from the population and compute the corresponding sample statistic. The sample statistic is used as an *estimate* of the population parameter.

TABLE 4.1 RELATIONSHIP BETWEEN POPULATION PARAMETERS AND SAMPLE STATISTICS

	Parameter (Population)		*Statistic (Sample)*
Mean	μ	←———	$\bar{Y}$
Standard deviation	σ	←———	s

EXAMPLE 4.3 An investigator wished to determine the mean height of all nurses in military service. A random sample of 50 military nurses was selected and their mean height calculated. In this example the parameter of interest is the mean height (μ) of *all* military nurses. The statistic, calculated from a sample of 50 nurses, is the sample mean ($\bar{Y}$). $\bar{Y}$ is used as an estimate of the true μ.

EXAMPLE 4.4 An investigator interested in determining the mean birth weight for *all* infants with Disorder X, calculated the birth weight of 25 infants born with Disorder X. The parameter of interest is μ, the mean of the entire population of infants with Disorder X. The sample statistic is $\bar{Y}$, the mean of the sample of 25 infants with Disorder X. $\bar{Y}$ is an estimate of μ.

EXERCISE 4.1

An investigator wished to study anxiety in mothers who are visiting their sick children for the first time. Each of 10 randomly selected mothers was given a standardized anxiety exam immediately before she visited her child and the mean score was determined. What is the parameter of interest? The statistic?

EXERCISE 4.2

In a research project to determine the effect of backrubs on the blood pressure of geriatric patients, 30 patients in a nursing home were given a 10-minute back rub and the change, if any, in their blood pressure was recorded. The average change in blood pressure was computed for the 30 patients. What is the parameter of interest? The statistic?

It was stated earlier that sample statistics are used to *estimate* population parameters. The value of a statistic will rarely, if ever, be exactly equal to a population parameter. However, if we repeatedly select random samples from the population, the value of the sample statistic will cluster around the value of the population parameter. In Example 4.3, suppose we selected 50 military nurses at random and found that their mean height was 5′4″. If we select another 50 nurses at random, we may find their mean height to be 5′6″ and another sample may have a mean height of 5′5″. All we know is that the sample values tend to cluster around the true (unknown) population mean. The failure of a sample statistic to be exactly equal to a population parameter results among other things from *sampling error*, which plays an important role in the statistical decision-making process.

Theoretical and Empirical Distributions

In Chapter 2 the concept of frequency distributions (frequency tables) for sets of data was introduced. When a frequency distribution is tabulated from a set of sample measurements, it is called an *empirical distribution*; the frequency distribution for all the measurements in the entire population is called a *theoretical distribution*. As the name implies, a theoretical frequency distribution is rarely obtainable in practice since it is rarely feasible to measure all elements in the population. Instead, an empirical frequency distribution is obtained for the sample measurements and is used as an *approximation* of the theoretical frequency distribution of all population measurements.

In most nonexperimental research designs where the purpose of the study is mainly descriptive, knowledge of the theoretical population frequency distribution is not essential. However, when it is of interest to make inferences about a population based on sample measurements, a knowledge of the properties of the theoretical distribution of population measurements is needed.*

The nature of the variable of interest in a study (i.e., whether it is discrete or continuous) is an important characteristic of a theoretical frequency distribution. It is possible to have either discrete theoretical distributions (the variable of interest is a discrete variable) or continuous theoretical distributions. We will now discuss one of the most widely used of all continuous theoretical frequency distributions, the normal distribution.

The Normal Distribution

Recall the birth-weight example in Chapter 2 (page 22). Suppose, in this example, the 35 birth weights are for infants who were born with a particular genetic disorder (Disorder X). The investigator tabulates the frequency distribution for the 35 infants in the sample (Table 2.7) and draws the histogram (Figure 2.3). The frequency distribution shown in Table 2.7 is the *empirical* frequency distribution since it is based on the 35 sample measurements. However, from the point of view of the research, the 35 infants do not constitute the population of interest. The investigator wishes to infer something about birth weight for any infant with Disorder X based on the results from the sample of 35 infants with Disorder X. In order to make this inference, it is necessary to have some idea of the shape of the frequency distribution of birth weights for *all* infants with the disorder.

Fortunately, based on much experience, it is known that the population frequency distributions (theoretical distribution) of many naturally occurring phenomena (e.g., height, weight, blood pressure,

* A rapidly growing area of statistics is nonparametric statistics. Using nonparametric techniques to make inferences, the investigator need not necessarily know the nature of the population distribution. Nonparametric techniques are discussed in Chapter 12.

cholesterol levels, scores on standardized exams, etc.) possess the familiar bell-shape characteristic of the "normal distribution."

To illustrate, consider Figure 4.1. The midpoints of the birth weight intervals (Figure 2.3) have been connected by a smooth curve. Conceptually, the smooth curve is a histogram whose interval widths become smaller and smaller as more observations are included. As the number of observations approaches the total number of observations in the population, the interval widths become infinitely small and form a smooth curve. Thus, the smooth curve is the shape of the frequency distribution if the entire population had been measured. We may then use the properties of the theoretical distribution (the normal, in this case) to make inferences about the population of interest.

Let us now consider some of the characteristics of the normal curve. In Chapter 6 we will see how these properties will be utilized in making statistical inferences.

PROPERTIES OF THE NORMAL CURVE

Returning to Figure 4.1, we may infer some of the important characteristics of the normal curve.

Consider the shape of the curve. It has the familiar bell shape and is symmetric about its center point, the mean (denoted by μ). Looking at the curve we see also that values that are very large or very small in relation to the mean occur infrequently. The closer the values are to the

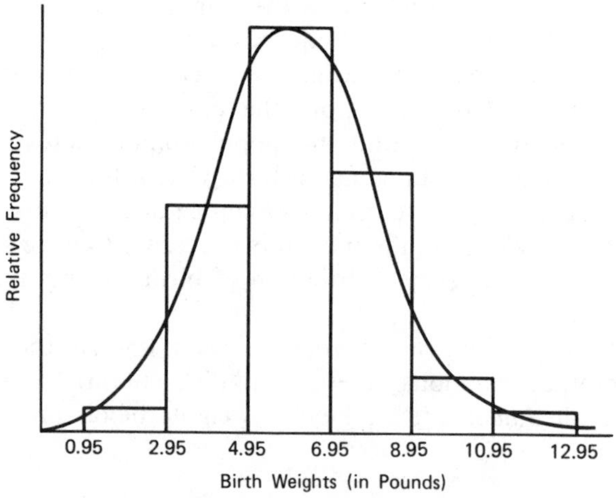

FIGURE 4.1 FREQUENCY DISTRIBUTION OF BIRTH WEIGHTS OF INFANTS WITH DISORDER X

mean the more frequently they occur. When we speak of an exam as being graded "on the normal curve," we mean that most of the scores will be around the average, or mean value, with less frequent high and low scores occurring approximately the same in either direction of the mean score.

The center point or location of a normal curve depends on the value of the mean and the "spread" depends upon the value of the standard deviation. Refer to Exercise 3.6. We said that histogram (*a*) has the largest standard deviation since the values are more spread out from the mean, while in (*b*) they are more closely clumped around the mean. Analogously, for distributions of measurements with a small standard deviation (σ), the curve is tall and thin while for distributions with large standard deviations the curve is short and fat. Figure 4.2 illustrates normal curves with different mean values and different standard deviations. In (*a*) we have two different normal curves with equal standard deviation (σ) but different mean values (μ). The frequency distributions for the two populations are identical in shape, with the mean of the distribution for females moved slightly to the right of the distribution of

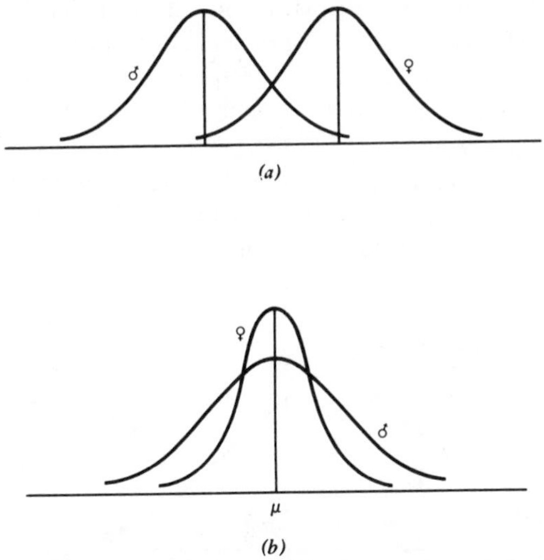

FIGURE 4.2 ILLUSTRATION OF EFFECT OF μ AND σ ON LOCATION AND SPREAD OF NORMAL CURVE (*A*) SCORES ON A STANDARDIZED NURSING EXAM SHOWING A HIGHER MEAN SCORE FOR FEMALES WITH A STANDARD DEVIATION OF SCORES THE SAME (*B*) LEVEL OF A PARTICULAR HORMONE SHOWING THE SAME MEAN VALUE FOR MALES AND FEMALES WITH A MUCH WIDER SPREAD OF VALUES (σ) FOR MALES

values for males. In (*b*) the standard deviation for the distribution of hormone measurements for males is larger than for females, with identical mean values for both groups of measurements. Thus, the two frequency curves are centered on the same mean value with the curve for males flatter and more spread out than for females.

For a population of measurements known to be normally distributed, if μ and σ are known, then the picture of the frequency distribution (normal curve) can be drawn. For the population frequency distribution (smooth curve), the area under the curve between two points is the *probability* that an observation from the population will fall in the interval specified by the two points. For example, referring to Figure 4.1, the probability that an individual from the specified population will have a birth weight in the range 4.95 to 6.95 pounds is obtained by finding the area under the smooth curve between these two points. These areas for the smooth curve may be obtained by the use of a table of normal distribution values. The procedure will be illustrated later in the chapter. Returning to Figure 4.1, we may observe that the *relative frequency* of observations in the *sample* falling in the interval 4.95 to 6.95 is given empirically by the height of the bar 4.95–6.95 in the histogram. If the population from which the sample of birth weights was drawn is indeed normally distributed, then we expect the relative frequency of observations in a specified interval obtained from the histogram (empirical frequency distribution) to be "close" to the probability obtained from the smooth curve (theoretical frequency distribution).

Refer to Figure 4.3. The percent values 68%, 95%, and 99% give the approximate number of observations falling within various standard deviations of the mean μ. That is, 68% of all the measurements will lie

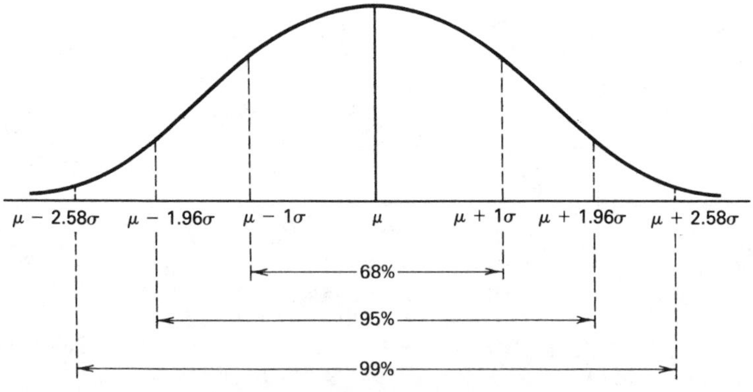

FIGURE 4.3 A NORMAL DISTRIBUTION SHOWING THE PERCENTAGE OF VALUES FOUND IN VARIOUS RANGES ABOUT THE MEAN

within 1 standard deviation above and below the mean; 95% of all measurements will lie within 1.96 standard deviations above and below the mean; and 99% of all measurements will fall within 2.58 standard deviations above and below the mean. (These proportions are properties of the normal curve that are derived from mathematical manipulations.) The significance of these particular values will become apparent when we discuss the process of making a statistical inference. For now, let us simply illustrate what the above percentages mean in terms of actual values.

EXAMPLE 4.5 From experience, it is known that blood pressure readings are distributed normally (i.e., the shape of the frequency distribution is that of the normal curve). Suppose, from a book of clinical norms, we learn that the mean systolic blood pressure for all males is 120 mmHg with a standard deviation of 20 mmHg. These values displayed on the normal curve are shown below.

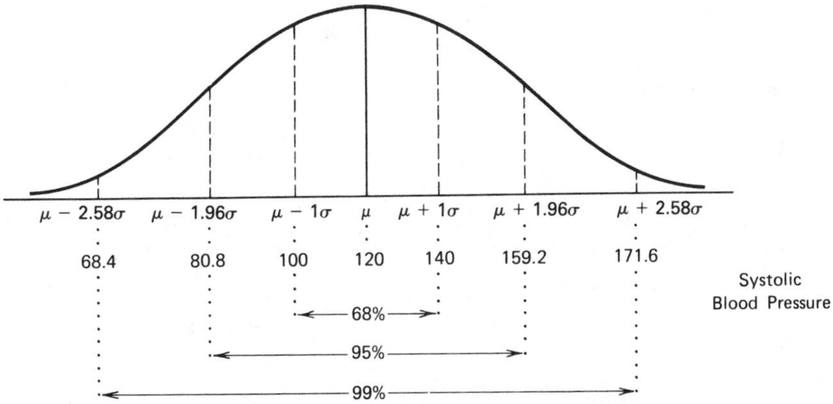

This tells us that we expect 68% of all males to have systolic blood pressures between 100 and 140; 95% between 80.8 and 159.2; and 99% between 68.4 and 171.6 mmHg.

EXERCISE 4.3

Scores on a particular personality inventory scale are assumed to be normally distributed with mean score equal to 50 and standard deviation equal to 10. Between what two scores would we expect 68% of the scores to fall? Between what two do we expect 95% to fall? 99%?

EXAMPLE 4.6 In a hypertensive screening program, health officials determined that the mean systolic blood pressure for normal males in a particular ethnic group is 140 mmHg with a standard deviation σ of 10 mmHg. It was decided that any individual in this group who had a systolic blood pressure two standard deviations above the mean normal systolic pressure would be considered hypertensive and referred for further diagnosis. A male from this group was tested and found to have a systolic blood pressure of 170 mmHg. How would he be classified?

SOLUTION The frequency distribution for the entire population of blood pressure readings (the *theoretical* frequency distribution) is assumed to have a normal distribution. It is further assumed that the mean μ of the population is 140 mmHg and the standard deviation (σ) is 10. These values are shown below.

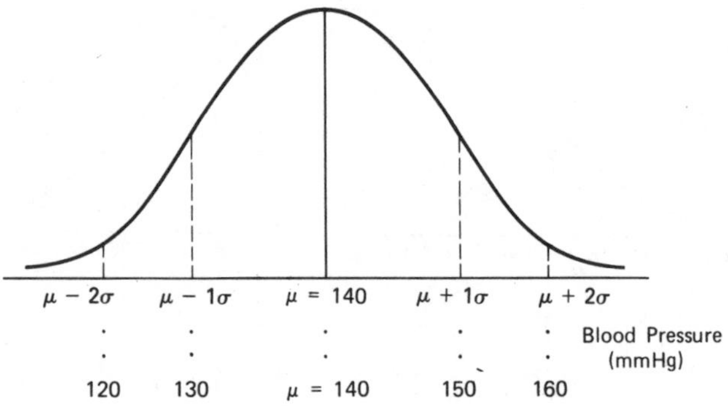

An individual with a systolic blood pressure reading of 170 would lie above the 2σ mark on the curve. More precisely he would lie three standard deviations above the mean systolic blood pressure for normal males in this group. This value is found by obtaining the difference between the mean μ and the observed value $Y = 170$ and then dividing by σ, the standard deviation; that is,

$$z = \frac{170 - 140}{10} = 3$$

where z is the number of standard deviations any observed value Y lies away from the mean μ. Since a blood pressure of 170 mmHg lies more than two standard deviations above the mean, the individual would be classified as hypertensive and referred for further diagnosis.

EXERCISE 4.4

In a study on infant mortality in South Carolina, 1,500 mortality records were selected at random from the files of a particular county. What is the target population for this study? The population sampled?

EXERCISE 4.5

The optimum mean weight for a particular height and body frame for females is 130 pounds with a standard deviation of 5 pounds. In a screening test for obesity, it was decided that all females whose weight is two standard deviations above the mean optimum weight would be classified obese. A female with the particular height and body frame weighed 143 pounds. How would she be classified?

EXERCISE 4.6

For Exercise 4.5, between what two weight measurements would 99% of all the weights lie?

EXERCISE 4.7

On a standardized nursing proficiency exam the mean score is 70 with a standard deviation of 10. Student ratings on the exam are classified as follows:

2–3 standard deviations below the mean—very poor
1–2 standard deviations below the mean—poor
0–1 standard deviations below the mean—average
0–1 standard deviations above the mean—average
1–2 standard deviations above the mean—good
2–3 standard deviations above the mean—excellent

Classify the following students with regard to their nursing proficiency. The scores for five students are:

a. 95
b. 65
c. 72
d. 52
e. 82

EXERCISE 4.8

For Exercise 4.7, between what two scores do we expect 95% of all the scores to lie? 99%?

†FINDING PROBABILITIES ASSOCIATED WITH THE NORMAL CURVE USING THE STANDARD NORMAL TABLE

In the previous section we noted that the probability that an observation will have a value in the interval a to b is obtained by finding the area under the smooth curve between the points a and b. We will now illustrate how these areas and, hence, probabilities can be obtained by using a standard normal table.

Suppose in the birth weight example cited earlier, we know from previous experience or from the literature that the mean of the population of birth weights μ is 5.95 pounds, and the standard deviation of the population of birth weights is 2 pounds. We are also assuming that the frequency distribution of the population of birth weights is the normal distribution. We wish to find the probability that an infant in this population will have a birth weight in the range 4.95 to 6.95 pounds. Symbolically, we write this probability as

$$\Pr(4.95 \leq Y \leq 6.95)$$

where Y represents birth weight of infants. The symbolic notation is translated as, "the probability that a birth weight Y lies in the interval 4.95 to 6.95 pounds." To obtain this probability from the table we must first convert the values 4.95 and 6.95 pounds to *standardized* units. That is, we convert our units (4.95, 6.95) taken from a normal population with μ = 5.95, $\sigma = 2$ to *equivalent* standardized units from a population with $\mu = 0$, $\sigma = 1$. This is done because otherwise we must have a different normal table for every different μ and σ combination, resulting in an infinite number of possible tables. By standardizing all units, we need only the *standard normal table* having $\mu = 0$, $\sigma = 1$. To standardize an observation Y, we first determine the distance between the mean μ and the observation Y and then convert to a standard score by dividing by σ, the standard deviation of Y. The resulting standardized value, called a *z-score,* is given symbolically by

$$z = \frac{\text{distance between } Y \text{ and } \mu}{\text{standard deviation of } Y} = \frac{Y - \mu}{\sigma}$$

† Sections marked by a dagger cover more difficult or theoretical topics.

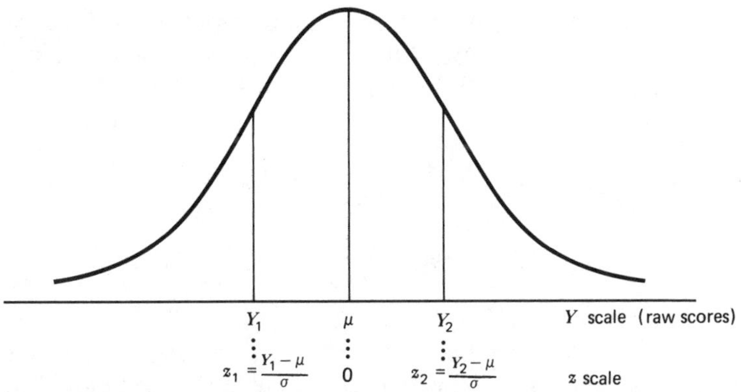

FIGURE 4.4 PICTORIAL REPRESENTATION OF EQUIVALENCE OF RAW SCORES (Y) AND z SCORES

In Figure 4.4, we see that for every Y value on the horizontal axis, there corresponds an equivalent z value. Since there is an exact equivalence between the raw scores (Y_1, Y_2) and the corresponding standardized scores (z_1, z_2), we may write

$$\Pr(Y_1 \leq Y \leq Y_2) = \Pr(z_1 < z < z_2)$$

That is, the area under the curve between Y_1 and Y_2 on the Y (raw score) scale is the same as the area between z_1 and z_2 on the corresponding standardized z scale. We obtain the probability that an observation falls between Y_1 and Y_2 by finding the area between z_1 and z_2 using the standard normal table (Table B in Appendix 3). We illustrate by returning to the birth weight example shown pictorially in Figure 4.5.

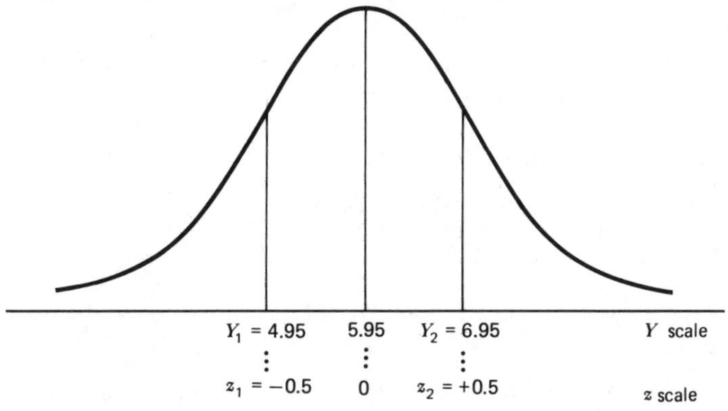

FIGURE 4.5 PICTORIAL REPRESENTATION OF Y SCORES AND z SCORES FOR THE BIRTH WEIGHT EXAMPLE

We wish to find the probability that an infant from this population will have a birth weight in the range 4.95 to 6.95, that is, Pr $(4.95 \leq Y \leq 6.95)$. We begin by converting the raw scores $Y_1 = 4.95$ and $Y_2 = 6.95$ to standardized scores.

$$z_1 = \frac{Y_1 - \mu}{\sigma} = \frac{4.95 - 5.95}{2} = -0.5$$

$$z_2 = \frac{Y_2 - \mu}{\sigma} = \frac{6.95 - 5.95}{2} = +0.5$$

Next, we must determine what information is provided by the standard normal table, Table B, in Appendix 3. In Table B the standard normal score z, tabulated to one decimal place, is given in the leftmost column; the second decimal place for z is given across the top of the table. The values given in the body of the table (four-digit numbers) are the areas under the standard normal curve between the mean ($\mu = 0$) and any observation z. Two features of the standard normal distribution will be utilized: (1) The standard normal curve is symmetric about 0 (i.e., the area between 0 and z_1 is the same as the area between 0 and $-z_1$); (2) 50% of the area of the curve lies above the mean and 50% of the area lies below the mean; that is, total area above 0 equals 0.5 and total area below 0 equals 0.5.

For the birth weight example depicted in Figure 4.5, we wish to find the area between $z_1 = -0.5$ and $z_2 = 0.5$. From Table B we see that the area between 0 (the mean) and 0.5 is .1915. We obtain this value by going down the first column labeled z until we reach $z = .5$. Since the second decimal place is 0, we moved over to the column headed 00 and read the value at the intersection, .1915. Due to symmetry, we know that the area between the mean 0 and $z_1 = -.5$ is also .1915. These areas are shown in Figure 4.6

We are interested in the total area between $z_1 = -.5$ and $z_2 = +.5$. Hence

$$\begin{aligned} \Pr(4.95 \leq Y \leq 6.95) &= \Pr(-.5 \leq z \leq .5) \\ &= .1915 + .1915 \\ &= .3830 \end{aligned}$$

That is, the probability that an infant from this population will have a birth weight between 4.95 and 6.95 pounds is .3830 or 38.3%.

EXAMPLE 4.6 For the birth weight example, find the probability that an infant from this population will have a birth weight greater than or equal to 8 pounds, that is, $\Pr(Y \geq 8)$.

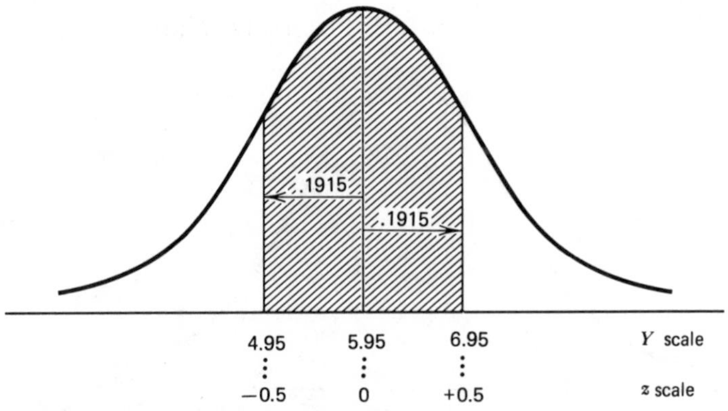

FIGURE 4.6 PICTORIAL REPRESENTATION OF AREAS UNDER THE NORMAL CURVE FOR THE BIRTH WEIGHT EXAMPLE

SOLUTION *Step 1*. Covert raw scores Y to standardized scores z.

$$z_1 = \frac{Y_1 - \mu}{\sigma} = \frac{8 - 5.95}{2} = 1.025$$

(We round to 1.03 since Table B is to two decimal places.)
Step 2. Illustrate the area pictorially.

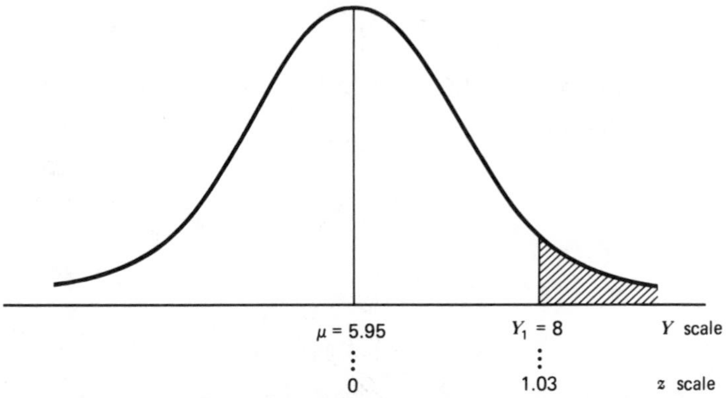

The area of interest is the shaded area above $Y_1 = 8$ or, equivalently, above $z_1 = 1.03$.
Step 3. Use Table B to obtain area. From Table B, we go down column 1 until we reach 1.0 and across to column .03. The value at the intersection, .3485, is the area be-

tween 0 and $z_1 = 1.03$. Pictorially, we have

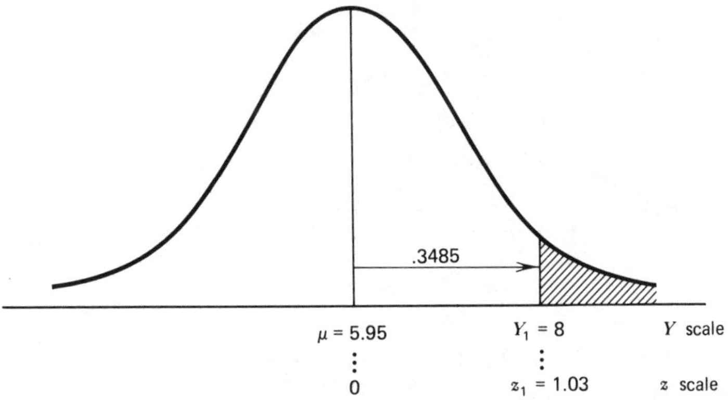

We wish to find the area *above* $z_1 = 1.03$ (shaded). To find this area, we recall that the *total area* above $z = 0$ (that is, the area from 0 to $+\infty$) is 0.5. Therefore, we obtain the area of the shaded region by subtracting the area from 0 to $z_1 = 1.03$ (.3485) from the total area above 0. We have

$$\begin{aligned} \Pr(Y_1 \geq 8) &= \Pr(z_1 \geq 1.03) \\ &= .5 - .3485 \\ &= .1515 \end{aligned}$$

EXAMPLE 4.7 In Example 4.6, find $\Pr(Y_1 \leq 8)$.

SOLUTION Pictorially, we have

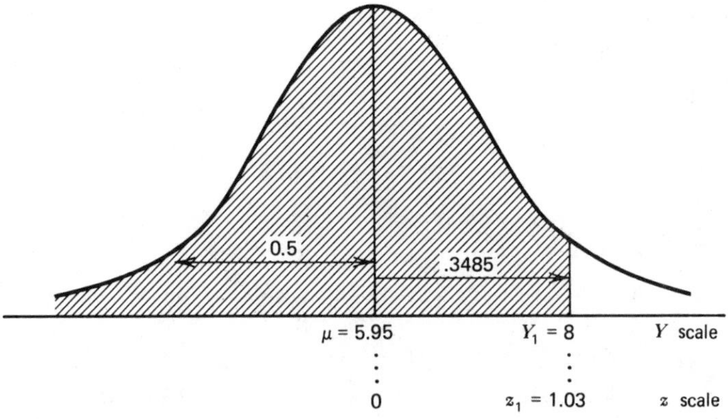

The area of interest is the area below $Y_1 = 8$ or, equivalently, *below* $z_1 = 1.03$. As before, the area between 0 and $z_1 = 1.03$ is .3485. The total area *below* 0 (from $\mu = 0$ to $-\infty$) is 0.5. Therefore,

$$\Pr(Y_1 \leq 8) = \Pr(z_1 \leq 1.03)$$
$$= .5 + .3485$$
$$= .8485$$

EXAMPLE 4.8 For the birth weight example, find the probability that an infant in this population will have a birth weight between 3.95 and 4.95 pounds, that is, $\Pr(3.95 \leq Y \leq 4.95)$.

SOLUTION *Step 1.* Convert raw scores Y to standardized scores.

$$z_1 = \frac{Y_1 - \mu}{\sigma} = \frac{3.95 - 5.95}{2} = -1.0$$

$$z_2 = \frac{Y_2 - \mu}{\sigma} = \frac{4.95 - 5.95}{2} = -0.5$$

Step 2. Illustrate pictorially.

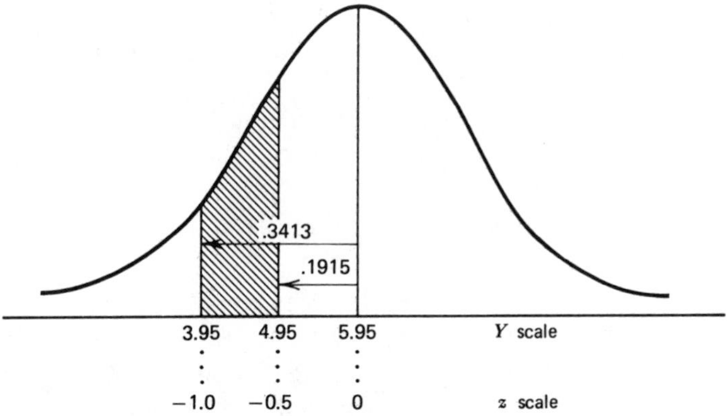

Step 3. Use Table B to find areas. Looking up $z_1 = -1.0$, we find the value .3413, the area between 0 and $z_1 = -1.0$. Similarly, the area between 0 and $z_2 = -0.5$ is .1915.

$$\Pr(3.95 \leq Y \leq 4.95) = \Pr(-1 \leq z \leq -.5)$$
$$= .3413 - .1915$$
$$= .1498$$

EXERCISE 4.9

For the birth weight example find

a. $\Pr(Y \geq 5.95)$
b. $\Pr(6.95 \leq Y \leq 7.95)$
c. $\Pr(Y \leq 3)$
d. $\Pr(3 \leq Y \leq 6)$

EXERCISE 4.10

Suppose a clinically accepted value for mean systolic blood pressure in males aged 20–24 is $\mu = 120$ mmHg, and the standard deviation is $\sigma = 20$ mmHg.

a. If a 22-year-old male is selected at random from the population, what is the probability that his systolic blood pressure is equal to or less than 150 mmHg? Equal to or less than 110?
b. The systolic blood pressure of a 20-year-old male selected at random from the population was 160 mmHg. How many standard deviations above the mean is this value? (Hint: z values are in units of standard deviations.)
c. Between what two blood pressure readings will 95% of all systolic blood pressure readings for 20 to 24-year-old males lie? Between what two values will 90% of the readings lie?
d. What proportion of the readings lie within the range 100–140 mmHg?
e. What proportion will be *outside* the range 60–180 mmHg?

†Sampling Distributions and the Central Limit Theorem*

DERIVED DISTRIBUTIONS

In previous sections the relationship between population parameters and sample statistics has been presented. A statistic is computed from a sample and is used as an estimate of the population parameter. Not only are statistics rarely if ever equal to their corresponding population parameters, but upon repeated random sampling under identical conditions the computed statistics will vary among themselves.

Suppose, for example, that an estimate of the average length of stay in the hospital of a certain type of hospitalized patient was desired. One

* This section is taken from Duncan, Knapp, and Miller, *Introductory Biostatistics*, 2nd ed., pp. 65–67. Copyright © 1977, John Wiley & Sons, Inc. Reprinted by permission of John Wiley & Sons, Inc.

approach would be to sample a certain number of hospital wards for the past calendar year and use the average of the sample values as the estimate. A second sample of the same size as the first would, in all likelihood, yield a different average value, as would further samples. Thus, the series of mean values from repeated samples of the same size would have a distribution of their own, a distribution derived from the original parent distribution.

The derived distribution of sample means (sometimes called the sampling distribution of the mean) is of unique importance in statistics. In order to gain insight into the meaning and utility of sampling distributions, let us first consider how the distribution of sample means for a given sample size n could be obtained. It is important to note that this will simply be an illustrative exercise; a sampling distribution is rarely generated in practice but a knowledge of its properties is essential for making statistical inferences discussed in the next chapter.

The steps for generating a sampling distribution of $\bar{Y}$'s are as follows:

1. Construct a population by recording population values on slips of paper and placing the slips in a container.
2. Select a sample of size n from the population and compute its mean $\bar{Y}$.
3. Replace these n observations in the population.
4. Continue steps 2 and 3 until a large number of samples of size n have been drawn.
5. Record the frequency of occurrence of each of the $\bar{Y}$'s. The resulting frequency distribution approximates the distribution of sample mean values, or more technically, the *sampling distribution of means*, for a given n.

As you may have perceived, the process of generating a complete sampling distribution becomes an impossible task for parent populations having more than a few observations. Fortunately, one does not have to rely on such formidable techniques to obtain information about the characteristics of every sampling distribution. In general, the information of value—the mean, the standard deviation, and the shape of the distribution when graphed—may be obtained mathematically. The mathematical derivation of these properties is beyond the scope of this text. The interested reader is referred to any theoretically oriented statistical text. We will simply state the mathematical conclusions as they are needed.

PROPERTIES OF THE DISTRIBUTION OF SAMPLE MEANS

The properties of the sampling distribution of the mean may be summarized as follows:

1. The mean $\mu_{\bar{y}}$ of the distribution of $\bar{Y}$ is equal to μ, the mean of the parent population from which the samples were drawn.
2. The standard deviation, $\sigma_{\bar{y}}$, of the sampling distribution of $\bar{Y}$ is equal to $\sigma/\sqrt{n}$, the standard deviation of the parent population divided by the square root of the number in each sample. The standard deviation of the distribution of $\bar{Y}$, called the *standard error of the mean*, is given in terms of the standard deviation of the individual values of the parent distribution. This allows the calculation of an estimate of the standard deviation of mean values from a single sample of individual observations.
3. The shape of the sampling distribution is that of the normal curve when sampling is from a parent population whose distribution is normal. The sampling distribution is *approximately* normal when sampling is from a nonnormal parent population. As the sample size increases, the approximation to normality becomes closer and closer.

This third property of the sampling distribution of $\bar{Y}$ is called the *Central Limit Theorem* and is of central importance to statistical methodology. Stated more formally, *the Central Limit Theorem is: given any parent population, not necessarily normal, having mean μ and standard deviation σ, the sampling distribution for fixed n which is generated from this population will be approximately normally distributed with mean, μ, and standard deviation $\sigma/\sqrt{n}$.* It is around this assumption of normality for the distribution of $\bar{Y}$ that much of statistical inference is built.

Utilizing the information provided by the Central Limit Theorem, we may now draw the curve depicting the distribution of sample means ($\bar{Y}$) of size n from a given population (Figure 4.7). You will note that the percentages given in Figure 4.7 are identical with those in Figure 4.3. For the frequency distribution of sample means ($\bar{Y}$), a unit of standard deviation is equal to $\sigma/\sqrt{n}$ rather than σ as for the parent population of measurements shown in Figure 4.3.

EXAMPLE 4.9 Refer to Example 4.6. Describe the frequency distribution of all possible sample means of size $n = 100$ from the

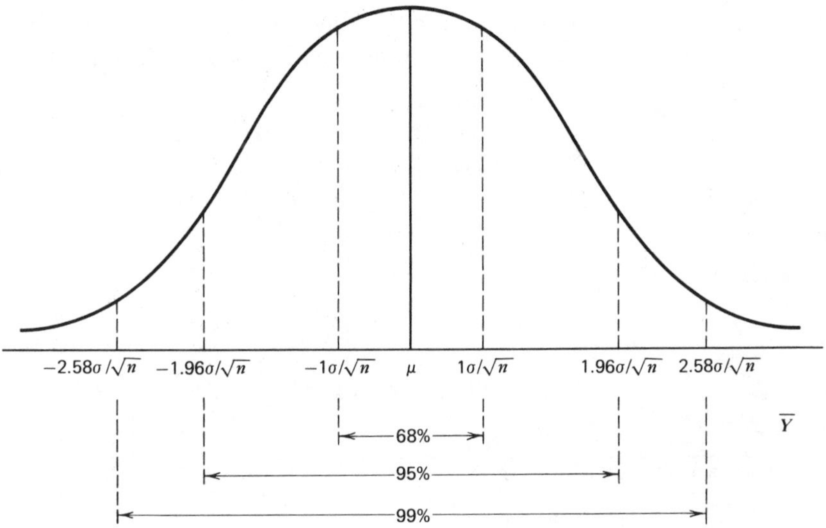

FIGURE 4.7 DISTRIBUTION OF SAMPLE MEANS $\bar{Y}$ OF SIZE n FROM A GIVEN POPULATION

given population of systolic blood pressure measurements.

SOLUTION The parent population of systolic blood pressure measurements as described in Example 4.6 has a mean μ equal to 140 mmHg and a standard deviation σ equal to 10 mmHg. According to properties of sampling distributions and the Central Limit Theorem, the population of sample means ($\bar{Y}$) of size $n = 100$ that could be drawn from this parent population is normally distributed with mean $\mu_{\bar{Y}}$ equal to the mean of the parent population and standard deviation $\sigma_{\bar{Y}}$ equal to the standard deviation σ of the parent population divided by the square root of the sample size. Hence,

$$\mu_{\bar{Y}} = \mu = 140 \text{ mmHg}$$

$$\sigma_{\bar{Y}} = \sigma/\sqrt{n} = 10/\sqrt{100} = 1 \text{ mmHg}$$

EXAMPLE 4.10 Refer to Example 4.9. Between what two sample *mean* values would we expect 95% of the $\bar{Y}$'s to fall? 99%?

SOLUTION The curve depicting the proportion of values found in the various ranges for the distribution of $\overline{Y}$ is shown below.

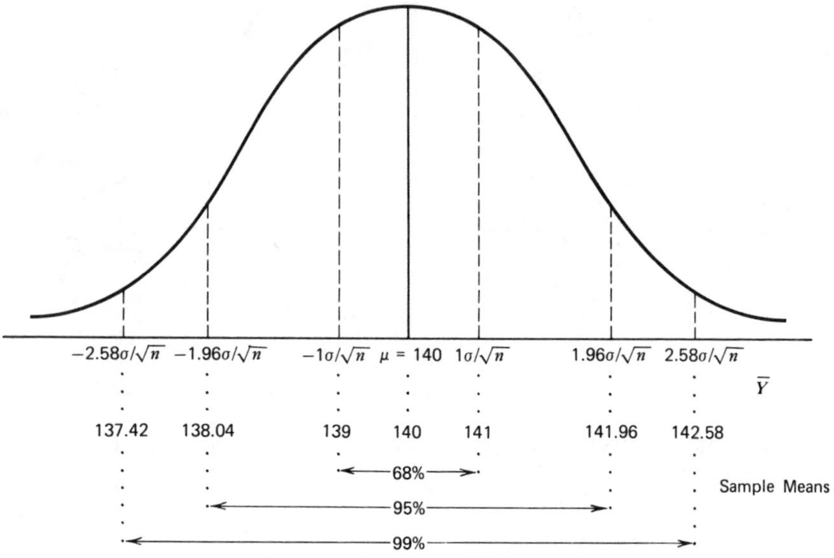

From the above curve, we see that 95% of the sample means of size $n = 100$ drawn from the given population would have values between 138.04 and 141.96 mmHg, that is:

$$\mu + 1.96\sigma/\sqrt{n} = 140 + (1.96)\ 10/\sqrt{100} = 141.96$$

and

$$\mu - 1.96\sigma/\sqrt{n} = 140 - (1.96)\ 10/\sqrt{100} = 138.04$$

Similarly, we would expect 99% of the sample means of size $n = 100$ to have values between 137.42 and 142.58 mmHg.

The concept of sampling distributions is often a difficult one to grasp especially for students in an introductory course. One application of the properties of sampling distributions will be presented in the next chapter on confidence intervals. Further applications will become apparent in the discussion of "Tests of Significance" in Chapters 6 through 12.

EXERCISE 4.11

As part of a screening project, the systolic blood pressures of 100 nursing students were taken. Assume that the population of blood pressures from which this sample was drawn is normally distributed with mean $\mu = 120$ mmHg and standard deviation $\sigma = 20$ mmHg. The mean systolic blood pressure for the sample of 100 students was 126 mmHg.

a. What is the probability that the *average* systolic blood pressure of a sample of $n = 100$ nursing students will be equal to or greater than 126 mmHg?
b. How many standard deviations above the true population mean μ is the observed sample mean $\overline{Y}$?
c. Between what two mean values would we expect 95% of all sample means (of size $n = 100$) to lie?
d. Above what value would 10% of the sample means (of size $n = 100$) lie?
e. What is the probability of selecting a sample of size $n = 100$ with mean $\overline{Y}$ equal to or less than 115 mmHg from a population with true mean systolic blood pressure $\mu = 120$ mmHg?

SUMMARY OF CRITICAL CONCEPTS IN CHAPTER 4

The following concepts were presented in Chapter 4:

1. Numerical quantities computed from a sample:
   a. may describe the sample itself (descriptive statistics).
   b. may provide inferences about the characteristics of the population from which the sample was collected (inferential statistics).
2. *Population* is the set of all measurements of interest to the investigator.
3. *Sample* is a subset of measurements selected from a given population.
4. *Target population* is the population about which one would like to make inferences.
5. *Population sampled* is the population from which the sample was taken. It is the population about which one can *statistically* make inferences.
6. Generalization from the population sampled to the target

population is subjective and open to controversy. The investigator must be assured that the characteristics of the two populations are identical.

7. A sample of measurements from a population rather than a census is usually taken because it is rarely feasible to measure every element of the population.
8. The objective in selecting a sample is to choose measurements that are "representative" of all measurements in the population.
9. *Empirical distribution* is a frequency distribution tabulated from a set of sample measurements.
10. Some knowledge of *theoretical distribution* is necessary for most classical statistical techniques. Theoretical distribution:
    a. is the frequency distribution for all measurements in the population.
    b. is rarely obtainable in practice.
    c. is approximated by the empirical distribution.
    d. depends on nature of variable being measured, i.e., may be continuous or discrete.
11. *Discrete data* are data in the form of integers resulting from a count of some sort. (Variable is referred to as a discrete variable.)
12. *Continuous data* are data which result from measuring a particular characteristic such as height, weight, etc., and which may take any value along a continuum.
13. Normal distribution:
    a. is continuous theoretical frequency distribution.
    b. represents the theoretical distribution of most naturally occurring phenomena, e.g., height, weight, blood pressure, cholesterol level, etc.
    c. has the following characteristics:
        (1) bell-shaped.
        (2) continuous measurements.
        (3) symmetric around its mean (μ).
    d. has its location of center determined by value of μ, spread determined by σ.
        (1) for small values of σ, curve is tall and thin.
        (2) for large σ, curve is short and fat.
    e. has 68% of values within 1 standard deviation of the mean; has 95% of values within 1.96 standard deviations of mean; has 99% of values within 2.58 standard deviations of mean.

14. The properties of the normal distribution provide the foundations and rationale for performing t tests and many other "tests of hypothesis."
15. *Parameter* is a numerical measure computed from the entire population.
    a. usually denoted by Greek symbols.
    b. μ = population mean; σ = population standard deviation.
    c. rarely obtainable in practice.
16. *Statistic* is a numerical measure computed from a set of sample measurements.
    a. usually denoted by English letters.
    b. $\overline{Y}$ = sample mean; s = sample standard deviation.
    c. is estimate of corresponding population parameter, e.g., $\overline{Y}$ estimates μ, s estimates σ.
17. The value of a statistic is rarely, if ever, exactly equal to the value of the corresponding parameter that it estimates.
18. Sampling error is the failure of a statistic to be exactly equal to its corresponding population parameter.
19. If many samples are taken and the statistic $\overline{Y}$ calculated for each sample, the values of $\overline{Y}$ will "cluster" around the true value of the parameter, μ.

†20. To find probabilities associated with the normal curve, we must first convert the raw scores Y to standard scores z (standard normal) by the formula

$$z = \frac{Y - \mu}{\sigma}$$

†21. Areas under the standard normal curve are found in Table B in Appendix 3.

†22. The Central Limit Theorem states: Given any parent population, not necessarily normal, having mean μ and standard deviation σ, the *sampling distribution* of $\overline{Y}$ for fixed n that is generated from this population will be normally distributed, or approximately so, with mean μ and standard deviation $\sigma/\sqrt{n}$.

†23. The standard deviation of the population of $\overline{Y}$'s is $\sigma/\sqrt{n}$, the *standard error of the mean*. The estimate of $\sigma/\sqrt{n}$ is $s/\sqrt{n}$ obtained from the sample.

CHAPTER 5

ESTIMATING POPULATION PARAMETERS

OVERVIEW

Inferential statistics, as defined in Chapter 1, is a set of procedures used to draw conclusions about a large body of data, called a *population,* based on a smaller set of data, a *sample,* taken from the population. There are two general areas of statistical inference: *estimation* and *hypothesis testing*. Procedures for estimating population parameters from information in a sample are discussed in this chapter. Statistical hypothesis testing is examined in Chapters 6–12.

OBJECTIVES

Upon completion of this chapter, the student will be able to:

1. Compute and interpret confidence intervals on a single population mean μ for
   a. population standard deviation σ known.
   b. population standard deviation σ unknown.
2. Compute and interpret confidence intervals on the difference in two population means $\mu_1 - \mu_2$ for
   a. population standard deviations known.
   b. population standard deviations unknown.
3. Compute and interpret confidence intervals on the true mean difference D in a paired experiment.
4. Compute and interpret confidence intervals on a true population proportion p.
5. Compute and interpret confidence intervals on the true difference in population proportions $p_1 - p_2$.

DISCUSSION

Confidence Interval for a Single Population Mean μ: Population Standard Deviation Known

A common problem encountered in most fields involving statistical investigation is that of estimating a population mean μ. For example, a nurse in nutritional research may wish to study the effect of a new diet in terms of weight gain for patients with a certain debilitating disease. The population of interest in this study is *all* patients with the given disease who are alive today, who may have lived in the past, and who may exist in the future. Because of the infinite nature of this population, it is impossible to examine the effect of diet for each member. Even for populations that are finite (that is, having bounds that may be specified, such as all patients in a particular hospital at a given time), the complete examination of all members may be prohibitive. To investigate the weight gain for patients on the diet, the investigator selects a sample of patients from the population. The information obtained from this subset of patients is used to estimate the parameter of interest, μ, the average weight gain for *all* patients in the population.

For example, suppose nine patients with a certain disorder are placed on the new diet and their weight gains measured at the end of a six-week period. At the end of this period, the investigator finds that the mean weight gain for the nine patients is 10.5 pounds. The sample mean, $\overline{Y}=$ 10.5 pounds, provides the best single guess as to the value of the true mean weight gain μ for all patients with the given disorder. In this case $\overline{Y} = 10.5$ is called a point estimate of μ, since it consists of a single value. It is known, however, from the discussion of sampling variation and sampling error that $\overline{Y}$ is rarely, if ever, equal to the true population mean μ. We use the sampling distribution of $\overline{Y}$, discussed in the preceeding chapter, to obtain an *interval* consisting of two numbers between which we expect the true population parameter μ to exist with a certain confidence or assurance level. This *interval estimate* is called a *confidence interval* for the population parameter.

From Figure 4.3 in Chapter 4, we know that 95% of all $\overline{Y}$'s for a given sample size will lie within a distance of $(1.96\sigma/\sqrt{n})$ on either side of μ. For the nutritional example discussed in the preceding paragraph, suppose it is reasonable to believe, based on previous research, that the true population standard deviation of weight gain measurements is $\sigma = 12$ pounds. Thus, 95% of the $\overline{Y}$'s of size $n = 9$ will have values within $1.96(\sigma/\sqrt{n}) = 1.96(12/\sqrt{9}) = 7.84$ units on either side of the true population mean μ, whatever its value. The values are shown on the curve in Figure 5.1.

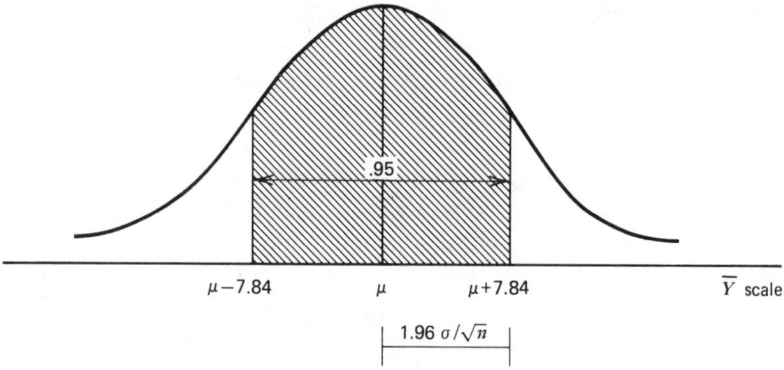

FIGURE 5.1 INTERVAL OF VALUES IN WHICH 95% OF WEIGHT GAIN MEAN VALUES (OF SIZE $n = 9$) ARE EXPECTED TO FALL

It must be emphasized that the parameter μ, although *unknown*, has fixed *constant* value. It is not subject to variation. The statistic $\overline{Y}$, on the other hand, may vary from sample to sample and hence is a random *variable*.

Consider the distance $(1.96\sigma/\sqrt{n})$ represented by the lower bar in Figure 5.1. If in the sampling process, a $\overline{Y}$ is obtained whose value falls on the horizontal axis under the shaded area, say at point 1 in Figure 5.2, and if we go a distance equivalent to the lower bar $(1.96\sigma/\sqrt{n})$ on either side of this $\overline{Y}$, then the resulting interval, specified by $\overline{Y}_1 \pm 1.96\sigma/\sqrt{n}$, will cover the true population mean μ. This is true since any sample mean that falls under the shaded area in Figure 5.2 will be less than $1.96\sigma/\sqrt{n}$ units from μ and the interval $\overline{Y} - 1.96\sigma/\sqrt{n}$ to $Y + 1.96\sigma/\sqrt{n}$ will contain μ. Similarly, if by chance our $\overline{Y}$ falls at point 2 in Figure 5.2, then the interval $\overline{Y}_2 \pm 1.96\sigma/\sqrt{n}$ will cover the true μ. However, if we obtain a $\overline{Y}$ on the horizontal axis under the *nonshaded* area, say at point 3, then the resulting interval $\overline{Y}_3 \pm 1.96\sigma/\sqrt{n}$ will *not* cover the true μ. Keep in mind that we do not know μ, hence we do not know, for any given sample, where on the horizontal axis of Figure 5.2 our particular $\overline{Y}$ falls. However, we do know that in repeated sampling, 95% of all $\overline{Y}$'s will fall on the horizontal axis under the shaded area. For example, if we take 100 samples of a given size n, we can expect approximately 95 of these 100 samples to yield $\overline{Y}$'s that do fall on the horizontal axis under the shaded area. Thus, if we take 100 samples and for each $\overline{Y}$ we obtain the interval $\overline{Y} \pm 1.96(\sigma/\sqrt{n})$, then we can expect approximately 95 of these 100 intervals to cover the true mean μ. Stated more generally, if we repeatedly sample from a normally distributed population, 95% of the intervals, $\overline{Y} \pm 1.96\sigma/\sqrt{n}$, will, in the long run, contain the true population parameter μ. In reality, however, we do not take multiple samples. We take a single sample and calculate a single interval $\overline{Y} \pm 1.96\sigma/\sqrt{n}$. In practical terms, we may extrapolate from the multiple sample case to conclude that, for the *single sample,* we are 95% confident that the single computed interval, $\overline{Y} \pm 1.96\sigma/\sqrt{n}$, contains the population mean μ.

The general form of a 95% confidence interval on μ is

$$\overline{Y} \pm 1.96\sigma/\sqrt{n}$$

In the nutrition example, the 95% confidence interval on the true population weight gain for all patients with the disorder is

$$\overline{Y} \pm 1.96\sigma/\sqrt{n}$$

$$10.5 \pm 1.96(12/\sqrt{9})$$

$$10.5 \pm 7.84$$

$$(2.66 \text{ to } 18.34)$$

Interpretation of Confidence Intervals on μ

In summary, confidence intervals may be interpreted in two ways. The first way is more theoretical in nature and involves a consideration of what happens in the long run. If a series of samples were obtained and 95% confidence intervals on the true mean μ constructed for each sample, then, in the long run, the relative frequency with which the computed intervals actually cover μ is 95%. A second, more useful interpretation, is that, for the single sample obtained in practice, we are 95% confident that

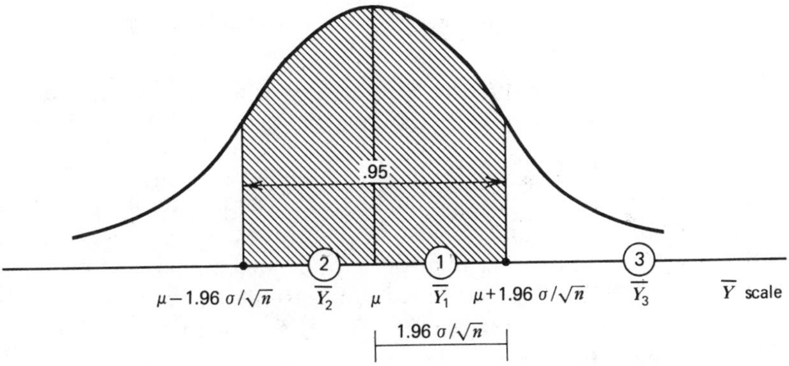

FIGURE 5.2 PICTORIAL DESCRIPTION OF CONFIDENCE INTERVALS

the interval $\overline{Y} \pm 1.96\sigma/\sqrt{n}$ computed from that particular sample covers the true population mean μ.

In the nutrition example, we are 95% certain that the interval (2.66 to 18.34) contains the true population weight gain for all patients with the given disorder who are on the special diet.

Other Confidence Intervals on μ

In general, any confidence interval on a population parameter may be written as follows:

(Estimate of parameter) ± (Confidence coefficient) ×
(Standard error of estimate)

In the preceeding discussion, the parameter of interest is μ, the population mean. The point estimate of μ is $\overline{Y}$, the sample mean. The standard error of the estimate $\overline{Y}$ is $\sigma/\sqrt{n}$, and, finally, the confidence coefficient for a 95% confidence interval is 1.96. If we are interested in a 99% confidence interval or a 90% confidence interval, the confidence coefficient would be 2.58 and 1.645, respectively. All other values in the expression are unchanged. Any desired confidence coefficient may be obtained from Table B in Appendix 3 by dividing the desired confidence level by 2 and locating this number in the body of the table. For example, to obtain the 95% confidence coefficient, we locate .95/2 = .4750 in the body of Table B. This value is located opposite the row 1.9 (under the first column heading z and the column headed .06). The values located down the first column labeled z give the confidence coefficients (z values) to one decimal place. The second decimal place (hundredth place) is across the remaining columns. Thus, since .4750 lies in the intersection of row 1.9 and column .06, the corresponding confidence coefficient is 1.96. Verify that 2.58 and 1.645 are the confidence coefficients for 99% and 90% confidence intervals, respectively.

Choice of a particular confidence interval is arbitrary and depends on the level of confidence desired by the investigator in a given situation.

EXAMPLE 5.1 In a study to determine the effect on pulse rate of a particular drug, each of 25 healthy men aged 24 to 38 years was given a single oral dose of 200 mg of the drug under study. Two hours later their pulse rates in beats per minute were taken, and the mean pulse rate for the group was found to be 68 beats per minute. The clinically accepted standard deviation σ of pulse rates is 10 beats per minute. Place a 99% confidence interval on the true mean pulse rate for all men aged 24 to 38 years who are on the medication. Interpret the interval.

SOLUTION The 99% confidence interval on the true population mean μ is given by

$$\overline{Y} \pm 2.58\sigma/\sqrt{n}$$

$$68 \pm 2.58 \times 10/\sqrt{25}$$

$$68 \pm 5.16$$

$$(62.84 \text{ to } 73.16)$$

Interpretation: We are 99% confident that the interval (62.84 to 73.16) contains the true mean pulse rate μ for all men aged 24 to 38 years who are on the medication.

EXERCISE 5.1

In a study to determine the effect of a statistics exam on stress of senior nursing students, the systolic blood pressures of 16 randomly selected students were taken immediately after a statistics midterm exam. The mean systolic blood pressure for the 16 students was found to be 150 mmHg. Assume that the standard deviation σ of systolic blood pressures is equal to 20 mmHg. Place a 95% confidence interval on the true mean systolic blood pressure for all senior nursing students following the statistics exam. Interpret this interval.

Confidence Intervals on a Single Population Mean μ Using the t-Distribution: Population Standard Deviation Unknown

The confidence intervals discussed in the previous section involved the population standard deviation σ. However, in practice the value of σ is rarely known. The most logical solution to the problem of an unknown σ is to use the estimate of the standard deviation s calculated from the sample. The obvious problem with the substitution of s for σ is that in addition to the variability of $\overline{Y}$, s now also varies with each sample. This additional variability in s is taken into account by use of the Student-t distribution (in common usage simply the t-distribution). (The values 1.96, 2.58, etc. used in the formula for confidence intervals in the previous section are called z values and are obtained from the standard normal or z-distribution.)

The t-distribution is a symmetric bell-shaped distribution centered on a mean of 0. In addition, the exact shape of the t-distribution depends on a value called degrees of freedom. For the case of estimating a single population mean μ, degrees of freedom for the t-distribution is defined to be $n - 1$. We use the concept of degrees of freedom for the t-distribution in obtaining values of the confidence coefficient from the table of t-values given in Table A of Appendix 3.

The rationale, methodology, and interpretation of confidence intervals are the same for both the t- and z-distributions. For a confidence interval on μ using t, the estimate s is substituted for σ and an appropriate value for t is obtained from a table of t values (Table A). The formula is

$$\bar{Y} \pm t(s/\sqrt{n})$$

As before, $\bar{Y}$ is the point estimate of μ as specified in the general formula for confidence intervals. However, for this case, the standard error of the estimate $\sigma/\sqrt{n}$ is approximated by $s/\sqrt{n}$, and the confidence coefficient is represented by the value t. To obtain a value for t in Table A for a 95% confidence interval on μ, we locate the value at the intersection of the column $(1 - .95)/2 = .025$ and the row labeled $df = n - 1$, where n is the size of the sample used in the study. Pictorially, the values in the body of the t table are points on the horizontal axis above which (1 − confidence level)/ 2 × 100% of the area of the curve lies. For a 95% confidence interval on μ when a sample of size 10 has been drawn, the confidence coefficient is $t = 2.2622$.

EXAMPLE 5.2 A study was conducted to evaluate the stress level of senior nursing students at a particular college of nursing. Ten students were selected at random from the senior class, and their stress level was monitored by attaching an electrode to the frontalis muscle (forehead). For the ten students the average EMG (electromyogram) activity was found to be 35.8 microvolts. In addition, for the ten students, the standard deviation of the EMG readings was found to be 2.5 microvolts. Place a 99% confidence interval on true mean EMG activity for all seniors in the class.

SOLUTION The population standard deviation σ is not known; hence, the t-distribution is appropriate for the interval

$$\bar{Y} \pm t(s/\sqrt{n})$$

The 99% confidence coefficient, obtained from Table A, column $(1 - .99)/2 = .005$, row $n - 1 = 9$, is 3.2498. The 99% confidence interval is then given by

$$35.8 \pm 3.2498(2.5/\sqrt{10})$$

Interpretation: We are 99% confident that the interval 33.2 to 38.4 covers the true mean EMG activity for the entire class.

EXERCISE 5.2

In a study to determine the effect, if any, of the mother's diet on the birth weight of infants, an investigator randomly selected nine women who were in their third month of pregnancy and placed them on a strict diet. At birth, the weights of the nine infants were recorded as

5.5, 6.0, 7.0, 5.5, 8.0, 7.0, 8.0, 5.5, 6.0

Place a 95% confidence interval on the true mean birth weight of infants for the population of mothers on the special diet. Interpret the interval.

Confidence Intervals on the Difference in Two Population Means: Population Standard Deviations Known

The previous sections of this chapter have described the rationale and procedure for placing a confidence interval on a single population mean μ. In nursing, the situation often arises in which we are interested in estimating the true *difference* in *two* population means $\mu_1 - \mu_2$. For example, a study carried out by Barsevick and Llevellyn (1982)* evaluates the relaxation effects of a towel bath versus a conventional bed bath for a group of patients experiencing unrelieved pain. A group of 26 patients randomly assigned to the towel bath group had a postbath mean score of 32.56 on the State-Trait Anxiety Inventory (STAI), while the 27 patients randomly assigned to the conventional bath procedure had a postbath mean score of 37.89. As a nurse researcher you may be interested in estimating the average difference in postbath anxiety score for the two types of baths for the whole population of patients experiencing unrelieved pain based on the information you have obtained for the 53 patients in the study.

In this example, the population parameter of interest is the difference, $\mu_1 - \mu_2$, where μ_1 is the true mean for the first population and μ_2 is the true mean for the second population. The statistic that serves as a point estimate of the population difference $\mu_1 - \mu_2$ is the difference in *sample* means $\overline{Y}_1 - \overline{Y}_2$. Without proof, we state that if we were to draw all possible samples of size n_1 from the first population whose size is N_1 and all possible samples of size n_2 from the second population whose size is N_2 and from these samples compute all possible differences $\overline{Y}_1 - \overline{Y}_2$, then

* Barsevick, A. and Llevellyn, J. A comparison of the anxiety-reducing potential of two techniques of bathing. *Nursing Research,* 1982, Jan–Feb; *31*(1): 22–27.

the frequency distribution of these differences is that of the normal distribution with mean $\mu_1 - \mu_2$ and standard deviation $\sqrt{(\sigma_1^2/n_1) + (\sigma_2^2/n_2)}$. Stated more formally, the *sampling distribution* of the differences $\overline{Y}_1 - \overline{Y}_2$ of independently drawn samples is normally distributed with mean $\mu_1 - \mu_2$ and standard deviation $\sqrt{(\sigma_1^2/n_1) + (\sigma_2^2/n_2)}$. This definition applies in a strict sense only when each of the two populations from which we are sampling has a normal distribution. However, if the two sampled populations are not normally distributed, or if the form of the frequency distributions of the populations is not known, the sampling distribution of the differences $\overline{Y}_1 - \overline{Y}_2$ is at least approximately normally distributed with mean $\mu_1 - \mu_2$ and standard deviation $\sqrt{(\sigma_1^2/n_1) + (\sigma_2^2/n_2)}$ for sample sizes n_1 and n_2 *large*. In this latter case, when the sample sizes n_1 and n_2 are large, procedures described for normally distributed populations are applicable.

Recall from the previous section that the general form of a confidence interval on a population parameter is as follows:

(Estimate of parameter) ± (Confidence coefficient) ×
(Standard error of estimate)

The parameter to be estimated is the population difference $\mu_1 - \mu_2$. The *estimate* of this parameter is the sample difference $\overline{Y}_1 - \overline{Y}_2$. By definition, the *standard error* of the estimate $\overline{Y}_1 - \overline{Y}_2$ is the standard deviation of the sampling distribution of $\overline{Y}_1 - \overline{Y}_2$, that is, $\sqrt{(\sigma_1^2/n_1) + (\sigma_2^2/n_2)}$. As was the case for confidence intervals on single population means, the confidence coefficient is a value from the standard normal table (Table B) corresponding to the desired level of confidence for the interval estimate. Hence, the 95% and 99% confidence intervals on the true population difference $\mu_1 - \mu_2$ are, respectively,

$$(\overline{Y}_1 - \overline{Y}_2) \pm 1.96\sqrt{\frac{\sigma_1^2}{n_1} + \frac{\sigma_2^2}{n_2}}$$

$$(\overline{Y}_1 - \overline{Y}_2) \pm 2.58\sqrt{\frac{\sigma_1^2}{n_1} + \frac{\sigma_2^2}{n_2}}$$

Interpretation: We may be 95% (or 99%) confident that the interval covers the true difference in population means $\mu_1 - \mu_2$.

EXAMPLE 5.3 In the example cited earlier in this section, a study was carried out to compare the effects of a towel bath versus a conventional bed bath for a group of patients experiencing unrelieved pain. For the 26 patients randomly assigned to the towel bath group, the mean postbath score on the STAI was 32.56, and for the 27 patients in the conventional bath group, the mean post-bath STAI score was 37.89. Suppose, based on previous knowledge of STAI scores for this type of situation, it was known (hypothetically) that the standard deviations $\sigma_1 = 10$ and $\sigma_2 = 12$. (Later, we will discuss the procedure to be used if the population parameters σ_1, σ_2 are not known). One objective of the investigation is to estimate the *true* difference in mean postscore for all patients of this type receiving towel bath and conventional bath.

SOLUTION The 95% confidence interval is an interval estimate of the true population difference in mean post-STAI scores for the two groups.

$$(\overline{Y}_1 - \overline{Y}_2) \pm 1.96\sqrt{\frac{\sigma_1{}^2}{n_1} + \frac{\sigma_2{}^2}{n_2}}$$

$$(32.56 - 37.89) \pm 1.96\sqrt{\frac{(10)^2}{26} + \frac{(12)^2}{27}}$$

$$-5.33 \pm 1.96(3.03)$$

$$(-11.27 \text{ to } 0.61)$$

Interpretation: We are 95% confident that the interval (−11.27, 0.61) covers the true difference in mean postbath STAI scores for the two populations studied.

EXERCISE 5.3

The serum indirect bilirubin levels were determined for five premature and five healthy full-term infants, as follows:

Premature	*Full-term*
1.0	2.0
2.0	4.0
3.0	6.0
2.0	4.0
2.0	4.0

It is reasonable to assume that the two populations have equal standard deviations, that is, $\sigma_1 = \sigma_2 = 1.0$ mg/100 cc. Place a 99% confidence interval on the true difference in mean serum indirect bilirubin levels for premature and full-term infants for the population studied. Interpret the interval.

Confidence Intervals on the Difference in Two Population Means: Population Standard Deviations Unknown

In the previous section describing confidence intervals on differences in population means, one major assumption that is made is that the population standard deviations σ_1, σ_2 are *known*. Rarely in the real world do we have knowledge of such population values. Usually, if the population means μ_1 and μ_2 are unknown, so are the population standard deviations σ_1 and σ_2. In the absence of knowledge of the true values of σ_1 and σ_2, the investigator must resort to use of their *estimators*, s_1 and s_2, the sample standard deviations. Recall that the sample standard deviation is calculated by the formula

$$s = \sqrt{\frac{\sum_{i=1}^{n} Y_i^2 - \frac{\left(\sum_{i=1}^{n} Y_i\right)^2}{n}}{n - 1}}$$

As was described for the single sample case, when the estimates s_1 and s_2 are used in place of the actual parameter values σ_1 and σ_2, this additional uncertainty must be taken into account by using the t-distribution (Table A) rather than the standard normal distribution (Table B) for finding the value of the confidence coefficient. Additionally, in the two sample case, another assumption must be made. We must now require that the values σ_1 and σ_2, although unknown, be *equal in value*.* This assumption is not unreasonably restrictive, since, in many situations, populations may differ with respect to their averages yet have essentially the same inherent variability among observations within the populations.

The general form of a confidence interval on the difference in two population means $\mu_1 - \mu_2$ is as before:

(Estimate of parameter) ± (Confidence coefficient) ×
(Standard error of estimate)

* There are statistical tests available for testing the assumption of equality of two population variances. Readers are referred to R. G. Steel and J. H. Torrie, *Principles and Procedures of Statistics,* 2nd Ed. (New York: McGraw Hill, 1980); and Wayne Daniel *Biostatistics: A Foundation for Analysis in the Health Sciences,* 2nd Ed. (New York: John Wiley and Sons, 1978).

The estimate of the parameter $\mu_1 - \mu_2$ is again the difference in sample means $\bar{Y}_1 - \bar{Y}_2$. The standard error of the estimate, using the estimates s_1 and s_2 in place of σ_1 and σ_2 and using the previous section as a guide, would be

$$\sqrt{\frac{s_1^2}{n_1} + \frac{s_2^2}{n_2}}$$

However, we have made the assumption that $\sigma_1 = \sigma_2 = \sigma$. Hence both s_1 and s_2 estimate the common unknown population standard deviation σ. Rather than use either s_1 or s_2 as the estimate of σ, we can "pool" information gathered from both samples to obtain an improved estimate of the common σ. This "pooled" estimate of σ is denoted by s_p and is computed by the formula

$$s_p = \sqrt{\frac{(n_1 - 1)s_1^2 + (n_2 - 1)s_2^2}{n_1 + n_2 - 2}}$$

While it is not easy to see by looking at this formula, s_p is actually a kind of "weighted" average of the two estimates s_1 and s_2.

The standard error of the estimate needed in the general formula for a confidence interval is $\sqrt{(s_p^2/n_1) + (s_p^2/n_2)}$, where s_p replaces both s_1 and s_2. This may be rewritten as $s_p\sqrt{(1/n_1) + (1/n_2)}$.

The confidence coefficient is obtained from Table A in the same manner as described for the one sample case. Calculation of degrees of freedom (df) for the two sample cases is a straightforward extension of the one sample case, that is

$$df = (n_1 - 1) + (n_2 - 1) = n_1 + n_2 - 2$$

The appropriate value t from Table A is located at the intersection of column (1 − confidence level)/2 and row $df = n_1 + n_2 - 2$. Thus, a 95% confidence interval for the difference $\mu_1 - \mu_2$ when two samples of size $n_1 = 10$ and $n_2 = 12$ are used is given by

$$(\bar{Y}_1 - \bar{Y}_2) \pm t\left(s_p\sqrt{\frac{1}{n_1} + \frac{1}{n_2}}\right)$$

$$(\bar{Y}_1 - \bar{Y}_2) \pm 2.086\ s_p\sqrt{\frac{1}{10} + \frac{1}{12}}$$

where $\bar{Y}_1$, $\bar{Y}_2$, s_1, s_2, and s_p are computed from the samples.

EXAMPLE 5.4 A study was carried out to determine the effect of parental participation in care on the anxiety level of hospitalized children. Eighteen hospitalized children aged 6 to 10 years were selected. Nine children were randomly assigned to the experimental group and nine to the control group. The parents of children in the experimental group were allowed to participate in the routine care of the child, while for the control group the parents were allowed to visit as usual but all care of the children was carried out by the nurse. At the end of a one-week period, all of the children were given a standardized 100-point anxiety exam. The mean anxiety score for the experimental group was 45 and the standard deviation s_1 was 20. The mean anxiety score for the control group was 55 and the standard deviation s_2 was 25. The investigator wishes to estimate, with 99% certainty, the true population mean difference in anxiety scores for the experimental and control groups.

SOLUTION First, the quantity s_p must be calculated where

$$s_p = \sqrt{\frac{(n_1 - 1)s_1^2 + (n_2 - 1)s_2^2}{n_1 + n_2 - 2}}$$

$$= \sqrt{\frac{8(20)^2 + 8(25)^2}{9 + 9 - 2}}$$

$$= 22.6$$

The 99% confidence interval on the true mean difference in scores is given by

$$(\bar{Y}_1 - \bar{Y}_2) \pm t\left(s_p\sqrt{\frac{1}{n_1} + \frac{1}{n_2}}\right)$$

$$(45 - 55) \pm 2.9208(22.6)\sqrt{\frac{1}{9} + \frac{1}{9}}$$

$$(-41.12 \text{ to } 21.12)$$

Interpretation: We are 99% confident that the interval (−41.12 to 21.12) covers the true population difference in mean anxiety scores for the experimental and control groups. Notice that the interval ranges from negative to positive values. Thus the study does not provide us with strong evidence that one group performed on the average better than the other group.

EXERCISE 5.4

A public health nurse was interested in determining the effect, if any, of pesticide exposure on blood pressure in adults. From a list of people known to have been exposed to pesticides, a random sample of 100 people was selected. Similarly, from a group of people for whom no exposure could be detected, a random sample of 100 was selected. The mean systolic blood pressure for the pesticide group was found to be 145 mmHg with a standard deviation s of 20 mmHg. The mean systolic blood pressure for the nonexposed group was found to be 120 mmHg with a standard deviation of 15 mmHg. Place a 90% confidence interval on the true difference in mean systolic blood pressure for the exposed and nonexposed groups for the populations studied. Interpret the interval.

Confidence Interval on Differences in Population Means for a Paired Study

The *paired* study is a research design often used in nursing. Pairing can be accomplished in three ways: (1) by giving an individual one treatment (or no treatment) on one occasion and another treatment on another occasion; (2) by using twins or animal litter mates and giving one twin or litter mate one treatment and the other twin or litter mate the second treatment; or (3) by artificially matching two individuals with respect to variables suspected of having an effect on the study variable and assigning one member of the pair to treatment 1 and the other to treatment 2. Pairing in a research design is used to control extraneous variables that effect the response being measured. The paired study is considered a two-group design even though in situation (1) above *one* group of individuals receives two treatments (or treatment and control). More will be said on the paired design in Chapters 7 and 8.

In a paired study, it is meaningful to consider the differences d_i of the two measurements for each of the individuals in the study. Furthermore, we are often interested in estimating the true mean D of all the differences, $d_i = Y_{1i} - Y_{2i}$, in the population being studied. The confidence interval approach may again be used. The general form of this confidence interval is

$$\bar{d} \pm t(s_d/\sqrt{n})$$

where $\bar{d}$ is the sample estimate of the true population difference D, t is the confidence coefficient from Table A ($df = n - 1$), s_d is the standard deviation of the n differences $Y_{1i} - Y_{2i}$ obtained from the sample, and $s_d/\sqrt{n}$ is the standard error of the estimate. In actuality, this confidence interval is identical to the one sample confidence interval on a population mean μ, except in this case the n observations in the sample are the n differences, $d_i = Y_{1i} - Y_{2i}$.

EXAMPLE 5.5 In a study to determine the effectiveness of a back rub in lowering the systolic blood pressure of geriatric patients, nine nursing home patients were selected. The systolic blood pressures of the nine patients were taken before a back rub and then again immediately after a 10-minute back rub. The investigator wishes to estimate the true average difference in before and after blood pressure measurements for the population of patients being studied. The data are as follows:

Systolic Blood Pressure (in mmHg)

Patient	*Before Back Rub*	*After Back Rub*	*Differences*
1	140	130	−10
2	160	150	−10
3	180	180	0
4	140	150	+10
5	160	140	−20
6	180	160	−20
7	150	150	0
8	170	130	−40
9	180	180	0

$$\bar{d} = \frac{-90}{9} = -10$$

The standard deviation s_d is

$$s_d = \sqrt{\frac{2700 - \dfrac{(90)^2}{9}}{8}} = 15$$

SOLUTION The 95% confidence interval on the true population difference D is then

$$\bar{d} + t(s_d/\sqrt{n})$$

$$-10 \pm 2.306(15/\sqrt{9})$$

$$(-21.53 \text{ to } 1.53)$$

Interpretation: With 95% assurance, we may state that the interval (−21.53 to 1.53) mmHg covers the true mean difference in before and after blood pressure measurements for the population of patients being studied. Further, since this interval contains the value 0, we do not have evidence to suggest that there is a statistically significant difference in the two sets of measurements. (More will be said about the interpretation of the term *statistically significant* in Chapter 8.)

EXERCISE 5.6

In order to determine the effect of a certain oral contraceptive on weight gain, nine healthy women were weighed before the start of the medication and again at the end of a three-month period. The weights are listed below.

	Weight (in pounds)	
Subject	*Initial*	*3 Months*
1	120	123
2	141	143
3	130	140
4	150	145
5	135	140
6	140	143
7	120	118
8	140	141
9	130	132

Place a 95% confidence interval on the true mean difference in weight gain before and at the end of the three-month period for the population studied. Interpret.

Confidence Interval on a Single Population Proportion

Often the nurse researcher is interested in estimating the proportion of patients who possess a particular characteristic, for example, the proportion of patients who have hospital-induced infections, the proportion of patients dying from a given illness, or the proportion of patients who experience relief when receiving a particular nursing procedure or medication. More generally, when the level of measurement of the response variable being studied is nominal, one of the few statistics that may be reported is proportion of individuals falling in various categories of the variable. For example, for the nominal variable marital status, we may report the proportion of married, single, widowed, or divorced individuals in our sample. Similarly, for ordinal data or for discrete or continuous equal-interval data that have been categorized, the proportion of individuals in the various categories is often of interest. For example, for the continuous variable systolic blood pressure, we may be interested in the proportion of individuals who have systolic blood pressure readings greater than 160 mmHg. Further, we may be interested in estimating, with a specified level of confidence, the true proportion of individuals having a certain characteristic in the population being studied. To do this, we employ a confidence interval of the form

$$\hat{p} \pm (\text{confidence coefficient}) \sqrt{\frac{\hat{p}(1 - \hat{p})}{n}}$$

The value $\hat{p}$ is an estimate of the population proportion p, and $\sqrt{\hat{p}(1 - \hat{p})/n}$ is an estimate of the standard error of $\hat{p}$. Without proof, we state that $\hat{p}$ has a sampling distribution that is approximately normally distributed with mean p and standard error $\sqrt{p(1 - p)/n}$ when both np and $n(1 - p)$ are larger than 5. Note that, in this section, sample statistics wear "hats" (ˆ), for example, $\hat{p}$, and population parameters wear no "hats," for example, p. Since the sampling distribution of $\hat{p}$ is approximately normally distributed, we may obtain the confidence coefficient from the table of standard normal values (Table B).

EXAMPLE 5.7 A student nurse carried out a study to estimate drug usage among nursing students in a particular college of nursing. She randomly selected 50 student nurses and asked them to answer anonymously the question "Have you ever experimented with drugs during your time at this college?" 15 of the 50 students in the sample responded "Yes" to the question.

SOLUTION A 95% confidence interval on the true proportion of nursing students at the given college who have experimented with drugs is given by

$$\hat{p} \pm 1.96\sqrt{\frac{\hat{p}(1-\hat{p})}{n}}$$

$$\frac{15}{50} \pm 1.96\sqrt{\frac{\left(\frac{15}{50}\right)\left(\frac{35}{50}\right)}{50}}$$

(.17 to .43)

Interpretation: We are 95% confidence that the true proportion of students who have experimented with drugs at this institution is between 17% and 43%.

EXERCISE 5.7

A nurse involved in the planning of an alternate birthing center for a community wished to estimate the proportion of women in the community who would use such a facility for the birth of a child. A survey was sent to 200 randomly selected women in the community, and, of these, 90 responded that they would use the facility. Place a 95% confidence interval on the true proportion of women in the community who would respond in favor of the center. Interpret the interval.

Confidence Interval for the Difference in Population Proportions

A nurse researcher may be interested in estimating the difference in the five-year survival rates for patients having two different types of cancer, or in estimating the difference in proportion of patients who experience relief from back pain for two different back rub procedures. Both examples may employ a confidence interval on the difference in population proportions having the form

$$(\hat{p}_1 - \hat{p}_2) \pm (\text{Confidence coefficient}) \sqrt{\frac{\hat{p}_1(1-\hat{p}_1)}{n_1} + \frac{\hat{p}_2(1-\hat{p}_2)}{n_2}}$$

So long as the true population proportions p_1 and p_2 are not close to 0 or 1 and the sample sizes n_1 and n_2 are large, we may obtain the confidence coefficient from Table B, the standard normal table.

EXAMPLE 5.8 In a study carried out to compare the proportion of patients experiencing relief from back pain for two different types of procedures, 80 out of 100 patients receiving Method 1 experienced relief while 50 out of 100 patients receiving Method 2 experienced relief. The investigator wished to estimate with 99% confidence the true difference in proportion of patients experiencing relief for the two methods.

SOLUTION The 99% confidence interval on the true difference in proportions is

$$(\hat{p}_1 - \hat{p}_2) \pm 2.58\sqrt{\frac{\hat{p}_1(1 - \hat{p}_1)}{n_1} + \frac{\hat{p}_2(1 - \hat{p}_2)}{n_2}}$$

$$(.8 - .5) \pm 2.58\sqrt{\frac{(.8)(.2)}{100} + \frac{(.5)(.5)}{100}}$$

$$(.13 \text{ to } .47)$$

Interpretation: We are 99% confident that the interval (13% to 47%) covers the true difference in proportion of patients experiencing relief for the two methods studied.

EXERCISE 5.8

A nurse wished to investigate the effects of an audiovisual teaching program designed to train diabetics to administer self-medication. Two groups of patients were randomly selected to participate in the study. The experimental group was asked to participate in the two-hour special training program, while the control group did not participate. Each patient was then asked to administer his or her own medication with a self injection of insulin. A team of nurses who were unaware of the patients' study group status evaluated each patient's performance as either "satisfactory" or "nonsatisfactory." For the 100 patients in the experimental group, 75 had satisfactory responses while 50 out of 90 patients in the control group had satisfactory responses. Place a 95% confidence interval on the true difference in proportion of satisfactory responses for the two groups. Interpret.

SUMMARY OF CRITICAL CONCEPTS

1. There are two general areas of statistical inference:
   a. Estimation
   b. Hypothesis testing
2. One method of estimating population parameters using information in the sample is the *confidence interval* method.

3. A confidence interval may be interpreted in two ways:
   a. If a series of samples were obtained and 95% confidence intervals on the true parameter constructed for each sample, then, in the long run, the relative frequency with which the computed intervals actually cover the true parameter is 95%.
   b. For the single sample obtained in practice, using a 95% confidence interval, we are 95% confident that the confidence interval computed from that particular sample covers the true population parameter.
4. The general form of any confidence interval is as follows:

   (Estimate of parameter) ± (Confidence coefficient) × (Standard error of estimate)

5. A confidence interval on the true population mean μ when the population standard deviation σ is known is

   $$\overline{Y} \pm z(\sigma/\sqrt{n})$$

   where z, the confidence coefficient, is a standard normal table value (Table B).
6. A confidence interval on the true population mean μ when the population standard deviation σ is not known is

   $$\overline{Y} \pm t(s/\sqrt{n})$$

   where t, the confidence coefficient, is a value from the t-distribution (Table A).
7. A confidence interval on the true difference in population means $\mu_1 - \mu_2$ when the population standard deviations σ_1 and σ_2 are known is

   $$(\overline{Y}_1 - \overline{Y}_2) \pm z\sqrt{\frac{\sigma_1^2}{n_1} + \frac{\sigma_2^2}{n_2}}$$

8. A confidence interval on the true difference in population means $\mu_1 - \mu_2$ when the population standard deviations σ_1 and σ_2 are *not* known is

   $$(\overline{Y}_1 - \overline{Y}_2) \pm t\left(s_p\sqrt{\frac{1}{n_1} + \frac{1}{n_2}}\right)$$

where s_p is the pooled standard deviation

$$s_p = \sqrt{\frac{(n_1 - 1)s_1^2 + (n_2 - 1)s_2^2}{n_1 + n_2 - 2}}$$

9. A confidence interval on the true mean difference D in a *paired* or matched study is

$$\bar{d} \pm t(s_d/\sqrt{n})$$

where $\bar{d}$ and s_d are the average and standard deviation of the sample differences, respectively.

10. A confidence interval on the true proportion of individuals in a population having a specified characteristic is

$$\hat{p}_1 \pm z\sqrt{\frac{\hat{p}(1 - \hat{p})}{n}}$$

where $\hat{p}$ is the proportion in the sample having the characteristic and the confidence coefficient z is from the standard normal table (Table B).

11. A confidence interval on the true difference in proportions of individuals with a specified characteristic for two populations is

$$(\hat{p}_1 - \hat{p}_2) \pm z\sqrt{\frac{\hat{p}_1(1 - \hat{p}_1)}{n_1} + \frac{\hat{p}_2(1 - \hat{p}_2)}{n_2}}$$

SUPPLEMENTARY EXERCISES

EXERCISE 5.9

A nurse midwife wished to compare the mean age of women who retained intrauterine devices (IUD's) and the mean age of those who expelled the devices. Fifteen subjects were randomly selected from among a group who had retained IUD's and 10 subjects from among those who had expelled IUD's. The data are shown below:

Expelled	*Retained*
$\bar{Y}_E = 28$	$\bar{Y}_R = 33$
$s_E = 10$	$s_R = 12$
$n_E = 10$	$n_R = 15$

Place a 95% confidence interval on the true difference in mean ages for the group who expelled IUD's and the group who retained IUD's for the populations studied. Interpret the interval.

EXERCISE 5.10

Nine randomly selected senior nursing students were given a standardized pretest on medical-surgical nursing knowledge prior to their taking a special clinical elective. After the six-week course, the students were given a posttest covering the same material. The pre- and posttest scores are given below.

Student	*Pretest*	*Posttest*
1	80	85
2	75	90
3	85	85
4	60	75
5	95	98
6	70	75
7	65	70
8	75	85
9	90	80

Place a 99% confidence interval on the true mean difference in pre- and posttest scores for the population of nursing students studied. Interpret the interval.

EXERCISE 5.11

In a study to estimate the effect on infants who are held by their parents immediately after birth, a nurse recorded the rectal temperatures of 100 infants immediately after they had been held for five minutes by their parents. The mean rectal temperature for the sample was found to be 98.02° F and the standard deviation was found to be .9° F. Place a 95% confidence interval on the true mean rectal temperature of the population of infants studied. Interpret the interval.

EXERCISE 5.12

Using a standard nursing audit instrument, nursing performance was evaluated in two hospitals. In both hospitals 15 patients were selected at random and their nursing care was evaluated by an audit committee. The mean nursing performance score for hospital 1 was found to be 85 with a standard deviation of 10, and the mean performance score for hospital 2 was 70 with a standard deviation of 10. Place a 99% confidence interval on the true difference in mean audit performance scores for the two hospitals studied. Interpret the interval.

EXERCISE 5.13

A study was carried out to compare smoking habits of freshmen and senior nursing students at a particular college of nursing. Random samples of 20 freshmen and 20 senior nursing students were selected. Each student participant was asked "Are you a smoker (of cigarettes or cigars) or nonsmoker?" Twelve of the freshmen were smokers while eight of the seniors were smokers. (a) Place a 95% confidence interval on the true proportion of smokers in the freshman class at this college of nursing. (b) Place a 95% confidence interval on the true proportion of smokers in the senior class at this college. (c) Place a 95% confidence interval on the true difference in proportion of smokers in the freshman and senior classes at this college. Interpret all three intervals.

CHAPTER 6

STATISTICAL INFERENCE: AN OVERVIEW

OVERVIEW

All of us are faced with decisions that must be made in both our professional and our personal lives. At home, we must decide which detergent washes more dishes, which oil cooks "grease-free," or which candidate to vote for in an election. In making these decisions, hopefully we "look at the relevant facts" (which statisticians refer to as observations, or data) and weigh the alternatives before making an inference or conclusion about the outcome of the decision. As a professional, you are called upon to make decisions that affect patients' welfare. The mechanism for reaching a decision about which nursing procedure is most beneficial or which drug is more effective also involves looking at the facts (data), weighing the alternatives, and arriving at a conclusion. However, experience has shown that most people are unable to scan large amounts of data, mentally weigh all relevant information, and reach a sound conclusion simply by relying on their own internal decision-making devices. The statistical decision-making process, called statistical inference, enables an investigator to view a decision-making situation in terms of the probability of making a correct or an incorrect decision and then to proceed accordingly. In this chapter, the general steps for testing a statistical hypothesis will be presented.

OBJECTIVES

Upon completion of this chapter, the student will be able to:

1. List the elements of a statistical test and describe in words what is meant by:
    a. Null hypothesis
    b. Test statistic

   c. Level of significance
   d. Rejection region
   e. Decision or conclusion
2. Identify the above elements in a given research situation.

DISCUSSION

Statistical Inference: A Test of Hypothesis

One process of statistical inference is called *hypothesis testing*. Hypothesis testing is concerned with making a decision about the value of a parameter by following a set of commonly accepted decision guides.

The reasoning behind a statistical test of hypothesis is basic: For some reason, we think that a population parameter has a specific value; we observe a sample of measurements from the population and calculate the corresponding statistic that estimates the parameter. If the difference between the "hypothesized" value of the parameter and the value calculated for the statistic is "large," then what we have observed in the sample contradicts what we assumed was true of the population. In this case, it must be concluded that the initial assumption about the value of the population parameter was incorrect. Stated more simply, to make an inference about the value of a population parameter, we (1) hypothesize that the population parameter of interest has a certain value; (2) observe a sample of measurements from the population and calculate the corresponding statistic; and (3) make a decision about the value of the parameter based on the discrepancy between the hypothesized value of the parameter and the computed value of the statistic. For "large" discrepancies we conclude that our initial assumption about the parameter is in error. The actual statistical techniques for deciding whether the observed discrepancy is large enough to warrant rejection of the hypothesis will be developed in subsequent chapters.

The following are examples in which statistical hypothesis testing would be used:

EXAMPLE 6.1 A public health nurse interested in determining the effect, if any, of pesticide exposure on blood pressure in adults selects a random sample of individuals known to have been exposed to pesticides and a random sample of individuals for whom no exposure could be detected. It is of interest to determine whether the mean blood pressure for all exposed individuals is different from that for nonexposed individuals.

EXAMPLE 6.2 A nurse midwife selected 100 female IUD users from the records of an Ob-Gyn clinic. These individuals were classified according to uterine sounding size and whether or not the IUD had been expelled. The investigator wished to determine whether there was an association between uterine size and whether or not an IUD was expelled.

EXAMPLE 6.3 In order to evaluate the curriculum of a particular nursing school, a study was carried out to determine whether a significant relationship existed between final college grade point average and performance on a state board examination.

EXAMPLE 6.4 A hospital administrator wished to determine whether the mean nursing costs per patient per day were different for two particular intensive care units.

While the specific methodologies for the above situations are different, the underlying principles and general procedure are the same for all. The purpose of this chapter is to introduce these general concepts of hypothesis testing through their application in a specific situation. The principles of hypothesis testing will be illustrated by considering the case of tests of hypothesis about the mean μ of a single population with known standard deviation σ. (Note: The purpose of this chapter is *not* to develop specific methodology for the z-test but rather to illustrate general concepts of hypothesis tests through their application in the special case.)

A Test of Hypothesis About a Mean μ, σ Known

EXAMPLE 6.5 Refer to Example 4.4. It was stated that the parameter of interest is the mean birth weight of all infants with Disorder X. Suppose that the investigator knows that the mean birth weight for all normal infants is 7 pounds with a standard deviation σ of 2 pounds, and wishes to determine whether infants with Disorder X have the same mean birth weight as normal infants. For a random sample of 25 infants, suppose the mean birth weight is calculated to be 6.2 pounds.

The question which the investigator must answer is "Does the population of infants with Disorder X have the same mean birth weight as the population of normal infants?" To answer this question by using statistical inference, we must first hypothesize that the birth weights of infants with Disorder X do not differ from the birth weight of normal infants; that is, we hypothesize that there is *no difference* in the mean birth weight of normal infants and those with Disorder X. This initial assumption about the value of a population parameter is called the *null hypothesis* and is denoted symbolically by H_0, read "H nought." In general, the investigator is seeking to discredit the null hypothesis.

The null hypothesis for Example 6.5 is that there is no difference in the mean birth weight of the population of normal infants and the mean birth weight of the population of infants with Disorder X. Written symbolically, this is:

$$H_0: \quad \mu_x = 7 \text{ pounds}$$

where the symbol μ_x represents the true population mean birth weight for all infants with Disorder X. The investigator seeks to discredit this statement in favor of the alternative that the mean birth weight for infants with Disorder X is different from the mean birth weight of normal infants. This last statement is called the *alternate hypothesis* and is written symbolically as:

$$H_A: \quad \mu_x \neq 7$$

(The symbol $\neq$ is read "not equal to.")

In general, the two types of hypotheses of interest to a researcher are *research hypotheses* and *statistical hypotheses*. The research hypothesis is the question to be answered by the research stated in a declarative form. For example, a nurse investigator may believe that patients' anxiety levels and hence blood pressure will be lower after a 10-minute back rub. The research hypothesis for this study would be stated "blood pressures of patients will be lowered (relative to pre-back rub levels) after a 10-minute back rub procedure." The statement reflects what the researcher suspects will result from the administration of the experimental treatment. It is the supposition to be tested by the study.

Statistical hypotheses, the *null* and *alternate* hypotheses, follow directly from research hypotheses. The statistical hypotheses have a special form dictated by the principles of statistical hypotheses testing. In general, the *alternate* statistical hypothesis reflects the research hypothesis. It is the supposition posited by the investigator stated in statistical terminology or symbolism. Thus, the choice of alternate hypothesis in a statistical analysis depends upon how the investigator has stated the research hypothesis of the study. The null hypothesis, on the other hand, is usually a hypothesis of "no difference." Simply speaking, the purpose of the null hypothesis is to play the role of "devil's advocate" relative to the research, or alternate, hypothesis. In the birth weight example being discussed, the research hypothesis is that birth weights of infants with Disorder X are different from birth weights of nondiseased infants. Hence the alternate hypothesis is stated as "the mean birth weight of infants with Disorder X is *not the same* as for nondiseased infants" ($\mu_x \neq 7$). The rationale of statistical hypothesis testing says, first, let us suppose (hypothesize) the opposite of the research (alternate) hypothesis is true, that is, "there is no difference in mean birth weight for infants with Disorder X and nondiseased infants" (the null hypothesis). Then we will attempt to discredit this statement in favor of the alternate hypothesis based on information gathered from the sample of infants with Disorder X. At this point it may not be clear to the student why we must test the null hypothesis rather than dealing directly with the hypothesis of interest, the alternate hypothesis. As we proceed through the remaining chapters on hypothesis testing, we hope the rationale will become apparent.

The particular research hypothesis about the effect of Disorder X on birth weight indicates that, given the current state of knowledge of the disorder, the investigator did not wish to speculate as to whether it increased or decreased birth weight relative to nondiseased infants. In some cases, however, at the outset of the study the investigator is interested in a directional hypothesis. For example, when studying the effects of back rub on patients' blood pressure levels, the investigator's research hypothesis is that blood pressures after a 10-minute back rub are *lower* than blood pressures before the back rub. The null hypothesis for this example is "there is *no difference* in mean blood pressures of patients before and after a 10-minute back rub" (H_0: $\mu_{\text{after}} = \mu_{\text{before}}$). The alternate hypothesis is "the mean blood pressure of patients after a 10-minute back rub is lower than that before the back rub" (H_a: $\mu_{\text{after}} < \mu_{\text{before}}$). When the alternate hypothesis has a direction, as in the back rub example, the statistical analysis is modified to take it into account. The statistical test is called a *one-tailed test*, as opposed to a *two-tailed test*, appropriate when nondirectional alternate hypotheses are used. One-tailed statistical tests are described in detail in Chapter 8.

It must be emphasized that both research and statistical hypotheses are stated *before* the collection of any data. We do not look at the results of our sample to decide the direction of the research and hence alternate statistical hypothesis. This decision is made based on the investigator's knowledge of the problem being studied and the nature of the question to be answered.

After the statistical hypotheses have been stated, the next step in a statistical test of hypothesis is to obtain a sample of measurements and to compute the sample statistic which corresponds to the parameter of interest in the study. For example, if the hypothesis concerns the value of a population mean μ, the corresponding sample mean $\overline{Y}$ is computed. Based on the principles of sampling error, it is known that the value of $\overline{Y}$ is rarely, if ever, equal to the population mean μ. In Example 6.5, we have hypothesized that the mean birth weight of the population of infants with Disorder X is equal to 7 pounds; the value of $\overline{Y}$, the mean birth weight of a random sample of 25 infants with Disorder X, was found to be 6.2 pounds. This discrepancy between the hypothesized value $\mu_x = 7$ and the observed $\overline{Y} = 6.2$ may be explained by one of two statements:

1. The sample mean $\overline{Y}$ is not equal to the hypothesized μ because of random chance (sampling error).
2. The sample mean $\overline{Y}$ is not equal to the hypothesized μ because the *true* population μ for infants with Disorder X *is different from* the population birth weight for normal infants and the sample reflects this difference.

Values of $\overline{Y}$ that are "close to" the hypothesized μ suggest that the first statement is the correct explanation; while values of $\overline{Y}$ that are "far away" from μ suggest that the second is the appropriate reason.

To decide which of the above two alternatives is the "most likely" explanation of the difference between the sample mean $\overline{Y}$ and the hypothesized mean μ, we calculate a mathematical quantity called a *test statistic*. Values for the test statistic quantify the magnitude of the discrepancy between the hypothesized population parameter and the corresponding statistic computed from the sample. "Large" values of the test statistic suggest that Statement 2 is the more likely explanation of the discrepancy between the hypothesized parameter and the calculated statistic, while "small" values favor Statement 1 as the rationale for the difference.

The test statistic that is appropriate for a test of hypothesis about the mean μ of a single population with known standard deviation σ, as in Example 6.5, is defined as:

$$z = \frac{\overline{Y} - \mu}{\sigma/\sqrt{n}}$$

Substituting the appropriate values from Example 6.5, we have:

$$z = \frac{6.2 - 7.0}{2/\sqrt{25}}$$

$$= \frac{-.8}{.4} = -2.0$$

This value obtained for the test statistic z tells us that the calculated sample mean $\overline{Y} = 6.2$ lies two standard deviations below the hypothesized mean $\mu_x = 7$ pounds. We must now decide whether this value of the test statistic ($z = -2$) is "large" enough to warrant discrediting the null hypothesis (that is, whether Statement 2 is the most likely explanation of the difference between $\overline{Y}$ and μ).

To make this decision regarding what constitutes "sufficiently large" values of the test statistic, we will specify a *rejection region for the test statistic*. When the value of the test statistic is "small," sampling error is a likely explanation for the discrepancy between the hypothesized population parameter and the calculated sample statistic. As the discrepancy between the observed sample statistic and the hypothesized parameter becomes larger, i.e., as the test statistic becomes larger, sampling error becomes a less likely explanation for the difference. The rejection region specifies those values beyond which sampling error is no longer accepted as the explanation for the discrepancy between the calculated sample statistic and the hypothesized parameter. Thus, values of the test statistic which fall into the rejection region imply that the true mean of the population is different from that specified by H_0 (Statement 2 above). Alternatively, values of the test statistic that do not fall into the rejection region suggest that sampling error or random chance is the explanation for the difference, and the null hypothesis cannot be rejected.

All possible values for the test statistic for the case where H_0 is true may be pictured as points on the horizontal axis of the curve of the frequency distribution of the test statistic.

Different types of test statistics have different frequency distributions. The test statistic z is known to have a normal frequency distribution (with mean $\mu = 0$ and $\sigma = 1$). The values of z on the horizontal axis that fall under the shaded areas of Figure 6.1 constitute the rejection region for the test of hypothesis. The values of the test statistic z that fall on the horizontal axis under the shaded areas of the curve are those values that are *less likely* to occur if the null hypothesis is true. Thus, when the calculated value of the test statistic falls into the rejection region we will conclude that it is unlikely that the null hypothesis is true. If the value of the test statistic is on the horizontal axis under the nonshaded area, we will conclude that sampling error (random chance) is a likely explanation of the discrepancy between the hypothesized population parameter and the observed sample statistic. In this case, the null hypothesis will not be rejected. It should be noted that hypothesis testing procedures cannot lead to "proof" of a hypothesis. Procedures of statistical inference merely indicate whether the available data support or do not support a particular hypothesis. For this reason, the terminology "accept the null

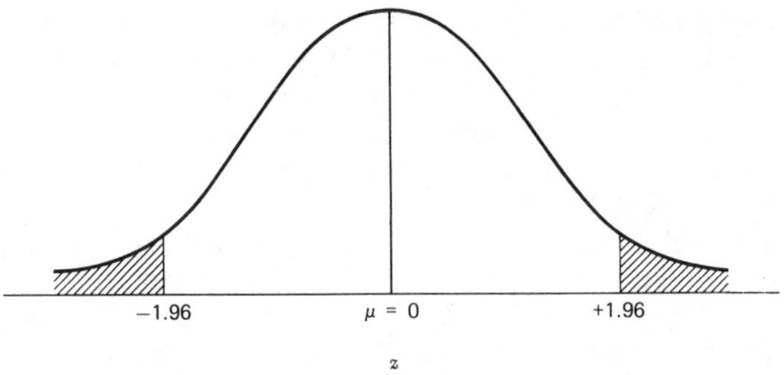

−1.96
μ = 0
+1.96
z

FIGURE 6.1 FREQUENCY DISTRIBUTION OF TEST STATISTIC Z

hypothesis'' is avoided since it implies proof that the hypothesis is true. Rather, we speak either of ''rejecting'' or ''not rejecting'' the null hypothesis.

In summary, the rejection region for a test of hypothesis is shown in Figure 6.2

The values of the test statistic that constitute the rejection region depend on the *level of significance of the test* chosen by the investigator. The level of significance, denoted by the symbol α (''alpha''), specifies the likelihood of incorrectly rejecting the null hypothesis when, in fact, it is true. To minimize the chances of falsely rejecting H_0, we select for the level of significance small values such as .05 and .01. In Figure 6.1, a rejection region corresponding to an α = .05 level of significance is shown. The curve shows the frequency distribution of all possible values

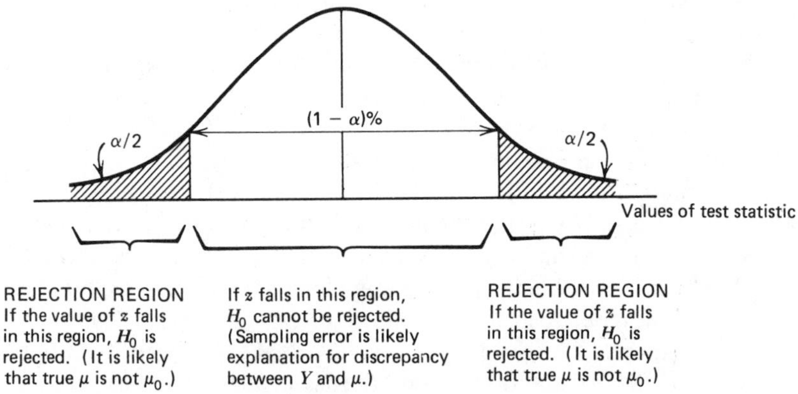

α/2
(1 − α)%
α/2
Values of test statistic
REJECTION REGION
If the value of z falls in this region, H0 is rejected. (It is likely that true μ is not μ0.)
If z falls in this region, H0 cannot be rejected. (Sampling error is likely explanation for discrepancy between Y and μ.)
REJECTION REGION
If the value of z falls in this region, H0 is rejected. (It is likely that true μ is not μ0.)

FIGURE 6.2 DIAGRAM SHOWING THE REJECTION REGIONS FOR A TEST OF HYPOTHESIS USING THE z STATISTIC

of the test statistic *given that* H_0 *is true*. The values that fall on the horizontal axis under the shaded portion of the curve have only a 5% chance of occurring in the population whose true mean is as specified by H_0. All values of the test statistic that fall under the shaded portion of the curve are regarded as contradictory to H_0 because they have a small likelihood of occurrence (5%) if H_0 is true. However, 5% of the possible values for the test statistic from the population specified by H_0 (H_0 true) will fall in the rejection region. When the decision is made to reject H_0, we are "playing the odds"; but "chances are" that the test statistic is from a population whose true mean μ is different from that specified by H_0 (H_0 false). Thus, when H_0 is rejected using a 5% ($\alpha = .05$) level of significance, there is a 5% risk of error. That is, there is a 5% chance that the null hypothesis has been rejected when in fact it is true.

An investigator may wish to risk only a 1% chance of rejecting the null hypothesis when it is true. The rejection region corresponding to an $\alpha = .01$ level of significance is shown in Figure 6.3. We would reject the null hypothesis for those values of the test statistic z that lie beyond $\pm$ 2.58 (under the shaded area).

The term level of significance accounts for the fact that tests of hypothesis are commonly referred to as *tests of significance*, and that a computed value of the test statistic is said to be "significant" if it falls in the rejection region.

The level of significance is an arbitrary choice depending on the implications of making an incorrect decision about H_0. In drug investigations where patient welfare may be affected by putting a new drug on the market, the level of significance should be very strict (at least .01). We want to be very sure that a correct decision is made. Often, when the

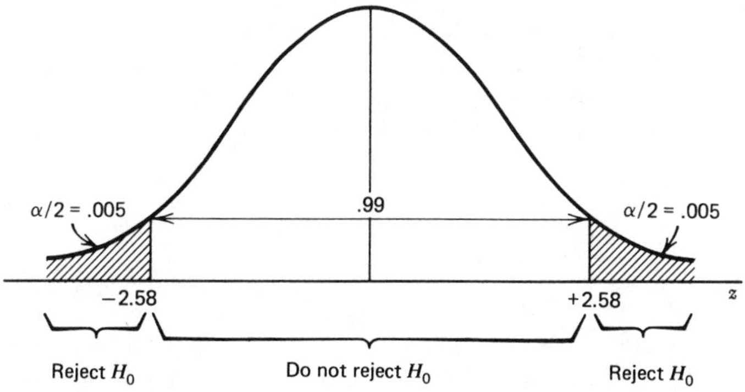

FIGURE 6.3 REJECTION REGION FOR z STATISTIC FOR A 1% LEVEL OF SIGNIFICANCE

consequences are not so dramatic, less strict values, say .05 or .10, may be selected. A more detailed discussion of choice of level of significance is given at the end of this chapter.

In the presentation of any research report, it is vital that the level of significance be reported. This allows the research consumer to weigh the odds of an incorrect decision and to decide about the validity of the conclusion. For example, a study reports that a new nursing procedure is significantly more effective in relieving decubitus ulcers than a currently used one with a level of significance of .20. This means that in saying that the new procedure is better, the investigator is willing to risk a 20% chance of being wrong. You may be unwilling to accept these odds and therefore decide against incorporating this new procedure into your care plan. Without knowing the level of significance, you do not have enough of the facts to make a decision about the reported work.

Returning to Example 6.5, suppose the investigator wished to test at the $\alpha = .05$ level of significance. The calculated value of the test statistic, $z = -2.0$, falls on the horizontal axis under the shaded portion of the curve in Figure 6.1. It is an *unlikely* value if H_0 is indeed true. The discrepancy between the hypothesized population mean birth weight $\mu_x =$ 7 pounds and the observed value of the sample mean birth weight $\overline{Y} = 6.2$ is large enough to warrant rejection of H_0; that is, it must be concluded that there is a difference in mean birth weight for normal infants and those with Disorder X ($\mu \neq 7$ pounds) at the $\alpha = .05$ level of significance. The $\alpha = .05$ level of significance states that we risk a 5% chance of error in rejecting the null hypothesis.

Up to this point, the discussion of rejection region for a statistical test has supposed a nondirectional alternate hypothesis, that is, one containing a "not equal to" ($\neq$) statement rather than the direction greater than (>) or less than (<). Figures 6.1 to 6.3 are appropriate for the former case, that is, for a *two-tailed* test of hypothesis. Suppose, in contrast, for the birth weight example the investigator suspected that Disorder X results in a lower birth weight for infants as compared to nondiseased infants. The statistical hypotheses in this case would be

$$H_0\text{: } \mu_x = 7$$

$$H_a\text{: } \mu_x < 7$$

Intuitively, we know that values of the sample mean $\overline{Y}$ much smaller than seven pounds would be evidence against the null hypothesis and in favor of the alternate hypothesis. Therefore, the rejection region for the test should consist of only *small* negative values of the test statistic, while the nonrejection region should consist of larger values of the test statistic.

Pictorially, the *one-tailed* rejection region for testing H_0: $\mu_x = 7$ versus H_a: $\mu_x < 7$ consists of all points on the horizontal axis below the shaded area in the lower tail of the frequency distribution of the test statistic as shown by the shaded area in Figure 6.4. Notice that for the one-tailed rejection region, the whole α (in this case $\alpha = .05$) is placed in the single tail and the cutoff point for the rejection region determined accordingly. The rejection region consists of all values of the test statistic less than -1.645. In contrast to Figure 6.4, the two-tailed rejection regions as shown in Figures 6.1 to 6.3 have both upper and lower tails reflecting the fact that either large *or* small values of the test statistic refute H_0: $\mu_x = 7$ in favor of H_a: $\mu_x \neq 7$. It is not important at this point that the student be able to obtain the cutoff points for the rejection region from the appropriate standard normal table. Accordingly, a detailed discussion of the methodology for obtaining cutoff values has not been given. As stated at the outset, the purpose of this chapter is to introduce the rationale and discuss general concepts of hypothesis testing. Detailed descriptions of how to carry out specific statistical tests of hypothesis and to recognize the situations in which they apply are given in subsequent chapters.

p Values

Often, researchers report the exact probability p of obtaining a sample value equal to or more extreme than that observed, assuming the null hypothesis is true. The symbol α is used to specify a predetermined level of significance for the test. Before the data in the study are collected, the investigator may state that he or she is willing to risk a 5% chance of erroneously rejecting H_0 when it is true. Here, $\alpha = .05$. The symbol p, on

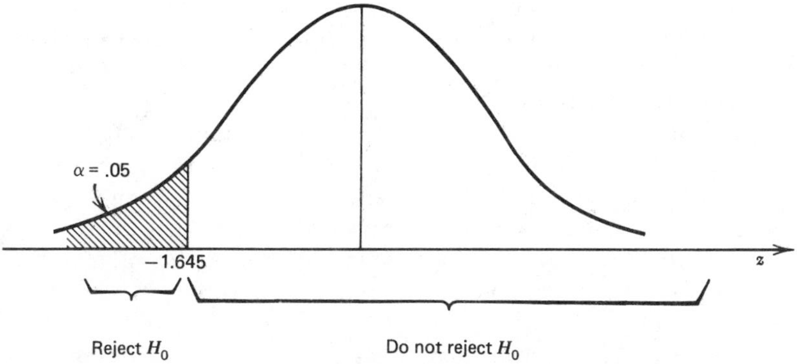

FIGURE 6.4 REJECTION REGION FOR A LOWER-TAIL TEST FOR THE z STATISTIC FOR A 5% LEVEL OF SIGNIFICANCE

the other hand, is the actual probability of obtaining the calculated value of the test statistic if H_0 is true. Thus, when $p \leq .01$ is reported in conjunction with a test of significance, it indicates that the probability of calculating the particular test statistic *given that* H_0 *is true* is less than 1%. Thus, the likelihood of getting the particular sample results from a population in which H_0 is true simply by random chance is less than 1%. Owing to lack of necessary tables, it is sometimes difficult to compute the exact value of p. In practice, researchers usually select an α level in advance of data collection, and, then, if the test statistic falls in the corresponding rejection region, report that $p \leq \alpha$.

Review

The general components of a test of hypothesis as illustrated by the above discussion are:

1. *Null hypothesis*. The null hypothesis is an initial claim about the value of a population parameter, generally stated in terms of "no difference" from some known population parameter. The hypothesis that is "accepted" as true when the null hypothesis is rejected is called the alternate hypothesis. The alternate hypothesis reflects the research hypothesis of interest in the study.
2. *Test statistic*. The test statistic is a measure of the magnitude of the disparity between the hypothesized value of a population parameter and the corresponding value of a statistic computed from a sample.
3. *Rejection region*. The rejection region for a test of hypothesis is a region whose values indicate contradiction of the null hypothesis. The size of the rejection region is determined by the desired *level of significance*, α.
4. *Decision or conclusion*. The null hypothesis is rejected at the given level of significance when the value of the test statistic is deemed to be "unlikely" when H_0 is true, that is, when it falls in the rejection region. Otherwise, we must state that the data do not provide enough evidence to discredit the null hypothesis. For values of the test statistic that do not fall in the rejection region, the null hypothesis is "not rejected." To "accept" the null hypothesis implies erroneously that we know the true value of a population parameter.

In Chapters 8, 9, 10, and 11, the most common tests of significance will be introduced. Each chapter will include:

1. Conditions under which the test is valid.
2. Overall description of test methodology.
3. Computational procedure, when meaningful.
4. Interpretation of results.

†Statistical Decisions and Their Outcomes*

Up to this point, the discussion of tests of hypotheses has supposed a preselected value of α, the level of significance of the test. In statistical decision making, the decision maker has the opportunity to select α. In doing so, he or she is determining how often an observation will be called unusual when in fact it is not unusual at all. Specifically, when a value for α, say 5 percent, is chosen, then in the long run, if the null hypothesis is true, it will be falsely rejected 5 percent of the time. For example, suppose we are testing the simple hypothesis

$$H_0: \quad \mu = \mu_0$$

versus

$$H_a: \quad \mu > \mu_0$$

where μ is the mean of a normally distributed population and σ^2 is known. The decision rule for the test for $\alpha = .05$ calls for rejection of the null hypothesis when the test statistic $z = (\bar{Y} - \mu_0)/(\sigma/\sqrt{n})$ is greater than 1.65 (Fig. 6.5).

When H_0 is true ($\mu = \mu_0$), we would expect 5 percent of the sample means ($\bar{Y}$), and, hence, 5 percent of the values of the test statistic, to fall on the horizontal axis in the rejection region (shaded area) for the test. Thus, we would expect to reject H_0 5 percent of the time when H_0 is really true. This type of incorrect decision, that is, rejecting a true H_0, is called a *Type I error*, and the probability of committing a Type I error is α. We have

$$\Pr(\text{reject } H_0 | H_0 \text{ true}) = \Pr(\text{Type I error}) = \alpha$$

The above is read "the probability that we reject H_0 given that H_0 is true"

* The following three sections are taken from Duncan, Knapp, and Miller, *Introductory Biostatistics For The Health Sciences*, 2nd Ed., pp. 102–111. Copyright © 1983, John Wiley & Sons, Inc. Reprinted by permission of John Wiley & Sons, Inc.

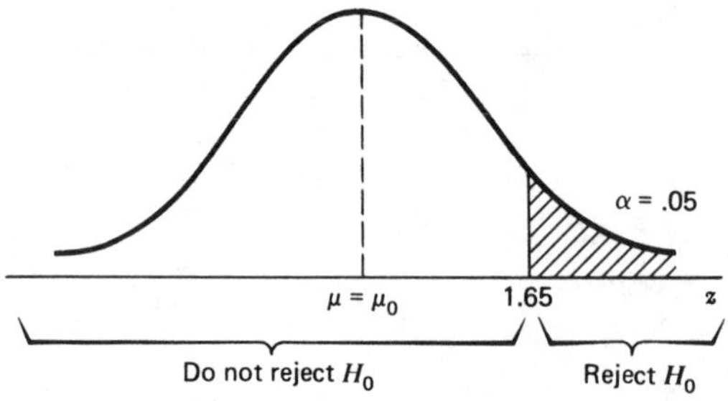

FIGURE 6.5 REJECTION REGION FOR α = .05 (ONE SIDED-TEST)

is α. An important concept that must be remembered is that we can never know the true "state of nature," that is, we cannot know whether H_0 is true or false. What we are saying is that "if H_0 is true," we risk falsely rejecting it (α) $\times$ 100 percent of the time.

If the null hypothesis is false and μ is not equal to μ_0 but rather is equal to some other value, say μ_1, then we make an incorrect decision, called a *Type II error,* if we fail to reject H_0. Consider Table 6.1. We see from this table that the probability of failing to reject a false null hypothesis is denoted by β.

$$\Pr(\text{do not reject } H_0 | H_0 \text{ false}) = \Pr(\text{Type II error}) = \beta$$

This probability is described pictorially in Figure 6.6*b*. If the null hypothesis is true, we are sampling from a population whose true mean is μ_0. The probability of rejecting H_0 if H_0 is true is α, the area of the shaded region in Figure 6.6*a*. If the null hypothesis is false, we are sampling from a

TABLE 6.1 OUTCOMES OF STATISTICAL DECISIONS

	True State of Nature	
	Data are from a population for which	
Statistical Decision	*H_0 True*	*H_0 False and H_a True*
Do not reject H_0	Correct decision	Incorrect decision: Type II error Pr(Type II error) = β
Reject H_0	Incorrect decision: Type I error Pr(Type I error) = α	Correct decision

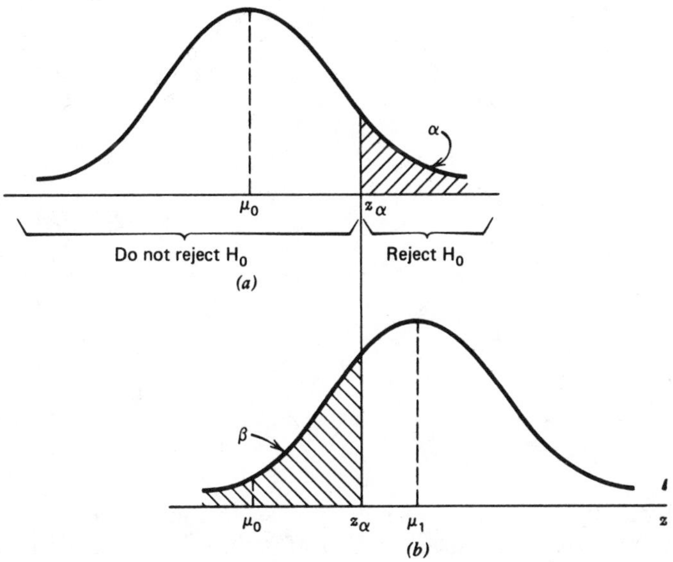

FIGURE 6.6 PICTORIAL DESCRIPTION OF α AND β; (*a* and *b*) FREQUENCY DISTRIBUTION OF TEST STATISTIC IF H_0 TRUE.

population whose true mean is not μ_0 but some other value μ_1. The test of hypothesis is carried out, however, using μ_0 only. That is, the decision is made to reject H_0 if the value of the test statistic falls in the rejection region (shaded area) of Figure 6.6*a*. We do not reject H_0 if the test statistic falls in the nonrejection region (nonshaded area) of Figure 6.6*a*. Thus the probability of not rejecting H_0 when H_0 is true is the area of the nonshaded region in Figure 6.6*a*. However, if H_0 is *not true* and we are sampling from a population whose mean is μ_1, then the probability of not rejecting H_0 is given by the shaded area in Figure 6.6*b*.

If we are able to specify an exact value for μ_1 in the alternate hypothesis, then, for a specified α level, we can calculate β, the probability of a Type II error. However, in practice, we usually test hypothesis of the form

$$H_0 : \mu = \mu_0$$

$$H_a : \mu > \mu_0 \quad \text{or} \quad H_a : \mu < \mu_0 \quad \text{or} \quad H_a : \mu \neq \mu_0$$

where the exact value of μ under H_a is not specified. In this case we cannot determine an exact value for β. We do know the following:

1. For a fixed sample size n, as α becomes smaller, β becomes

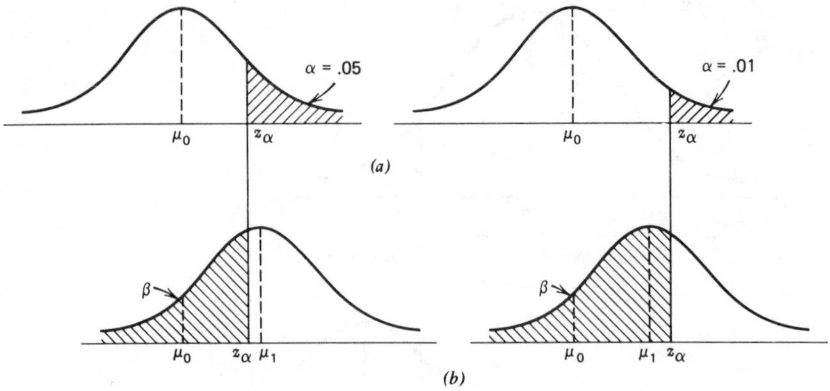

FIGURE 6.7 THE SIZE OF α AND β FOR α DECREASED TO .01 FROM .05 (ONE-SIDED TEST)

larger. In Figure 6.7 we have decreased α from .05 to .01. As can be seen, as α decreases, the cutoff point for the rejection region in Figure 6.7*a* moves farther to the right. Correspondingly, the area of the shaded region in Figure 6.7*b* (β) becomes larger.

2. As the difference between μ_0 and μ_1 increases, β decreases (Fig. 6.8).
3. For a fixed α, as the sample size is increased, β is decreased, since the standard error $\sigma/\sqrt{n}$ decreases as n increases. The following discussion of the dependence of power on sample size will illustrate this point.

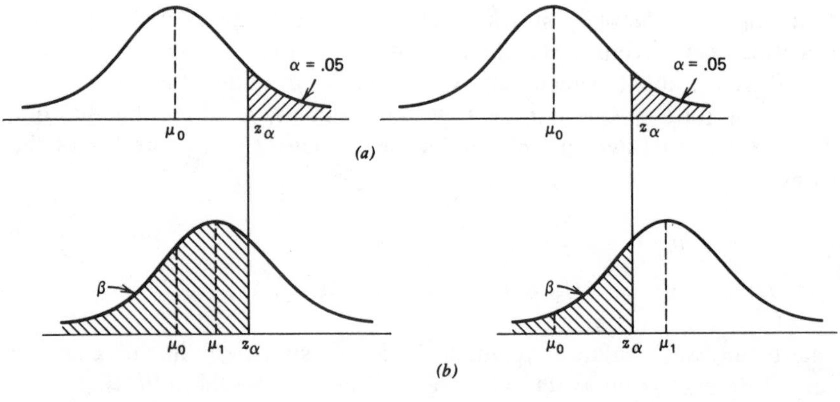

FIGURE 6.8 THE SIZE OF α AND β FOR μ_1 INCREASED (ONE-SIDED TEST)

†Power of the Test and Determination of Sample Size

In the preceding section we considered the probabilities of the two types of errors that can be made when making a statistical decision regarding the credibility of the null hypothesis. We stated

$$\Pr(\text{reject } H_0 | H_0 \text{ true}) = \alpha$$

$$\Pr(\text{do not reject } H_0 | H_0 \text{ false}) = \beta$$

A related concept is that of *power* of the statistical test. Power is defined as the ability of the test to reject the null hypothesis given that H_0 is false.

$$\text{Power} = \Pr(\text{reject } H_0 | H_0 \text{ false}) = 1 - \beta$$

The power of the test is the complement of the Type II error rate. Consider the following example.

Suppose a group of nurse researchers know, based on prior experience, that the mean of a clinical variable for a certain population is 50 with a known standard deviation, σ, of 15. The researchers plan to investigate the effect of a new treatment applied to a sample of individuals selected from this population. Further, they feel that if the treatment would increase the variable being studied by an average of 5 units, this would be a clinically meaningful result. Thus, they wish to test the hypothesis $H_0 : \mu = 50$ versus $H_a : \mu = 55$. How does the choice of sample size affect the probability of finding such an increase if the treatment is truly effective? In statistical terminology, what is the probability of rejecting the null hypothesis when the alternative hypothesis is true; that is, what is the power of the statistical test?

Suppose the researchers choose a sample of size $n = 9$ and for these nine individuals find the sample mean $\overline{Y}$ to be equal to 56. Thus, the value of the test statistic is

$$z = \frac{\overline{Y} - \mu_0}{\sigma/\sqrt{n}} = \frac{56 - 50}{15/\sqrt{9}} = 1.2$$

The rejection region for this test is given by the shaded area shown in Figure 6.9. Note that a value of 2.33 on the z scale corresponds to a value on the $\overline{Y}$ scale given by

$$\overline{Y} = z(\sigma/\sqrt{n}) + \mu_0 = 2.33(15/\sqrt{9}) + 50 = 61.65$$

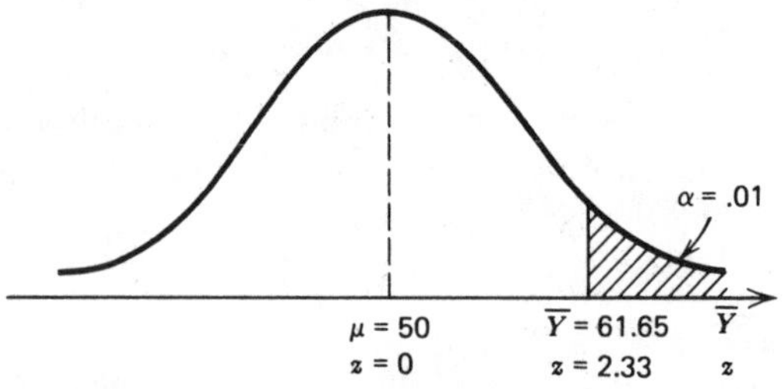

FIGURE 6.9 REJECTION REGION FOR $\alpha = .01$

Therefore, the decision rule for this test is to reject H_0 when the test statistic z is greater than 2.33. Correspondingly, in terms of $\overline{Y}$, we reject H_0 when the value of the sample mean $\overline{Y}$ is greater than 61.65. In this example, the test statistic (or $\overline{Y}$) falls in the nonrejection region. We cannot reject H_0 and therefore cannot conclude that $\mu = 55$.

The power of the test for this example is computed as follows:

$$\beta = \Pr(\text{do not reject } H_0 | H_0 \text{ false and } H_a \text{ true})$$

$$= \Pr(\overline{Y} \leq 61.65 | \mu = 55)$$

$$= \Pr\left(\frac{\overline{Y} - \mu}{\sigma/\sqrt{n}} \leq \frac{61.65 - 55}{15/\sqrt{9}}\right)$$

$$= \Pr(z \leq 1.33) = .9082$$

$$\text{Power} = 1 - \beta = .0918$$

Thus, with a sample of size 9, we have only a 9.18 percent chance of being able to reject a false null hypothesis. That is, if the true difference in means is as large as 5 units, with this particular test we have only a 9.18 percent chance of detecting such a difference.

If the study were done using $n = 100$ rather than $n = 9$, the reader may verify that the decision rule for $\alpha = .01$ is to reject H_0 when $\overline{Y}$ is greater than 53.495. Then

$$\beta = \Pr(\text{do not reject } H_0 | H_0 \text{ false and } H_a \text{ true})$$

$$= \Pr(\overline{Y} \leq 53.495 | \mu = 55)$$

$$= \Pr\left(z \leq \frac{53.495 - 55}{15/\sqrt{100}}\right) = \Pr(z < -1.00)$$

$$= .1587$$

and Power $= 1 - \beta = .8413$.

By increasing the sample size from $n = 9$ to $n = 100$, we have increased the power of the test from 9.18 percent to 84.13 percent. The probability of detecting a true difference of 5 units (rejecting H_0 in favor of H_a when H_a is true) is increased to 84 percent.

This discussion illustrates a very important concept in the interpretation of statistical tests; that is, "What does a statistically nonsignificant result really mean?" In the example of this section, the investigator feels that a difference of at least 5 units in the variable being measured would be of clinical or practical importance. Differences of less than 5 units are considered by those familiar with the clinical issues to be of no practical importance. Therefore, it is desirable to construct a test which we feel, with reasonable probability, can detect a difference of at least 5 units, if a difference of this magnitude really exists. Assume, in the first example, that $H_a: \mu = 55$ is true; that is, a true difference of 5 units does exist. Based on the data, however, we are unable to reject the null hypothesis $H_0: \mu = 50$ since $z_{cal} = 1.2$ is less than the rejection region cut-off value 2.33. How do we interpret this negative result? Many investigators are tempted to interpret a statistically nonsignificant result as demonstrating that the null hypothesis is "true." They proceed to declare that a difference of 5 units *does not exist* in the population. Consider, however, that the power of this test, when the sample consists of 9 observations, is only .0918. Therefore, even if a difference as large as 5 units does exist in the population, we have only a 9.18 percent chance of detecting it. To declare, then, that a difference of 5 units does not exist, based on the outcome of the statistical test, would indeed be misleading.

In the example with the sample size increased to $n = 100$, the power of the test has been increased to 84 percent. With this power, we know that if a difference of 5 units exists, then there is an 84 percent chance that the test will detect it. An investigator in this situation can have more faith that a statistically nonsignificant result does in fact suggest that a difference as large as 5 units does not exist in the population.

One important use of the idea of power of a statistical test is that of determination of a sample size, in advance, that will ensure a specified power for the test. The calculation of sample size, based on power considerations, requires that the investigator specify the following:

1. The size of the effect that is clinically worthwhile to detect
2. The probability of falsely rejecting a true null hypothesis (α)
3. The probability of failing to reject a false null hypothesis (β)
4. The standard deviation of the population being studied

The first three items above are under the control of the investigator. A value for the population standard deviation may be determined from previous work, work reported in the literature, a pilot study, or simply an educated guess.

A discussion of determination of sample size will be given in the chapters describing specific statistical tests.

Statistical Significance Versus Practical or Clinical Significance

An important concept in the interpretation of a statistical test is that of statistical significance versus practical significance. Consider an example in which an investigator is comparing the mean systolic blood pressure level of a group of patients being given an experimental antihypertensive drug with the mean systolic blood pressure of a control group. For the purpose of illustration, suppose that the true difference in the mean systolic blood pressure of the populations from which the two groups were selected is 1 mmHg ($\mu_T - \mu_C = 1$). As described earlier, we may choose the sample size n so that the statistical test has sufficient power to detect even this small a difference in population means. Thus, while we may declare a "statistically significant" difference in the mean blood pressure of the control and treatment groups, we may be unwilling to promote the use of a drug which, on the average, may reduce blood pressure by only 1 mmHg. The point to remember is that statistical significance does not imply necessarily that the true difference in population means is of sufficient magnitude to be of clinical or practical importance. For large sample sizes, the actual difference in the observed sample values ($\overline{Y}_1 - \overline{Y}_2$) is usually a good estimate of the actual difference in the means ($\mu_1 - \mu_2$) of the populations being studied. However, the decision as to whether a difference is of practical importance is not a statistical one, but rather must be made by those with expertise in the subject area being investigated.

EXERCISE 6.1

In Exercise 4.2, what is the relationship between the parameter μ and the statistic $\overline{Y}$? Will $\overline{Y}$ necessarily equal μ? Discuss sampling error in the context of this problem.

EXERCISE 6.2

A study was carried out to determine if pulse rate is elevated in male patients who are approached by a nurse carrying a hypodermic syringe. The mean pulse rate of 10 patients under this circumstance was found to be 90 beats/minute (normal is assumed to be 72 beats/minute). A test of hypothesis was carried out and the results indicated a significant increase in pulse rate for these patients at $\alpha =$

.01. In terms of this example, discuss:

a. Hypotheses
b. Test statistic
c. Level of significance (α) (what does it mean in terms of the conclusion?)
d. Rejection region

EXERCISE 6.3

Suppose that the mean urine chloride level in normal full-term infants is assumed to be 210 mEq/24 hr and the standard deviation is 20 mEq/24 hr. The mean urine chloride level for 25 premature infants was found to be 170 mEq/24 hr. A test of significance was carried out and it was reported that premature infants have a lower mean urine chloride level than the clinical norm at $\alpha = .01$.

1. List two possible statistical reasons why the mean urine chloride level for the sample of premature infants ($\bar{Y}$) is not equal to the mean value for normal full-term infants.
2. What would be the hypotheses in this study?
3. Discuss the conclusion in terms of test statistic, rejection region, and level of significance.

SUMMARY OF CRITICAL CONCEPTS IN CHAPTER 6

The following concepts were presented in Chapter 6:

1. Statistical inference—the process by which one draws conclusions about a population based upon information contained in the sample.
2. Hypothesis testing—one method of statistical inference.
3. To make an inference about the value of a population parameter using statistical hypothesis testing, we:
   a. Hypothesize that the population parameter has a specified value.
   b. Observe a sample of measurements from the population of interest and calculate the sample statistic corresponding to the hypothesized parameter.
   c. Calculate a test statistic that quantifies the discrepancy between the hypothesized population parameter and the calculated sample statistic.

d. Reject the initial claim about the value of the population parameter if the test statistic is "sufficiently large."

4. The initial claim about the value of a population parameter is called the *null hypothesis* and is denoted by the symbol H_0.
5. The *alternative hypothesis* is what will be accepted as true upon rejection of the null hypothesis.
6. The *rejection region* of a test of hypothesis specifies which values of the test statistic are "sufficiently large" to warrant rejection of the null hypothesis. The size of the rejection region depends on the level of significance for the test.
7. The *level of significance,* α, is the likelihood of rejecting the null hypothesis when it is true.
8. To reject a null hypothesis at an α level of significance implies that the investigator is willing to risk an α chance of error in so rejecting H_0 (e.g., $\alpha = .05$ specifies a 5% chance of rejecting H_0 when it is true).
9. The choice of level of significance is arbitrary and depends on the implications of making an incorrect decision. The most common choices of α are .05 and .01.
10. An α value refers to a predefined level of significance (before the data are collected). A p value is the probability of obtaining the particular value of the test statistic given that H_0 is true.

†11. A Type I error is committed when H_0 is rejected given that H_0 is really true.

†12. Pr (Type I error) $= \alpha$, the level of significance of the test.

†13. A Type II error is committed when H_0 is not rejected when in fact it is *false*.

†14. Pr (Type II error) $= \beta$.

†15. For a given sample size n, as the level of significance α becomes *smaller,* β, the probability of a Type II error, becomes *larger*.

†16. For a fixed α, as the sample size is *increased,* β is *decreased.*

†17. As the distance between μ_0, the value hypothesized under H_0 and μ_1, the value hypothesized under H_a increases, β decreases for fixed n.

†18. The *power* of a statistical test is the ability of the test to *reject* H_0 when in fact H_0 is *false.*

†19. Power $= 1 - \beta$.

†20. For statistical tests with low or undetermined power, failure to reject H_0 does not imply that H_0 is true. It says "if H_0 is false, my test, conducted under the given circumstances (e.g., sample size, variability of observations), was unable to detect it."

†21. Any difference may be statistically significant if the sample size is large enough and hence the power of the test sufficiently great to detect the difference.

†22. A difference, although statistically significant, may not be clinically important.

CHAPTER 7

SELECTION OF AN APPROPRIATE STATISTICAL TEST

OVERVIEW

After the research and statistical hypotheses have been formulated, one of the most difficult tasks facing the nurse researcher is that of selecting the appropriate statistical test of hypothesis. A long-standing problem of most introductory statistical methodology courses is that emphasis is placed on the mechanics of how to carry out the statistical tests with little or no emphasis on recognizing which tests should be employed in given research situations. It is the purpose of this chapter to outline a decision tree algorithm for choosing a statistical test based on the student's ability to determine the level of measurement of the response variable being studied and the type of research question being employed.

OBJECTIVES

Upon completion of this chapter, the student will be able to:

1. Use the decision tree presented in Figure 7.1 to select an appropriate statistical analysis.

DISCUSSION

There is often considerable confusion and disagreement over a proper taxonomy for describing types of research and types of research designs. For the purpose of this text we will classify nursing research into two broad categories: (1) experimental and quasiexperimental, and (2) nonexperimental, or observational, research. Research designs will be considered to be of three types: one group (historical) designs, two group designs, and more-than-two group designs. Both the two group designs and more-than-two group designs can be subdivided into independent group (nonmatched) designs and paired group (matched) designs. Each of these terms will now be briefly described. These descriptions are deliberately simplified since the purpose of this chapter is to provide a general framework for selecting a statistical test rather than to present a detailed and all-encompassing account of research design methodology. For the latter, the student is referred to the many excellent texts on research methodology, specifically to Campbell and Stanley's *Experimental and Quasi-Experimental Designs for Research* (1963) or to Polit and Hungler's *Nursing Research: Principles and Methods* (1978).

Experimental research and quasiexperimental research differ from observational, or nonexperimental, research in that, for the former, the investigator manipulates the independent variable in order to determine its effect on the dependent or response variable. For example, a nurse researcher conducting an experimental study to determine the effect of isolette temperature (independent variable) on body temperature of newborns (dependent variable) can randomly assign a group of newborns to one of three isolettes, each set at a different temperature. The temperature setting of the isolette (independent variable) is under the control of the investigator.

In observational studies, on the other hand, the investigator is only an observer in the sense that he or she does not exert control over the independent variable. For example, in a study of the relationship of lung cancer to smoking, an investigator does not assign individuals to be smokers or nonsmokers. She or he may instead observe a group of smokers and a group of nonsmokers for a period of years and then determine if there is an association between smoking and occurrence of lung cancer for the two groups.

Polit and Hungler (1978) state that a true experiment can be characterized by the following properties:

1. *Manipulation:* the experimenter does something to at least some of the subjects in the study.
2. *Control:* the experimenter introduces one or more controls over the experimental situation, including the use of control or comparison groups.
3. *Randomization:* the experimenter assigns subjects to a control or experimental group on a random basis.

Quasiexperimental designs, like true experimental designs, involve manipulation of the independent variable by the investigator. However, they differ from true experimental designs in that either the control group component (2) or the randomization (3), or both, is missing.

For observational research, we further classify studies as being *prospective, retrospective,* or *cross-sectional.* Often, due to their wide use in epidemiological studies, we refer to the independent variable as the risk factor and the dependent variable is usually the occurrence of a disease or condition. In *prospective* studies, we identify groups of individuals exposed and not exposed to the risk factor and follow them *forward in time* to determine the occurrence of disease. For *retrospective* studies, we identify groups of individuals with and without the disease and follow them *backward* in time to determine the presence or absence of the risk factor. A *cross-sectional* study involves looking at a group of individuals at a given point in time and determining occurrence of risk factor and disease.

It should be noted that all experimental studies are prospective in nature.

A question related to type of research is, "What type of question am I trying to answer with my study?" This question may be subdivided into two components: (1) "Am I interested in detecting *differences* between (or among) the groups being studied with respect to the dependent (response) variable being measured?"; or, (2) "Am I interested in measuring *associations* or describing the *nature of the relationship* between the dependent variable and independent variable?" Linton and Gallo (1975) state, "In experimental research you will ask about differences more often than you will ask about degree of relationship between variables." For observational studies, in contrast, we are most often interested in assessing degree of association between the independent and dependent variable. In the isolette example, an experimental study, the researcher wishes to answer the question "Is there a difference in rectal temperature of infants in the three isolettes having three different temperature settings?" In the observational study concerning lung cancer and smoking the question of interest is whether there is an association between the independent variable (smoking) and the dependent variable (lung cancer).

The rule relating questions of differences to experimental studies and associations to observational studies is by no means without exceptions. There may be cases in which experimental studies are designed to answer questions about associations between independent and dependent variables and observational studies designed to answer questions concerning differences in the dependent variable for several levels or factors of the independent variable. The important point with respect to selecting an appropriate statistical analysis is that you be able to distinguish the kind of question you are asking in your research. Accordingly, the type of research question, that is, whether differences or associations, is the first major branching point in the decision tree algorithm (Figure 7.1) for selecting a statistical procedure.

The second major branching point for our decision tree algorithm is the *level of measurement* of the response (dependent) variable. These levels are *nominal, ordinal,* and *equal-interval.* The reader is referred to Chapter 1 for definitions and examples of these terms. In a research study the data we collect on a qualitative variable measured on a nominal scale is *count,* or enumeration, data. For example, for the variable blood type, the response for each patient may be "Type A," "Type O," "Type B," or "Type AB." The data to be analyzed are the number (count) of individuals in the study falling in each of the categories. Similarly, for a group of patients receiving a treatment, we may record "improved" or "not improved," and the data to be analyzed are the numbers falling in each of these two categories. In the decision tree (Figure 7.1) for the branch labeled *nominal* level of measurement, we are concerned with analyzing data consisting of *counts.*

The data we collect on quantitative variables (having an ordinal or equal interval scale) are usually in terms of "amount" of the variable being measured as opposed to the *count* of individuals falling into the subcategories of the variable. Thus, the branches for ordinal level of measurement and equal-interval level of measurement require that the data to be analyzed represent amount as opposed to count. For example, suppose we are studying the variable "patient satisfaction." The data, measured on an ordinal scale, are recorded for each patient as 1 = very satisfied, 2 = satisfied, 3 = neutral, 4 = dissatisfied, and 5 = very dissatisfied. For each patient we are recording "amount" of satisfaction on a scale from 1 to 5. Similarly, for the equal-interval variable, systolic blood pressure, we record amount of blood pressure in mmHg for each study participant. It must be noted, however, that both ordinal and equal-interval data may be converted to counts. For example, with the variable "patient satisfaction" we may be interested in the number falling into the five subcategories of the scale or for "systolic blood pressure" we may have divided the scale into intervals and recorded the number falling into each of the intervals. (We do not advocate converting the actual measurements to counts in an interval because, in the process, we have lost information). *It is important to note that when ordinal or equal-interval data have been converted to counts, we must use the branch for nominal data to determine the appropriate statistical analysis.* The branches, ordinal and equal-interval as outlined for the decision process, require that the data be recorded as the amount of the variable being measured.

In Chapter 1, we further subdivided data obtained on a quantitative variable measured on an equal-interval scale into continuous data and discrete data. Theoretically, the statistical tests at the end of the branch labeled "equal-interval level of measurement" require that the data be *continuous*. However, in practice, most of the tests may be used for discrete data as long as there are enough categories to approximate a continuum.

The final branching point in our decision tree algorithm is *research design*. As we stated earlier, for the purpose of selecting a statistical test we classify research designs into three main groups: (1) the single group (historical) design, (2) the two group design, and (3) the more-than-two group design. Within the two group and the more-than-two group designs we further subdivide on the basis of whether we have independent, non-matched groups or paired (matched) groups. In addition, we may have combinations of independent and matched samples within a single study. This we call a *mixed* design.

The first design, the single group or historical design, involves the use of a single randomly selected sample of individuals. The response measured for the sample is compared to a clinical norm, past experience, or previous work cited in the literature; hence, the name *historical* design. Consider the following example of a one group experimental study.

EXAMPLE 7.1 In a study to determine the effect, if any, of the mother's diet on the birth weight of infants, an investigator randomly selected nine women who were in their third month of pregnancy and placed them on a strict diet. At birth, the weights of the nine infants were recorded. From experience, it is known that the average birth weight of all infants is approximately 7 pounds. The nurse investigator wished to answer the question, "Does the average birth weight of infants whose mothers are on the special diet differ from the known average birth weight for all infants?"

In this example, the birth weights of the *single sample* of nine infants is compared to the known, historical value of 7 pounds. When using historical controls to measure the efficacy of a new treatment or procedure, we must be extremely careful to avoid biases that invalidate the results of the study. These sources of possible bias are discussed, in detail, in Chapter 8. The one sample design, because of the possible occurrence of these inherent biases, is the *least* desirable of all the designs and is rarely used in practice. It is included in our discussion for the sake of completeness.

The two group and the more-than-two group designs overcome many of the shortcomings of a historical design by the use of *concurrent* sampling. With these designs, each treatment group studied is represented by a sample of subjects selected from the population of subjects who may theoretically receive each treatment. In the two sample (group) design, for example, one sample of patients may receive one treatment while another sample of patients receives the other treatment. Each of the samples is treated as alike as possible except for the treatment under study. The two sample and the more-than-two sample designs are further subdivided into independent (nonmatched) samples and paired (matched) samples. For the former, we are considering two or more *independently selected, nonrelated* groups of individuals. In the independent sample case in an experimental setting, the process of randomization is assumed to secure comparable pretreatment groups. That is, by randomly assigning study participants to the levels of the independent variable being studied (treatment groups), the investigator hopes, by the principles of randomization, to ensure that any extraneous variables, such as age, sex, occupation, or physical condition, will be spread evenly over the two (or more than two) study groups. For example, suppose 40 patients participate in a study of a patient teaching program designed to teach patients about a particular therapeutic procedure. Using the two sample, independent sample design, 20 of these patients may be randomly assigned to the treatment group that receives the special teaching and 20 patients randomly assigned to the control group that receives no special teaching. The researcher recognizes that age and educational level of the patients, among other variables, will affect the response being measured (level of knowledge of the hospital procedure). The researcher is counting on the randomization procedure to "cancel out" the effects of age and educational level. That is, it is the function of the randomization procedure to equalize the two groups with respect to such extraneous factors that affect the response being investigated. However, randomization *does not guarantee* that the groups will be equal, particularly when there is only a small number of subjects. Thus, just by chance, the treatment group could consist of many more individuals with higher levels of education than the control group, or vice versa, although it is unlikely. It is wise to consider the distribution of values for the recognized extraneous variables for each of the study groups to confirm that randomization has achieved its goal of equalization.

The teaching method example just cited applies to a two group, independent sample design. However, it is easy to visualize how the process of random assignment of subjects to independent groups can be extended to the more-than-two group designs as well.

In presenting our decision tree algorithm for selecting an appropriate statistical test, the second subdivision of the two group and the more-than-two group designs is the *paired* (matched) sample case. For the two sample design case, pairing, as we will define it, can be accomplished in one of three ways: (1) *natural pairing,* (2) *artificial pairing,* and (3) *self-pairing.* Natural pairing occurs when a study consists of pairs of human twins or animal litter mates, and one member of each pair is *randomly* assigned to each treatment being studied. In the case of artificial pairing, individuals are matched with respect to certain specified extraneous variables such as age, educational level, and so on. Again, one member of each pair is *randomly assigned* to each of the treatment groups. Finally, with self-pairing, each individual serves as his own control. This includes the "before-and-after" (or "pre-and-post") study or the case in which each individual may receive one treatment on one occasion and the other treatment on another occasion. The self-pairing case is sometimes referred to as a "repeated measures" design. The purpose of pairing, in the sense it is used in this text, is to control extraneous variables, thereby reducing variation in the response variable attributable to these sources and thus increasing the precision of the statistical test.

It is not on the basis of the presence or absence of *randomization* that the independent sample case and the paired sample case differ. Rather, the question of importance is whether there is a relationship between measurements on individuals receiving Treatment 1 and those receiving Treatment 2. In the independent sample case, the measurement made on Individual 1 receiving Treatment 1 is independent of the measurement made on Individual 1 receiving Treatment 2. In the paired case, the measurements are related either because they are both made on the same individual or because the individuals on whom the measurements are made are deliberately (or naturally) matched. For the paired case, we expect measurements *within* a pair to be more alike than measurements *between* pairs.

The preceeding discussion of paired or matched samples has been relevant for the two group design. We will expand the concept of matched samples to the more-than-two group case. Because of the nature of the statistical analyses appropriate for the three or more sample case, the factor (extraneous variable) on which the subjects are matched can be included in the study design as an independent variable. Consider an example in which a nurse wishes to compare the knowledge level of a particular therapeutic procedure for patients who are taught using three different teaching methods. Recognizing that educational level is an extraneous variable that may affect the response being measured, the nurse divides the patients in the study into four different educational levels: those who (1) did not complete elementary school, (2) completed elementary but did not complete high school, (3) completed high school, and (4) completed at least one year of college. Within each educational level, the researcher has available three patients, each of whom is then randomly assigned to the three treatment groups (teaching method). Schematically, we have

	Teaching Method		
Educational Level	*A*	*B*	*C*
1	X	X	X
2	X	X	X
3	X	X	X
4	X	X	X

In the table, X represents the measurement of knowledge level of each patient with respect to the therapeutic procedure. The variable on which the subjects are matched, educational level, is called the *blocking* factor, and the design has a special name, the *randomized block design*. It should be noted that this design can also be employed when there are only two groups.

In summary, for the more-than-two group designs, as for the two group designs, to choose a statistical analysis the investigator must be able to answer the question, "Were the study subjects *independently* assigned to the treatment groups or were the subjects *matched* on the basis of an identifiable extraneous variable?"

The final classification that appears under the two group and more-than-two group designs is the *mixed* design. This design, frequently employed in educational research, may be illustrated schematically, as

Treatment 1		*Treatment 2*	
Pre	*Post*	*Pre*	*Post*
X	X	X	X
X	X	X	X
X	X	X	X

In this design, we have both *matched* samples (pre and post) within a treatment and independent samples between treatments. The most common procedure for analyzing data obtained using this design is to use *difference,* or *gain,* scores between pre and post periods within each treatment and to analyze the difference scores as two independent samples from the two treatments. There are other and possibly more desirable methods of analysis, such as those using pre scores as covariates and then using a covariance analysis, or incorporating pre scores into the design as a *blocking* factor. Covariance analysis is an advanced technique and is beyond the scope of this text. For completeness, however, it is listed where appropriate in the decision tree.

The branch of the decision tree dealing with associations (II) is much simpler than that for differences (I). Because of the introductory nature of this text, we consider only the case in which the relationship of interest involves *two* variables. There are techniques available for the analysis of relationships among three or more variables. For these, the reader is referred to more advanced statistical texts. For branch II, questions 3 and 4 in the steps outlined below are not relevant for the procedures described in this text. The exception to the rule is for nominal data in which the X^2 procedure may be used for any number of groups. Branch IIA may be considered the same as IA.

On the following pages we present a schematic diagram of our decision tree for selecting an appropriate statistical test. To use the tree, the reader must be able to answer a series of questions relating to the previous discussions. These questions are as follows:

1. What type of research question am I asking? Am I asking
   I. Is there a *difference* in the dependent variable for different levels of the independent variable (treatment groups)?

   or

   II. Is there an *association* between the dependent variable and the independent variable(s)?
2. What is the level of measurement of the response variable being investigated? Is it
   A. Nominal (or count data)
   B. Ordinal
   C. Equal-interval
3. What is the research design employed in my study? Is it
   1. One group (historical) design
   2. Two group design
   3. More-than-two group design
4. How are my samples obtained. Are they
   a. Independent (nonmatched) samples
   b. Paired (matched) samples
   c. Mixed samples

We issue a word of warning to users of this decision tree algorithm. In order to create a situation in which all types of data fit our conceptualization of the decision-making process relative to choosing a statistical test, we have, perhaps, in some situations oversimplified or created an artificial framework that may offend those well-schooled in statistical methodology. The distinction between differences (I) and associations (II) may be particularly confusing when dealing with nominal data. Also, for this case, the one group, two group, and more-than-two group categorization may trouble the statistical purist. It must be kept in mind that the *only* purpose of the decision tree algorithm is to present a simplified and useable framework to aid the student in the task of selecting a statistical test. All specific information relevant to a given test, such as the conditions under which it is used, type of data, assumptions, and so on, should be obtained from the particular chapters in which the tests are discussed.

Before proceeding with specific examples of how to select a statistical test using our decision tree algorithm, we address the troublesome issue of nominal data. Let us consider an example. Suppose a nurse researcher is interested in comparing the efficacy of two treatments for decubitus ulcers in bedridden patients. For 60 patients receiving the new treatment, 40 improved and 20 did not improve. Similarly, for the standard treatment 25 improved and 15 did not improve. The data are shown in Table 7.1. In this example both variables, "type of treatment" and "improvement status," have two levels each. Under our decision tree scheme, this example would be classified as a two group design (new treatment versus standard treatment). Had there been three treatments (Treatment 1, Treatment 2, standard treatment), we would have classified it as a more-than-two group design regardless of the number of levels of improvement status.

In some situations in which nominal or count data is used, it is not clear which variable is the independent variable and which is the response variable. For example, in a study on the relationship of sleeping habits and performance in clinical work, 100 student nurses may be cross-classified as to sleeping habits ("poor," "good") and clinical performance

TABLE 7.1

	Improved	*Not Improved*	
New treatment	40	20	60
Standard treatment	25	15	40

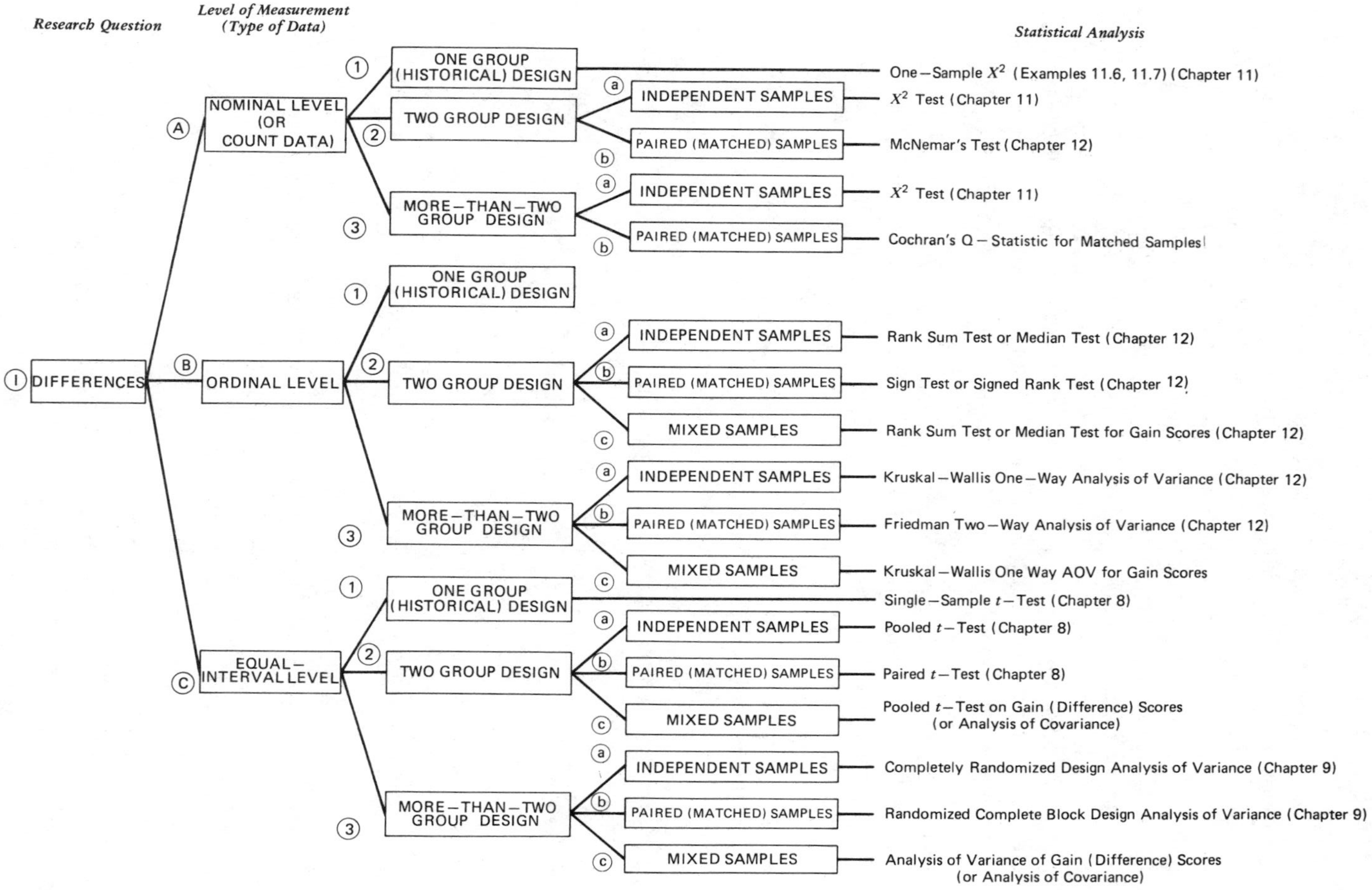
Research Question
Level of Measurement (Type of Data)
Statistical Analysis
I DIFFERENCES
A NOMINAL LEVEL (OR COUNT DATA)
1 ONE GROUP (HISTORICAL) DESIGN
One–Sample X^2 (Examples 11.6, 11.7) (Chapter 11)
2 TWO GROUP DESIGN
a INDEPENDENT SAMPLES
X^2 Test (Chapter 11)
b PAIRED (MATCHED) SAMPLES
McNemar's Test (Chapter 12)
3 MORE–THAN–TWO GROUP DESIGN
a INDEPENDENT SAMPLES
X^2 Test (Chapter 11)
b PAIRED (MATCHED) SAMPLES
Cochran's Q – Statistic for Matched Samples
B ORDINAL LEVEL
1 ONE GROUP (HISTORICAL) DESIGN
2 TWO GROUP DESIGN
a INDEPENDENT SAMPLES
Rank Sum Test or Median Test (Chapter 12)
b PAIRED (MATCHED) SAMPLES
Sign Test or Signed Rank Test (Chapter 12)
c MIXED SAMPLES
Rank Sum Test or Median Test for Gain Scores (Chapter 12)
3 MORE–THAN–TWO GROUP DESIGN
a INDEPENDENT SAMPLES
Kruskal–Wallis One–Way Analysis of Variance (Chapter 12)
b PAIRED (MATCHED) SAMPLES
Friedman Two–Way Analysis of Variance (Chapter 12)
c MIXED SAMPLES
Kruskal–Wallis One Way AOV for Gain Scores
C EQUAL–INTERVAL LEVEL
1 ONE GROUP (HISTORICAL) DESIGN
Single–Sample t–Test (Chapter 8)
2 TWO GROUP DESIGN
a INDEPENDENT SAMPLES
Pooled t–Test (Chapter 8)
b PAIRED (MATCHED) SAMPLES
Paired t–Test (Chapter 8)
c MIXED SAMPLES
Pooled t–Test on Gain (Difference) Scores (or Analysis of Covariance)
3 MORE–THAN–TWO GROUP DESIGN
a INDEPENDENT SAMPLES
Completely Randomized Design Analysis of Variance (Chapter 9)
b PAIRED (MATCHED) SAMPLES
Randomized Complete Block Design Analysis of Variance (Chapter 9)
c MIXED SAMPLES
Analysis of Variance of Gain (Difference) Scores (or Analysis of Covariance)

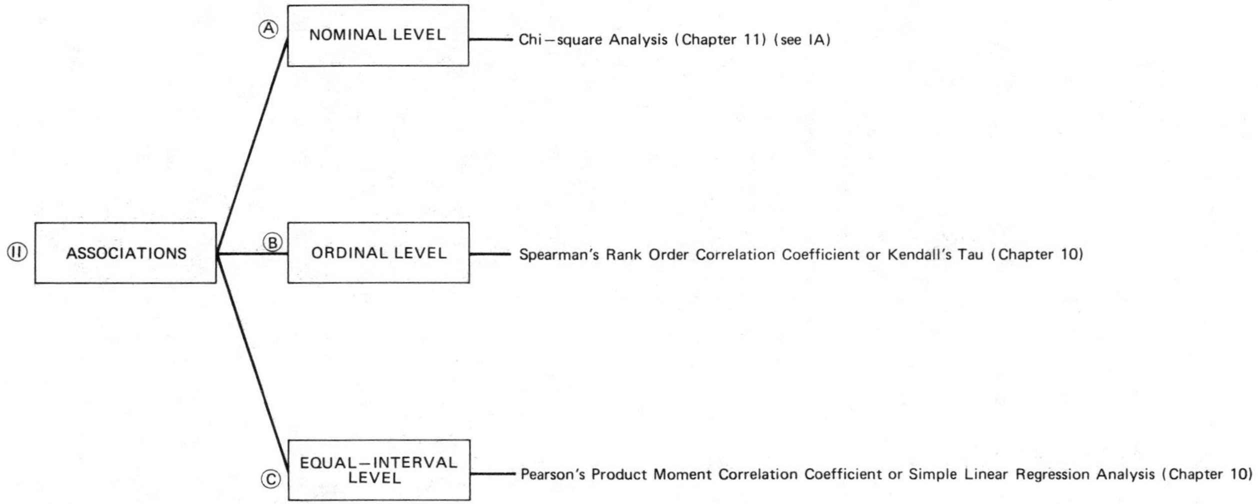

FIGURE 7.1 DECISION TREE ALGORITHM FOR SELECTING AN APPROPRIATE STATISTICAL ANALYSIS

TABLE 7.2 SLEEPING HABITS

Clinical Performance	*Poor*	*Good*
Poor		
Average		
Good		

("poor," "average," "good"). The data would consist of the numbers of students in each of the cells in Table 7.2.

In this example, it is difficult to answer the question, "How many groups are being compared?" In this case, we may think of both "sleeping habits" and "clinical performance" as independent variables, with the response variable being the number of individuals falling into each cell. For the nominal level (or count data) branch in the decision tree in Figure 7.1, the actual question of importance in choosing a statistical test is not number of treatment groups but rather whether the measurements are independent or matched. All branches labeled "independent" lead to a X^2 analysis. For the branches labeled "paired," the McNemar's test is appropriate when there are two variables each having only two categories. The more-than-two group paired branch requires a not commonly used analysis cited as a reference. Thus, for the example of sleeping habits versus clinical performance, even though we may be unsure of how to answer the question "How many groups?", we recognize that we have independent observations and hence a X^2 test is appropriate.

The other issue to be addressed with nominal data is the considerable overlap between the case of testing for differences (branch I) or detecting association (branch II). In fact, for nominal data we use the same analysis (e.g., X^2) to answer a question phrased in terms either of differences or of associations. In the example just given (Table 7.1), the question may be stated as (1) "Is the proportion of patients showing improvement different for the new and standard treatments?", or as (2) "Is there an association between treatment received and improvement status?" Since questions may be phrased either in terms of differences in proportions or in terms of associations between variables, we include essentially the same branches for nominal data under both the major headings *differences* (I) and *associations* (II).

We will complete our discussion of the decision tree by giving several examples and illustrating how to choose a statistical test by following the appropriate branches on the tree.

EXAMPLE 7.2 Nine student nurses were asked to instruct pediatric patients in proper dental hygiene techniques. The performance of each student was rated by each instructor on a scale from 1 to 10. Subsequently, the students participated in a lecture–slide demonstration on teaching proper dental hygiene techniques. They were again rated on teaching techniques. The researcher wished to determine if there is a difference in performance ratings before and after the demonstration. Specifically, the research hypothesis states "There will be a significant improvement in rating scores of students following the lecture–slide demonstration." What statistical test is appropriate?

SOLUTION *Question 1:* What type of research question is being asked?
Answer: The research question asks whether there are *differences* in the dependent variable (rating score) for two levels (pre and post) of independent variable (lecture demonstration). Thus, we begin at point I (differences).
Question 2: What is the level of measurement of the response variable?
Answer: Ordinal. We now go to IB.
Question 3: What is the research design? Is it one group, two groups, or more-than-two groups?
Answer: Although technically there is only one group of subjects on which two measurements were made, we classify before-and-after designs as two group designs. (Think of it as *two groups* of data.) We now proceed to IB2.
Question 4: Were the samples independently selected or were they paired (matched)?
Answer: This is an example of *self*-pairing. All before-and-after (pre versus post) studies are *paired* designs. Proceeding along the branch IB2b, we see that we may use either the *sign test* or *signed ranks test*. The chapter on nonparametric procedures should be consulted for specific details of these tests.

EXAMPLE 7.3 Suppose in Example 7.2, nine additional students were used as a control group. Each of these students had pre and post ratings made at the same time as the pre and post ratings for the treatment group. However, the control group did not attend the lecture–slide demonstration as did the treatment group. The researcher hypothesized that improvement (pre and post) scores for the treatment group would be higher than improvement (pre and post) scores for the control group. What statistical test is appropriate for testing this hypothesis?

SOLUTION *Question 1:* What type of research question is being asked? *Answer:* We are asking a question about *differences* in the performance of the control and treatment group. Specifically, we are asking whether the improvement scores for the treatment group are different from improvement scores for the control group. We begin at point I.
Question 2: What is the level of measurement of the response variable?
Answer: In this example, we consider the response variable for each individual in the study to be the difference in his or her pre- and posttest scores. Since we said in Example 7.2 that the pre- and posttest scores were ordinal data, then we also consider the improvement (difference or gain) scores to the *ordinal* data.
Question 3: What is the research design?
Answer: In this example we have two groups. *Within* each group we have a set of matched observations, and *between* the two groups we have independent observations. Since our purpose is to compare the treatment and control groups, the analysis is done on the improvement scores for the two groups, which are considered two sets of *independent* observations. We follow I to B to 2 to c. The appropriate analysis is a median test or rank sum test on the two sets of improvement scores.

EXAMPLE 7.4 A study was conducted to compare three different methods of administering an intravenous (IV) injection of heparin. Each of the 150 patients in the study received a heparin injection by one of the three methods. For each patient the amount of bruising around the injection site was recorded as "bruising" and "no bruising." The data were recorded as follows:

Method of Injection	*Bruising*	*No Bruising*	
A	30	20	50
B	10	40	50
C	25	25	50
	65	85	150

The investigator wished to determine if there were a *difference* in proportion of bruises for methods A, B, and C (or, identically, if there is a difference in proportion of no bruising). What statistical test is appropriate?

SOLUTION *Question 1:* What type of research question is being asked?
Answer: The research question is concerned with ascertaining *differences* in proportion of bruising (or no bruising) among the three methods of injection. We begin at I.
Question 2: What is the level of measurement of the response variable?
Answer: In this study each patient's response is recorded as "bruising" or "no bruising"; thus, we are dealing with nominal type data. We analyze the *count* or *number* of individuals falling in the cells of the table. (Enumeration or count data are classified as nominal data.) We proceed to IA.
Question 3: What is the research design?
Answer: We are interested in comparing *three* methods of injection; thus, we have three treatment groups. This is a *more-than-two group design*. We are located at IA3.
Question 4: Are the samples independent or matched?
Answer: An individual's response (bruising or no bruising) to a given method of injection in no way affects the next individual's response. The study consists of three separate, nonmatched (independent) groups receiving the three methods. (*Note:* Had each individual received all methods on different occasions, then we would no longer have independent samples.) We proceed to IA3a. We find that a X^2 test of proportions is appropriate.

EXAMPLE 7.5 A study was conducted to compare the effects of three different drugs on the alleviation of anxious depression in nonpsychotic patients. Twelve nonpsychotic patients, all suffering from moderate to severe depression and anxiety, were grouped according to initial level of severity. Patients in each severity level group were randomly assigned to the three treatments (drugs). At the end of the experimental period, a combined anxiety and depression score as determined by the MMPI and Taylor Manifest Anxiety Scale was recorded for each patient. The researcher hypothesizes that there is a difference in average anxiety/depression scores for the three drugs. The data are as follows:

Initial Severity Level	*Drug A*	*Drug B*	*Drug C*
1	35	30	25
2	40	25	20
3	25	25	20
4	30	25	25

What statistical test is appropriate for testing the researcher's hypothesis?

SOLUTION *Question 1:* What type of research question is being asked? *Answer:* The researcher has asked, "Is there a *difference* in average anxiety/depression scores for the three drugs?" We begin at location I.
Question 2: What is the level of measurement of the response variable?
Answer: The data consist of anxiety/depression *scores*, which are considered *continuous equal-interval* measurements. We are located at IC.
Question 3: What is the research design?
Answer: The researcher is interested in comparing three treatment groups; thus, we have a *more-than-two group design*. We proceed to IC3.
Question 4: Are the samples independent or matched?
Answer: Individuals were *matched* according to initial severity level. For example, within row 1 of the data table, the three individuals receiving the three drugs all had the same initial level 1. The design involved using initial level as a "blocking" factor. We proceed to IC3b and see that an analysis of variance for a randomized complete block design is the statistical analysis of choice.

EXAMPLE 7.6 A study was carried out to determine the degree of association between creativity of children and creativity of their parents. Six children and their parents were interviewed and ranked according to their creativity. The researcher hypothesizes that there is a high degree of positive association between creativity of children and creativity of their parents. The ranks are as follows:

Child	*Parent*
4	5
2	2
1	3
3	1
5	6
6	4

Choose an appropriate statistical procedure.

SOLUTION *Question 1:* What type of research question is being asked? *Answer:* This investigation is concerned with assessing *associations* between the two variables "creativity of child" and "creativity of parent." The researcher is *not* interested in whether there exists a difference in creativity of parents and their children. Instead, the investigator wishes to measure the magnitude of the relationship (association) between creativity of parents and children to determine if the association is statistically significant. We begin at point II.

Question 2: What is the level of measurement of the variables?

Answer: Both parents and children were *ranked* according to creativity level; thus, we have *ordinal* data. We proceed to IIB. The appropriate measure of association is the Spearman's rank order correlation coefficient accompanied by a test of its statistical significance.

EXAMPLE 7.7 A nurse educator, investigating the degree of relationship between student scores on a battery of personality tests and performance in nursing school, gave a random sample of 50 applicants accepted for admission to a nursing school a personality inventory evaluation. At the end of the first year, college grade point average (GPA) was also recorded for each of the 50 students in the study. What statistical procedure is appropriate for assessing degree of association between personality scores and freshman GPA?

SOLUTION *Question 1:* What type of research question is being asked? *Answer:* The investigator wishes to measure degree of association between the two variables "personality score" and "GPA." Here, it is not reasonable to be concerned with assessing *differences* in these two variables because they are indeed *different* by their nature. We begin at point II.

Question 2: What is the level of measurement of the variables?

Answer: Both variables are measured on a continuous equal-interval scale. We proceed to IIC and find that a Pearson's correlation coefficient is appropriate. If the researcher wished to develop an equation for predicting freshman GPA based on personality scores, then a simple linear regression analysis is appropriate.

SUMMARY

It has been the purpose of this chapter to acquaint the student with a simplified procedure for selecting a statistical analysis. It is hoped that as students progress through the remaining chapters that describe particular tests of hypotheses, they will return frequently to the decision tree to orient the test being described with its location in the overall decision process.

EXERCISES

Using the decision tree, select the appropriate analyses for the following problems.

EXERCISE 7.1

A nurse investigator carried out a study to determine the relationship, if any, between amount of caffeine ingested during pregnancy and birth weight of infants. She selected 100 patients in a prenatal clinic who were beginning their second trimester of pregnancy. Each participant kept a daily log of caffeine intake during the remainder of her pregnancy. The birth weights of the infants were also recorded for the study subjects. The researcher hypothesized that there is a significant positive relationship between amount of caffeine and birth weight.

EXERCISE 7.2

The mean urine chloride level in nondiseased, full-term infants is assumed, from previous knowledge, to be 210 mEg/24 hr. A nurse researcher, interested in investigating the effects of premature delivery on urine chloride level of infants, measured urine chloride levels on 25 randomly selected premature infants. The research hypothesis for this study is that premature infants have elevated urine chloride levels relative to nondiseased, full-term infants.

EXERCISE 7.3

In a large medical-school-affiliated hospital, all women from age 15 to 45 who were discharged with a diagnosis of idiopathic thromboembolism were identified and then classified as to type of contraceptive used. The data consisted of number of women using (1) oral contraceptive, (2) IUD, (3) diaphragm, and (4) other. The researcher wished to answer the question, "Is there a difference in proportion of women using the four types of contraception among women suffering from thromboembolic disease?"

EXERCISE 7.4

A nurse midwife wished to determine if the mean age of women who expel IUD's is different from the mean age of those who retain the devices. Fifteen subjects were randomly selected from among a group who had retained IUD's and 10 subjects from among those who had expelled IUD's. The ages of the participants were recorded.

EXERCISE 7.5

A group of nine randomly selected senior nursing students were given a standardized pretest on medical-surgical knowledge prior to their taking a special clinical elective. After the six-week course, the students were given a posttest covering the same material. The researcher wished to determine if scores following the elective were higher, on the average, than scores obtained prior to the elective.

EXERCISE 7.6

A nurse investigator wished to compare the elapsed time of three different methods of collecting information on patients in an emergency room. The patients participating in the study were randomly assigned to one of the three methods. For each participant, the time in minutes to collect the information was recorded.

EXERCISE 7.7

A nurse hypothesized that mothers who received counseling concerning their reactions to their children's illnesses would react more appropriately when greeting their hospitalized children for the first time than would mothers who received no counseling. Two groups of mothers were selected to participate in the study. In one group each mother met with a staff psychologist and the attending nurse. The nature of the child's illness was explained, and the mother was counseled concerning her reactions to the illness. The second group of mothers received no such counseling and served as the control group. A nurse then observed the initial encounter between each mother and child and assigned a greeting behavior rating score (on a scale from 1 to 10, with 10 most desirable). The question the researcher wished to answer was, "Are the greeting behavior ratings higher, on the average, for the counseled group than for the group who received no counseling?"

EXERCISE 7.8

One hundred institutionalized schizophrenic patients were classified according to the season of the year in which they were born. An investigator hypothesized that there is a difference in season of birth among institutionalized patients. The data were as follows:

Season of Birth Among 100 Schizophrenic Patients

Fall	20
Winter	35
Spring	20
Summer	25
	100

EXERCISE 7.9

Researchers reviewed the records of 100 patients with hypertension (cases) and of 100 patients without hypertension (controls) who were comparable in other respects to the cases. The history of the use of oral contraceptives was reviewed, and "used oral contraceptives" or "did not use oral contraceptives" was recorded for each participant. The researchers hypothesized that there is an association between oral contraceptive use and presence or absence of hypertension.

EXERCISE 7.10

Stroke patients at a VA hospital were evaluated at discharge with respect to certain physiological and psychological factors. This evaluation resulted in the assignment of a "functional ability score." The scores ranged from 0 to 100 points and were obtained by scoring and summing several components of the patients' ability to function outside the hospital environment. In addition, for each patient length of hospital stay was recorded. An investigator wished to assess the degree of association between functional score at discharge and length of hospital stay.

EXERCISE 7.11

Each student in a research/statistics course was ranked with respect to final course grade. In addition, each student was asked to rate the course on a scale from 1 (very poor) to 5 (very good). A researcher wished to determine whether there is an association between grade rankings and evaluation scores.

BIBLIOGRAPHY

Cambell, D. T., and Stanley, J. C. *Experimental and Quasi-Experimental Designs for Research.* Chicago: Rand-McNally, 1963.

Fleiss, J. L. *Statistical Methods for Rates and Proportions.* New York: John Wiley and Sons, 1973.

Linton, M., and Gallo, P. *The Practical Statistician: Simplified Handbook of Statistics.* Monterey: Brooks/Cole Publishing Co., 1975.

Polit, D., and Hungler, B. *Nursing Research: Principles and Methods.* Philadelphia: J.B. Lippincott, 1978.

CHAPTER 8

MAKING INFERENCES ABOUT MEANS: THE t-TEST

OVERVIEW

One of the most common methods of statistical inference is *hypothesis testing*, a technique that aids the researcher in making a decision about the values of a population parameter based on what is observed in a sample from that population. Examples of situations in which hypothesis testing may be used are: (1) decide whether a particular patient counseling program reduces presurgery anxiety; (2) determine which of two nursing procedures is most beneficial in reducing the number of decubitus ulcers in patients; and (3) decide which method of hygiene instruction results in retention of learning in children after they have left the hospital.

Hypothesis testing, as suggested in Chapter 6, follows a commonly accepted sequence of actions. For the remainder of this chapter, the procedure used to test a hypothesis will be presented in terms of six basic aspects.

1. The nature of the data.
2. The assumptions necessary for the test to be valid.
3. The hypothesis to be tested.
4. The calculation of the test statistic.
5. The formulation of a decision rule regarding rejection of the null hypothesis.
6. The decision or conclusion to be made.

It should be pointed out that while the above six steps are somewhat arbitrary, it was felt that presenting the material in this manner would aid the student in understanding and carrying out these techniques.

OBJECTIVES

Upon completion of this chapter, the student will be able to:

1. State the assumptions and nature of the data necessary for the use of a simple t-test.
2. State the assumptions and nature of the data necessary for the use of the pooled t-test and the paired t-test.
3. Carry out a test of hypothesis about a single mean using a simple t-test.
4. Interpret results of a simple t-test, pooled t-test, and paired t-test.
5. Carry out a test of hypothesis using a pooled t-test and a paired t-test.

DISCUSSION

Basic Steps of Hypothesis Testing

Before beginning a discussion of the t-test, we will first discuss briefly each of the six steps in any statistical test of hypothesis.

1. *Nature of the data.* The choice of a statistical test often depends on the nature of the variable(s) of interest in the investigation; that is, the choice of statistical analysis depends on the underlying scale of measurement of the variable(s) and whether or not the data are discrete or continuous in nature.
2. *Assumptions.* As in Step 1 the choice of a statistical test is dependent upon certain basic conditions that must be met in order for the test to be valid. Different tests often have different assumptions. These include whether the population from which the sample data were taken follows a normal distribution, the method in which the sample was drawn, and the equality of variances for two populations being compared. For the purposes of simplicity, not all the assumptions for each test will be presented. In practice, many of the assumptions are often relaxed; therefore, only the most basic will be discussed.
3. *Hypothesis to be tested.* The hypotheses to be tested usually consist of statements about the value of a parameter from one or more populations. The *null hypothesis,* set up for the purpose of

being discredited, is usually an hypothesis of "no difference." The alternate hypothesis, in general, corresponds to the research hypothesis of the study. Rejection or nonrejection of the null hypothesis may not be taken as "proof" of the hypothesis; it merely indicates whether the available data support or do not support the alternate, or research hypothesis.

4. *Test statistic*. The test statistic is a value that is calculated from the sample data. Its magnitude reflects the degree of the discrepancy between the hypothesized value of the parameter and the corresponding value calculated from the sample. In general, "large" values of the test statistic indicate a large discrepancy between the hypothesized and sample values, thus leading to rejection of the null hypothesis. In this manner, the test statistic functions as the decision-maker in a test of hypothesis.
5. *Decision rule regarding rejection of the null hypothesis*. All possible values that the test statistic can assume are divided into those values that are likely to occur if the null hypothesis is true and those values that have a small likelihood of occurrence if the null hypothesis is true *(rejection region)*. The desired *level of significance* of the test, α, dictates where the cut-off point lies for the rejection region. While choice of level of significance is up to the investigator, the most common values of α are .05 and .01. As the value of α becomes smaller, the cut-off values for the rejection region move further away from the hypothesized μ. For example, the test statistic must be much larger to reject at the .01 level than at the .10 level. This is a reflection of the fact that α is, in reality, the likelihood, or probability of rejecting the null hypothesis *when it is true*. Therefore, as the chance of incorrectly rejecting a true H_0 is reduced, the values of the cut-off points become larger, thus requiring larger discrepancies between the hypothesized and sample values in order to reject H_0.
6. *Decision or conclusion*. If the value of the test statistic (Step 4) falls in the rejection region (Step 5), the null hypothesis is rejected in favor of the alternate hypothesis. Otherwise it must be concluded that the data do not provide enough evidence to reject the null hypothesis. This is called the statistical conclusion. An investigator must now go one step further and translate the statistical conclusion into a decision formulated in terms of the investigation; that is, the outcome of the investigation needs to be stated in nonstatistical terminology.

Hypothesis Tests About the Mean of a Single Population: The Simple *t*-Test

In Chapter 6, a test of hypothesis about the true mean μ for a single population was discussed. It was assumed for this test that the true population standard deviation σ was known. In most situations encountered in practice, however, the true value of σ is rarely known. When σ is unknown, its estimate s must be calculated from the sample and used in place of σ. Because σ is not known and has been estimated from the sample, we can no longer calculate the z statistic and use its distribution (the normal) in testing hypotheses about the true mean μ of the population. Instead, a statistic called t, and its corresponding distribution, the t-distribution, is employed. The t-distribution (sometimes called the Student-t distribution) is similar in most respects to the z described previously. The frequency distribution of the t statistic, while bell-shaped and symmetric like the normal (the distribution of the z statistic), is somewhat more widely spread than the curve for the z distribution. Its exact shape depends on the sample size n. When the sample size is large, the frequency distributions for t and z are almost indistinguishable.

The general procedure for testing hypotheses about μ using the t-statistic is identical with that described for z except for the substitution of s for σ in the formula for the test statistic and the use of the curve for the t-distribution for specifying the cut-off points for the rejection region for the test.

The general steps for carrying out tests of hypotheses about population means using the t-statistic will now be presented in terms of the six steps of hypothesis testing.

1. *Data*. The data consist of *one* sample of measurements from a *single* population. In theory, the data must represent measurements on at least an interval scale made on a *continuous* variable. In practice, however the t-test is often applied to discrete data as long as there are enough categories to approximate a continuum.
2. *Assumptions*. The population of measurements from which the sample was taken is assumed to be normally distributed. This is a relatively safe assumption for most measurements of naturally occurring phenomena, e.g., blood pressure, cholesterol level, height, and weight.

    † When sampling is from a population which is not nor-

† Sections marked by a dagger (†) cover more difficult or theoretical topics.

mally distributed, results of the Central Limit Theorem allow the use of normal theory when the sample size is large enough. (See Chapter 4.)

3. *Hypothesis to be tested.* The null hypothesis is that the mean μ of the population of interest is equal to some specified value. Symbolically, the null hypothesis is:

$$H_0: \quad \mu = \mu_0$$

where μ_0 represents any value. (Later in this unit, the null hypothesis of the equality between two population means will be discussed. For the present, the discussion will be restricted to the one-sample or simple t-test.) The alternate hypothesis reflects the research hypothesis posed by the investigator. This hypothesis, for the single sample case, may take one of three forms:

$$H_a: \quad \mu \neq \mu_0$$

or

$$H_a: \quad \mu > \mu_0$$

or

$$H_a: \quad \mu < \mu_0$$

For example, a researcher may know from previous experience or from the literature that the mean systolic blood pressure in the general population is 120 mmHg. As part of a study of stress levels of nursing students, the researcher may formulate the research hypothesis, "the mean systolic blood pressure of nursing students is higher than that for the general population." The null hypothesis would be

$$H_0: \quad \mu_{NS} = 120$$

or, in words, "there is no difference in mean systolic blood pressure for nursing students and for the general population." The alternate hypothesis would be, "the mean systolic blood pressure of nursing students is higher than that for the general population." Symbolically,

$$H_a: \quad \mu_{NS} > 120$$

In this example, if the investigator had been interested in detecting a difference in mean systolic blood pressure of nursing students and the general population in either a positive or negative direction, then the alternate hypothesis would have been

$$H_a: \quad \mu_{NS} \neq 120$$

4. *Test statistic.* The value that estimates μ calculated from the sample is the sample mean $\overline{Y}$. The decision-maker, or test statistic, that quantifies the discrepancy between the hypothesized μ_0 and the calculated $\overline{Y}$ is called the *t*-statistic. The value of t is:

$$t = \frac{\overline{Y} - \mu_0}{s/\sqrt{n}}$$

where $\overline{Y}$ = sample mean value, μ_0 = hypothesized value of μ specified by H_0, s = standard deviation calculated from the sample (see Chapter 3):

$$s = \sqrt{\frac{\Sigma Y^2 - \frac{(\Sigma Y)^2}{n}}{n - 1}}$$

and n = number of observations (measurements) in the sample. Note that the formula for t is identical with that for z given in the previous chapter except for the substitution of s for σ in the denominator.

5. *Decision rule.* To determine the rejection region for the *t*-statistic, the level of significance α must first be specified. The cut-off value is then found from a table of *t*-values (Table A in the Appendix). In addition, to level of significance, a quantity called *degrees of freedom* (*df*) must be calculated in order to obtain a value from the table. Degrees of freedom (*df*) is equal to $n - 1$. Symbolically,

$$df = n - 1$$

where n is the sample size. When the alternate hypothesis being tested is of the form H_a: $\mu \neq \mu_0$, we obtain a value from the table, by finding the column headed by the desired level of significance divided by 2, i.e., $\alpha/2$ and the row with the desired df. The appropriate cut-off point for the rejection region is the intersection of this column and this row. For example, suppose there are 10 measurements in the sample and we wish to test at the .05 level of significance. We proceed down the column headed $\alpha/2 = .025$ until we reach the row labeled $df = 9(df = n - 1 = 10 - 1 = 9)$. The value at the intersection is 2.2622. Pictorially, the rejection region consists of all values of the test statistic that fall on the horizontal axis under the shaded areas shown below. (Note that the rejection region is located in both tails of the curve.) Table A gives only positive values. The lower tail value is found simply by affixing a minus sign to the positive value.

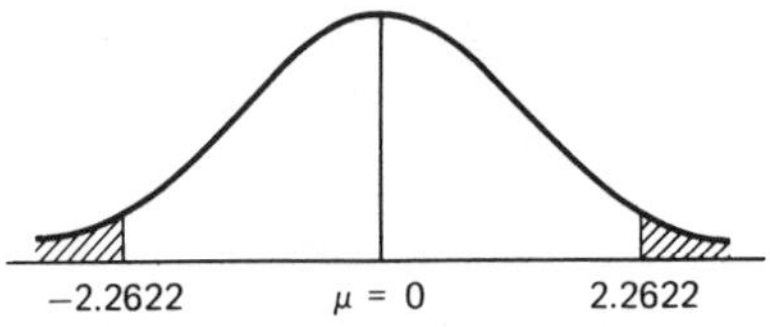

The procedure for finding one-tailed rejection regions, corresponding to $H_a: \mu < \mu_0$ or $H_a: \mu > \mu_0$, is given in a later section.

6. *Conclusion, or decision.* If the value calculated for the test statistic (Step 4) falls on the horizontal axis under the shaded areas, then H_0 is rejected. If the null hypothesis is rejected using an .05 level of significance, then there is a 5% chance that H_0 has been rejected when in fact it is true. If the value of the test statistic falls on the horizontal axis under the nonshaded area, then we must state that we do not have enough evidence to discredit the null hypothesis.

EXAMPLE 8.1 In a study to determine the effect, if any, of mother's diet on birth weight of infants, an investigator randomly selected nine women who were in their third month of pregnancy and placed them on a strict diet. At birth, the weights of the nine infants were recorded. From experience, it is known that the average birth weight for all infants is approximately 7 pounds. Based on the data shown below, can it be said that the birth weight of the infants born to mothers on this particular diet is different from the normal average birth weight? The birth weights of the nine infants (in pounds) are:

5.5, 6.0, 7.0, 5.5, 8.0, 7.0, 8.0, 5.5, 6.0

SOLUTION

1. *Data*. The data consist of birth-weight measurements made on nine infants. The measurements have at least an interval scale and are continuous in nature.
2. *Assumptions*. The sample of measurements was randomly selected from a population of birth weights that are assumed to be normally distributed.
3. *Hypotheses*. The investigator would like to determine if mean birth weight for infants whose mothers were on the special diet is different from the mean birth weight of normal infants. In null form, it is hypothesized that the birth weight of the infants is equal to the normal average birth weight, i.e., 7 pounds. Symbolically,

$$H_0: \quad \mu = 7$$

The alternative of interest is that birth weight of infants whose mothers were on a special diet is not equal to the normal average birth weight. Symbollically,

$$H_a: \quad \mu \neq 7$$

4. *Test statistic*. The values calculated from the sample are:

$$\overline{Y} = \frac{\Sigma Y}{n} = \frac{58.5}{9} = 6.5 \text{ pounds}$$

$$s = \sqrt{\frac{\Sigma Y^2 - \dfrac{(\Sigma Y)^2}{n}}{n-1}} = \sqrt{\frac{388.75 - \dfrac{(58.5)^2}{9}}{8}} = 1.03$$

The test statistic is:

$$t = \frac{\overline{Y} - \mu_0}{s/\sqrt{n}}$$

$$= \frac{6.5 - 7.0}{1.03/\sqrt{9}} = \frac{-.5}{.34} = -1.47$$

5. *Decision rule*. The determination of the rejection region depends on the specified level of significance.

Suppose the investigator wishes to test at the $\alpha = .05$ level of significance, thus risking a 5% chance of erroneously rejecting H_0. The cut-off point for the rejection region is then found by locating in Appendix Table A the intersection of the column headed $\alpha/2 = .025$ and the row labeled $df = n - 1 = 8$. The rejection region consists of values of the test statistic on the horizontal axis under the shaded areas shown below.

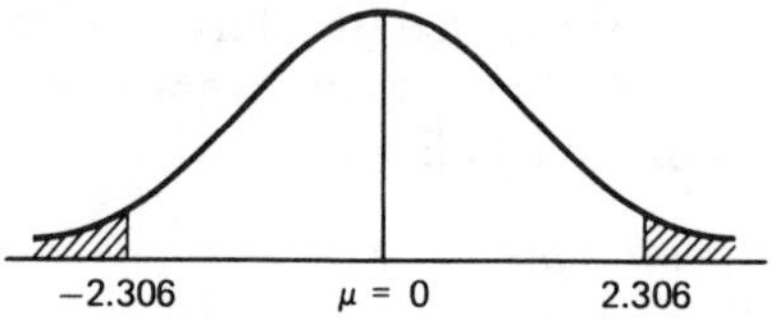

The rejection region has two tails because the researcher is interested in detecting differences in either a positive or negative direction. That is, either small negative or large positive values of the test statistic are evidence against H_0 and in favor of H_a.

6. *Conclusion*. Since the value of the test statistic (Step 4) does not fall on the horizontal axis under the shaded areas (Step 5), the investigator cannot reject the null hypothesis. It must be concluded that the data do not provide enough evidence to show that the birth weight of infants born to mothers on the special diet is significantly different from the average normal birth weight (at the .05 level of significance).

EXAMPLE 8.2 In a sociological study, it was determined that the average amount of money spent on food per week for a four member family in the United States is $50 (hypothetical study). A public health nurse wished to determine if families who had received instruction in nutrition would, on the average, spend an amount that was different from the amount spent by the population as a whole. Ten four-member families were selected at random and counseled for one month on good nutritional practices. For the following four weeks, the amount of money each family spent on groceries was recorded. The weekly averages (in dollars) for the 10 families were:

50, 70, 65, 45, 80, 75, 85, 70, 50, 70

SOLUTION

1. *Data*. The data, which consist of measurements of the average weekly amount spent for food by the 10 families, have at least an interval scale and are continuous in nature.
2. *Assumptions*. It is assumed that the 10 measurements were randomly selected from a population of measurements that is normally distributed.
3. *Hypotheses*. The null hypothesis is that the average amount spent weekly for food by the group receiving instruction is the same as for all four-member United States families. Symbolically,

$$H_0: \quad \mu = 50 \text{ dollars}$$

The alternate hypothesis is that the average amount spent weekly by this group is not the same as for all four-member United States families.

$$H_a: \quad \mu \neq 50 \text{ dollars}$$

4. *Test statistic*. From the data, we find that the average amount spent for food per week by families who have been instructed in nutrition is:

$$\overline{Y} = \frac{\Sigma Y}{n} = \frac{660}{10} = 66 \text{ dollars}$$

and the standard deviation, s, of this sample of measurements is:

$$s = \sqrt{\frac{\Sigma Y^2 - \frac{(\Sigma Y)^2}{n}}{n-1}} = \sqrt{\frac{45200 - \frac{(660)^2}{10}}{9}}$$

$$= 13.5$$

The test statistic is:

$$t = \frac{\overline{Y} - \mu_0}{s/\sqrt{n}}$$

$$= \frac{66 - 50}{13.5/\sqrt{10}} = 3.75$$

5. *Decision rule*. Assume that the investigator wishes to test at the .01 level of significance. The cut-off points for the rejection region are found by locating in Appendix Table A the intersection of the column headed $\alpha/2 = .005$ and the row labeled $df = n - 1 = 9$. The value is 3.2498. The rejection region consists of values of the test statistic on the horizontal axis under the shaded areas shown below.

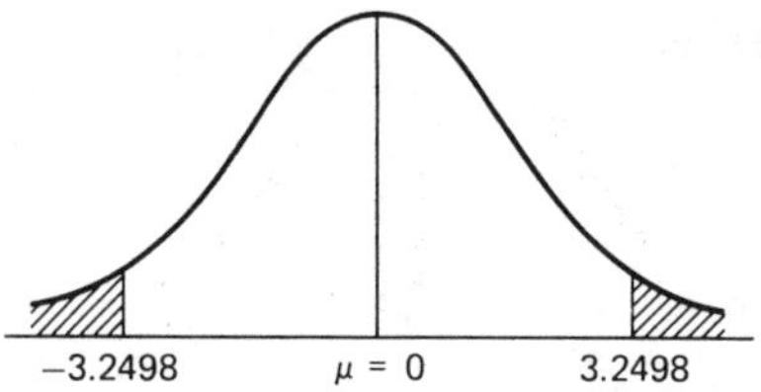

Since the alternate hypothesis specifies that the investigator wishes to detect a departure from the null hypothesis in either a positive or negative direction, the rejection region consists of both an upper and lower tail. Values of the test statistic that are either large positive or small negative values are evidence in favor of H_a, that is, favor rejection of H_0.

6. *Conclusion*. The value of the test statistic (Step 4) falls in the rejection region (shaded areas). Therefore, the investigator may reject H_0 and conclude that the average amount spent weekly for food by the nutritional group is not the same as for the United States population as a whole. Since the level of significance was specified to be .01, there is a 1% chance that the null hypothesis has been rejected when in fact it is true.

Review: *t*-Test for Single Samples

1. *Null hypothesis*

$$H_0: \quad \mu = \mu_0$$

Alternative (two-tailed)

$$\mu \neq \mu_0$$

2. *Test statistic*

$$t = \frac{\overline{Y} - \mu_0}{s/\sqrt{n}}$$

3. *Decision rule.* Cut-off point found at intersection of Appendix Table A column headed $\alpha/2$ (α = level of significance) and row labeled $df = n - 1$.
4. *Decision.* Reject if test statistic falls in rejection region (outer tails of curve). Otherwise, do not reject. If H_0 has been rejected at an α level of significance, then the chance that H_0 has been rejected when it is true is $\alpha\%$.

EXERCISE 8.1

A researcher hypothesized that production of glucose-6-dehydrogenase in the blood is different for alcoholics and normal "nondiseased" adults. For nine adult male patients newly admitted to an alcohol treatment center, the mean glucose-6-dehydrogenase level in the blood was found to be 450 units/10^9 cells and the standard deviation, s, of the sample was found to be 60 units/10^9 cells. The adult normal value for glucose-6-dehydrogenase is 375 units/10^9 cells. Do the data present sufficient evidence to support the researcher's hypothesis? (Use α = .05.) Follow the six steps for hypothesis testing outlined above.

1. *Data*
2. *Assumptions*
3. *Hypotheses*
4. *Test statistic*
5. *Decision rule*
6. *Conclusion*

EXERCISE 8.2

An investigator wishes to study the effect of fondling on the body temperature of infants. Nine infants were selected at random and were held for 30 minutes prior to having their temperature recorded.

The "normal" body temperature for infants is assumed to be 99.6° F. The temperatures recorded for the nine infants in the sample are:

100.8, 101, 99.3, 100.4, 101.1, 100.6, 99.6, 101, 99.8

Can it be said that fondling affects the body temperature of infants? Follow the six-step hypothesis testing procedure. (Use $\alpha = .05$.)

One-tailed Tests of Hypothesis

EXAMPLE 8.3 The national average on a standardized exam given to graduating seniors to test general nursing knowledge is 60. A group of 16 student nurses were randomly selected to participate in a special educational project. There were no formal classroom lectures; all teaching was in the form of clinical demonstrations, slides, and self-instructional texts. The investigator wished to determine if the experimental group performed better (had a higher mean score) than the national average. The mean score of the experimental group was 65 with a standard deviation, s, of 20. What conclusions can be drawn based on the data using an .05 level of significance?

In Example 8.3, notice that the investigator wished to determine if the mean score for the experimental group was *greater than* the norm for the exam. In all previous examples, we have been concerned with showing only that there is a difference (either greater than or less than) between the hypothesized mean and the sample mean. Tests of hypothesis in which we are concerned with the direction of the difference are called *one-tailed tests of hypothesis*, as opposed to the two-tailed tests in the previous examples. The name is taken from the fact that for two-tailed tests, the rejection region consists of two-tails (an upper and a lower). For one-tailed tests, the rejection region consists of only a single tail. When it is of interest to determine if the true value of μ is significantly *greater* than the hypothesized μ_0, then the rejection region is in the upper tail (righthand side). Alternatively, when it is of interest to determine if the true value of μ is significantly *less than* the hypothesized μ_0, then the rejection region is in the lower tail (lefthand side) (see Figure 8.1).

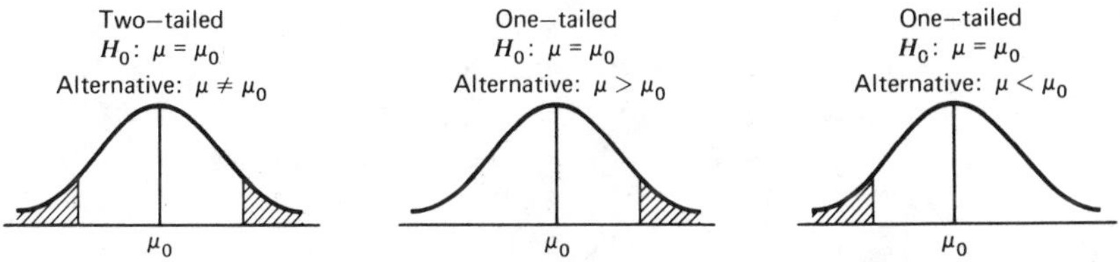

FIGURE 8.1 REJECTION REGIONS FOR TESTS OF HYPOTHESIS

SOLUTION Returning to Example 8.3, to carry out the test of hypothesis, we again follow the six-step procedure.

1. *Data.* The data, which consist of 16 scores on a standardized exam, are continuous in nature. (Score data will be treated as if the underlying scale of measurement is interval.)
2. *Assumptions.* It is assumed that the population of test scores from which the random sample of 16 scores was taken is normally distributed.
3. *Null hypothesis.* The investigator would like to determine whether the experimental group scored higher than the national average on the standardized exam. In null form, it is hypothesized that the mean for the experimental group is the same as the national average. Symbolically,

$$H_0: \quad \mu = 60$$
$$H_a: \quad \mu > 60$$

   where the alternate hypothesis, representing what will be accepted as true upon rejection of H_0, reflects the research hypothesis formulated by the investigator.
4. *Test statistic.* The test statistic is calculated as before.

$$t = \frac{\overline{Y} - \mu_0}{s/\sqrt{n}}$$

$$= \frac{65 - 60}{20/\sqrt{16}} = \frac{5}{5} = 1.0$$

5. *Decision rule.* The rejection region for a one-tailed test is found by placing the entire α in a single tail. (Recall that for two-tailed tests α was divided by 2 in order to locate the cut-off point in Appendix Table A.) To find the cut-off point, we find the intersection of the column headed $\alpha = .05$ (rather than $\alpha/2 = .025$) and the row $df = n - 1 = 15$. The rejection region consists of values of the test statistic on the horizontal axis under the shaded area shown below.

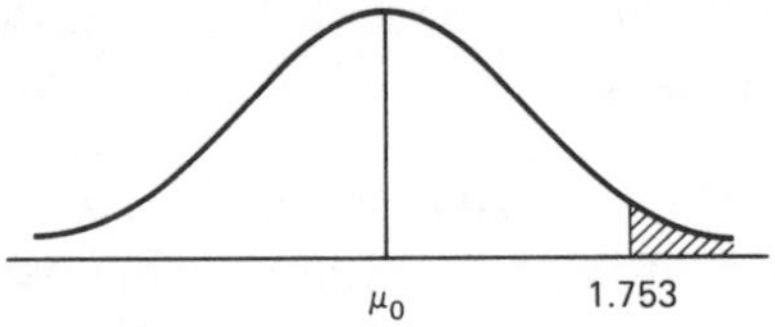

6. *Conclusion.* Since the value of the test statistic (Step 4) does not fall into the rejection region (Step 5), H_0 cannot be rejected. It must be concluded that the data do not provide enough evidence to show that the experimental group scored higher than the national average on the standardized nursing exam.

EXERCISE 8.3

A researcher hypothesizes that premature infants have lower urine chloride levels than the clinical norm for full-term infants. Suppose that the mean urine chloride level in normal full-term infants is assumed to be 210 mEq/24 hr. The mean urine chloride for nine randomly selected premature infants was found to be 205 mEq/24 hr with a standard deviation of 12 mEq/24 hr. Can it be concluded that premature infants have a lower urine chloride level than the clinical norm? (Use $\alpha = .01$.) Follow the six-step hypothesis testing procedure.

Further Interpretation of Single-sample Tests of Hypothesis

Either in carrying out a research project or evaluating the results of a project carried out by others, there are several important points about tests of hypothesis that must be considered.

All tests of hypothesis refer to a null hypothesis that must be either explicitly stated or clearly implied. A test of hypothesis determines whether the observed sample statistic is a likely or an unlikely occurrence if H_0 is true. Unlikely values of the test statistic lead to rejection of the null hypothesis at the specified level of significance (α). The specified level of significance states the risk of incorrectly rejecting the null hypothesis when it is true. Likely values of the test statistic (those which fall into the nonrejection region) do not constitute "proof" that the null hypothesis is actually true. For this reason, we never "accept" the null hypothesis, but rather we fail to reject H_0.

The level of significance (α) must *always* be clearly stated. The choice of a particular level of significance for a test specifies the chance of an incorrect decision regarding rejection of the null hypothesis that the investigator is willing to risk. One investigator may be willing to accept a 10% chance ($\alpha = .10$) of incorrectly rejecting while another investigator is willing only to reject H_0 when the risk of doing so falsely is only 1% ($\alpha = .01$).

As pointed out earlier, often tests of significance will be accompanied by p values. The p values report the actual probability of observing a given test statistic when H_0 is true. Thus, a statement that a given test is significant at $p \leq .01$ means that the chance of obtaining the particular value of the test statistic given that the null hypothesis is true is only 1 in 100, or 1%. A α level, on the other hand, is a value which is determined prior to data collection and which states the risk of error the investigator is willing to accept.

Another important consideration in the evaluation of results is that all tests of hypothesis are based on the assumption of random sampling from the population of interest. Unfortunately, in the nursing profession as in all medical areas, the use of strict random sampling is not always feasible. The method of sample selection should be evaluated in terms of how far it departs from strict random sampling, and how this departure may bias the representativeness of the sample in terms of the target population.

The last major point to consider in evaluating tests of significance is the adequacy of the design of the study. In most of the examples discussed previously, the sample values were compared to a clinical norm or to past experience. In such cases, the statistical analysis may imply that patients who have received a particular treatment or procedure are statistically different from patients who received an old procedure (the value specified under H_0). Even though the statistical analysis has been carried out correctly, we might be suspicious of the actual appropriateness of the design of the study. The patients who receive the new treatment are compared to past patients who did not receive the procedure. The apparent difference between the new group and the "historical" group may be attributed to several factors. Among these are: (1) techniques and facilities for patient care may have improved in general since patients were treated under the old procedure; in drug studies, there is always the possibility that the nature of the disease being treated has changed over time; (2) the psychological effect of being part of an experimental group may affect patient response; and (3) unconscious supportive therapy on the part of the investigator not present in the historical controls may contribute to the observed difference between the two groups.

Obviously, when using historical controls to measure the efficacy of a new treatment or procedure an investigator must be extremely careful to avoid the above effects, which may invalidate the results of the study. One method for overcoming many of the shortcomings of historical controls is to use *concurrent* sampling. Under this design, a sample of patients may receive one treatment while another sample of patients receive another treatment. Each sample is treated as alike as possible except for the treatment under study. The use of concurrent sampling, or sampling from two populations, will be discussed in the next section.

(Note: The term "treatment" refers to anything that is done to a study unit whether it is a drug, a teaching method, a nursing procedure, etc.)

EXERCISE 8.4

The following paragraph is an excerpt from a hypothetical research study:

> The investigator wished to determine whether a 20-minute counseling period with presurgery patients by a trained nurse psychologist would result in a lowering of blood pressure. (The mean systolic blood pressure for presurgery patients in this particular environment was known to be 150 mmHg). Five patients selected from among a group of presurgery patients were counseled for 20 minutes on the day prior to their surgery. A *t*-test performed on the blood pressure readings showed conclusively that counseling does not in fact lower blood pressure.

Critique the above excerpt with respect to null hypothesis, test statistic, and validity of conclusion.

EXERCISE 8.5

A curriculum committee wishes to evaluate the efficacy of a new teaching program in medical-surgical nursing. Under this new approach, students are given self-instructional materials, including slides and demonstrations covering all topics in the course. For a four-week period, the students spend eight hours per day in clinical experience with ample opportunity given for discussions with an instructor. At the end of the four-week period, the students are given a standardized proficiency exam on the course content. The national average for this exam is 70 (out of a total of 100 possible points).

a. Discuss briefly how you as a member of the curriculum committee would carry out this research project. Include a discussion of null hypothesis, sample selection, and method of analysis.
b. Suppose the study is carried out and it is reported that students participating in the program score significantly higher than the national average at the .20 level of significance. Discuss what this conclusion means in terms of significance level. Would you recommend that this new method be instituted? Why or why not?

Hypothesis Tests: The Difference Between Two Population Means

Hypothesis tests about two population means attempt to answer the question "if two groups are treated differently (i.e., receive different treatments or procedures), will there be a difference in their resulting behavior?" Essentially, we are attempting to determine if there is a "real" (statistically significant) difference in the mean response of two groups who are treated alike except for the treatment under investigation. From the previous unit, we know that differences in the calculated sample means for the two groups could be due to (1) sampling variation of mean values (sampling error), or (2) the sample mean values could have been obtained from two populations with different true mean values.

Two types of t-tests for testing significance of differences between means will be presented: the *pooled* t-test and the *paired* t-test. The distinction between these two lies in the method in which the sample is drawn. Both the pooled t-test and the paired t-test will be presented in terms of the six-step hypothesis testing procedure.

t-TEST FOR INDEPENDENT GROUPS

1. *Data.* The data usually consist of two sets of sample measurements that possess at least an interval scale and are continuous in nature.

2. *Assumptions*. One of the basic assumptions of the pooled t-test is that the data constitute two *independent random samples* drawn from populations that are normally distributed. One additional assumption is that the standard deviations of the two populations are equal, i.e., $\sigma_1 = \sigma_2$. It is usually not unreasonable to expect that the two populations contain roughly the same amount of variation among measurements.*

   † When the populations from which the two samples are drawn are not normally distributed, for sufficiently large sample sizes, normal theory is still applicable, based on results of the Central Limit Theorem.
3. *Hypotheses*. In the two sample cases, the question of interest is whether the mean of one population is equal to the mean of some other population. Samples have been drawn concurrently from the two populations. Symbolically, the null hypothesis is written:

$$H_0: \quad \mu_1 = \mu_2, \quad \text{or} \quad \mu_1 - \mu_2 = 0$$

   Rejection of the null hypothesis may be in favor of one of the following alternatives:

$$H_a: \quad \mu_1 \neq \mu_2, \quad \text{or} \quad \mu_1 - \mu_2 \neq 0$$

$$H_a: \quad \mu_1 > \mu_2, \quad \text{or} \quad \mu_1 - \mu_2 > 0$$

$$H_a: \quad \mu_1 < \mu_2, \quad \text{or} \quad \mu_1 - \mu_2 < 0$$

   Note that while the most common null hypothesis is that the difference between the two population means is zero (i.e., $\mu_1 = \mu_2$), we may also test whether the difference is equal to some value other than zero.
4. *Test statistic*. The test statistic provides a measure of the discrepancy between the hypothesized difference in the true means (usually zero) and the observed difference in the sample means. The test statistic or decision-maker for the pooled t-test is:

$$t = \frac{(\bar{Y}_1 - \bar{Y}_2) - (\mu_1 - \mu_2)}{s_p \sqrt{\dfrac{1}{n_1} + \dfrac{1}{n_2}}}$$

* The assumption $\sigma_1^2 = \sigma_2^2$ may be tested by carrying out a statistical test of hypothesis. The procedure for carrying out a test of H_0: $\sigma_1^2 = \sigma_2^2$ may be found in most intermediate-level statistics texts. If the assumption $\sigma_1^2 = \sigma_2^2$ is not valid, there is an approximate t-test that may be employed.

Since we are ordinarily testing whether $\mu_1 - \mu_2 = 0$, the test statistic simplifies to:

$$t = \frac{\overline{Y}_1 - \overline{Y}_2}{s_P\sqrt{\frac{1}{n_1} + \frac{1}{n_2}}}$$

where $\overline{Y}_1$ = mean of Sample 1, $\overline{Y}_2$ = mean of Sample 2, s_P = pooled or "average" standard deviation for the two samples, n_1 = number of measurements in Sample 1, and n_2 = number of measurements in Sample 2. The pooled or "average" standard deviation may be found by:

$$s_P = \sqrt{\frac{s_1{}^2(n_1 - 1) + s_2{}^2(n_2 - 1)}{n_1 + n_2 - 2}}$$

where s_1 = standard deviation of Sample 1 and s_2 = standard deviation of Sample 2. The pooled standard deviation is an estimate of the population standard deviation σ, common to both populations. (Recall that we have assumed $\sigma_1 = \sigma_2 = \sigma$.)

5. *Decision rule*. The rejection region for the pooled *t*-test is found by locating the cut-off point in Appendix Table A in the same manner as for the one-sample case. The only difference is that for the pooled case, degrees of freedom is defined as:

$$df = n_1 + n_2 - 2$$

For two-tailed alternatives (i.e., $\mu_1 - \mu_2 \neq 0$) the level of significance α is divided by 2 before locating the column in Appendix Table A. For one-tailed alternatives ($\mu_1 - \mu_2 > 0$, $\mu_1 - \mu_2 < 0$), the column headed by α is used. The rejection regions for the pooled *t*-test are summarized below.

H_0: $\mu_1 - \mu_2 = 0$	H_0: $\mu_1 - \mu_2 = 0$	H_0: $\mu_1 - \mu_2 = 0$
Alternative $\mu_1 - \mu_2 \neq 0$	Alternative $\mu_1 - \mu_2 > 0$	Alternative $\mu_1 - \mu_2 < 0$
Column: $\alpha/2$	Column: α	Column: α
Row: $df = n_1 + n_2 - 2$	Row: $df = n_1 + n_2 - 2$	Row: $df = n_1 + n_2 - 2$

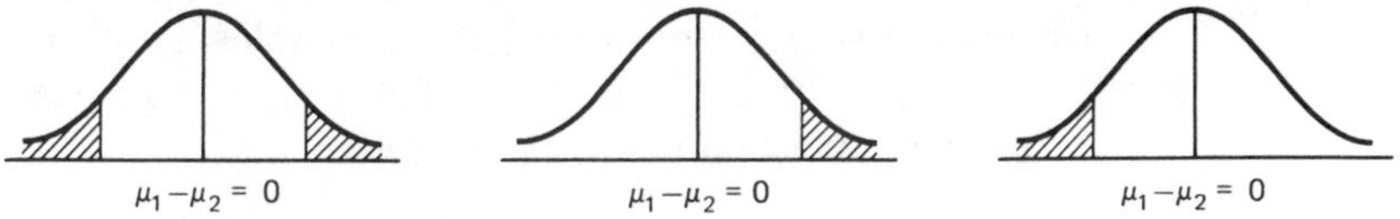

6. *Conclusion*. If the value of the test statistic (Step 4) falls in the rejection region (horizontal axis under the shaded areas), the null hypothesis of no difference between the population means is rejected at the specified α level. The level of significance specifies the risk of rejecting H_0 when it is true. If the value of the test statistic falls on the horizontal axis in the nonshaded regions, then we must conclude that the data do not provide enough evidence to show that the means for the two populations are different (i.e., we do not reject H_0).

EXAMPLE 8.4 A researcher hypothesized that hospitalized children whose parents participate in their care have lower anxiety levels than hospitalized children whose parents do not participate. Eighteen hospitalized children aged 6 to 10 were selected. Nine children were randomly assigned to the experimental group and nine to the control group. The parents of children in the experimental group were allowed to participate in the routine care of the child, while for the control group the parents were allowed to visit as usual but all care of the children was carried out by the nurse. At the end of a one-week period, all of the children were given a standardized 100-point anxiety exam. The mean anxiety score for the experimental group was 45 with a standard deviation, s, of 20. The mean anxiety score for the control group was 55 with a standard deviation, s, of 25. Can it be said that hospitalized children whose parents participate in their care have lower anxiety levels than hospitalized children whose parents do not participate? (Use an .05 level of significance.)

SOLUTION

1. *Data*. The data consist of nine anxiety scores for the experimental group and nine anxiety scores for the control group. The scores represent measurements on a continuous scale and will be treated as if they possess interval scaling.
2. *Assumptions*. In this study, there are two random samples: one representing the population of children whose parents participate in their care and one representing the population of children whose parents do not participate. It is further assumed that the populations of scores from which these samples were taken are normally distributed and that the standard deviations for the two populations are equal ($\sigma_1 = \sigma_2$).

3. *Hypotheses*. The investigator's research hypothesis is that the true population mean anxiety score for the experimental group is less than the true mean for the control group. Stated in null form, the hypothesis to be tested is that there is no difference between the two population means, i.e., the mean anxiety level for children whose parents participate in their care is no different than the true mean anxiety level for children whose parents do not participate. Symbolically, this is written:

$$H_0: \quad \mu_E - \mu_C = 0$$

Upon rejection of H_0, the alternative of interest is that the true mean anxiety score for the experimental group is *less than* the true mean anxiety score for the control group. Symbolically, the alternate hypothesis is:

$$H_a: \quad \mu_E - \mu_C < 0$$

4. *Test statistic*. The following information was calculated from the sample measurements:

Experimental	*Control*
$\overline{Y}_E = 45$	$\overline{Y}_C = 55$
$s_E = 20$	$s_C = 25$
$n_E = 9$	$n_C = 9$

To calculate the test statistic we must obtain an estimate of the pooled standard deviation for the two samples.

$$s_P = \sqrt{\frac{s_E{}^2(n_E - 1) + s_C{}^2(n_C - 1)}{n_E + n_C - 2}}$$

$$= \sqrt{\frac{(20)^2(8) + (25)^2(8)}{9 + 9 - 2}}$$

$$= 22.6$$

The test statistic is:

$$t = \frac{\overline{Y}_E - \overline{Y}_C}{s_P\sqrt{\dfrac{1}{n_E} + \dfrac{1}{n_C}}} = \frac{45 - 55}{22.6\sqrt{\dfrac{1}{9} + \dfrac{1}{9}}}$$

$$= -.94$$

5. *Decision rule.* Since the investigator is interested in showing that the mean of the experimental group is *less than* the mean of the control group, the rejection region is located in the lower tail of the curve (a one-tailed alternative). The chosen level of significance is $\alpha = .05$ and $df = n_E + n_C - 2 = 9 + 9 - 2 = 16$. The cut-off point for the rejection region is found in Appendix Table A at the intersection of the column headed $\alpha = .05$ and the row labeled $df = 16$. The rejection region is shown below.

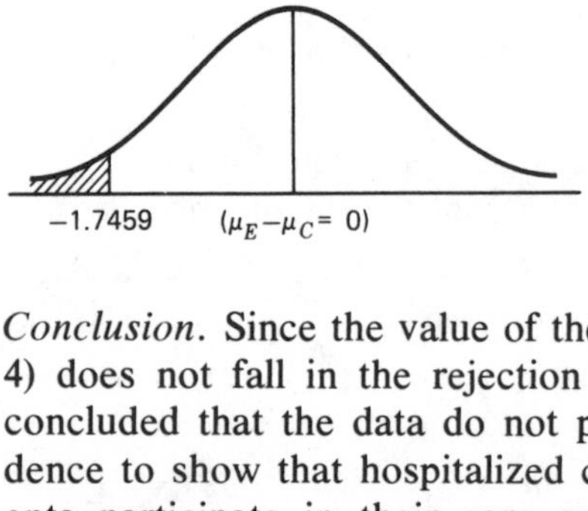

6. *Conclusion.* Since the value of the test statistic (Step 4) does not fall in the rejection region, it must be concluded that the data do not provide enough evidence to show that hospitalized children whose parents participate in their care are significantly less anxious than hospitalized children whose parents do not participate in their care at the $\alpha = .05$ level of significance.

EXAMPLE 8.5 A public health nurse was interested in determining the effect, if any, of pesticide exposure on blood pressure in adults. From a list of people known to have been exposed to pesticides, a random sample of 100 people was selected. Similarly, from a group of people for whom no exposure could be detected, a random sample of 100 people was selected. The mean systolic blood pressure for the pesticide group was found to be 145 mmHg with a standard deviation, s, of 20 mmHg. The mean systolic blood pressure for the not-exposed group was found to be 120 with a standard deviation of 15 mmHg. Can it be said that the true mean systolic blood pressure for the two groups is different (at the .01 level of significance)?

SOLUTION

1. *Data*. The data, which consist of systolic blood pressure readings on 100 pesticide-exposed individuals and on 100 individuals with no pesticide exposure, have at least an interval scale and are continuous in nature.
2. *Assumptions*. The two samples were independently and randomly selected from the two populations of interest. It is assumed that the underlying populations from which the samples were drawn are normally distributed with equal standard deviations.
3. *Hypotheses*. The null hypothesis is that the population mean systolic blood pressure for people exposed to pesticides is no different than the population mean systolic blood pressure of people not exposed to pesticides. Symbolically,

$$H_0: \quad \mu_1 - \mu_2 = 0$$

where 1 = exposed to pesticide and 2 = not exposed. The alternate hypothesis is that there is a difference in the mean systolic blood pressure for the two groups. (Notice that in this example the investigator is interested simply in showing that there is a difference, regardless of direction.) Symbolically, the alternative of interest is:

$$H_a: \quad \mu_1 - \mu_2 \neq 0. \qquad (\mu_1 \neq \mu_2)$$

4. *Test statistic*. The information obtained from the sample measurements is:

Exposed	*Not Exposed*
$\bar{Y}_1 = 145$	$\bar{Y}_2 = 120$
$s_1 = 20$	$s_2 = 15$
$n_1 = 100$	$n_2 = 100$

The pooled standard deviation is:

$$s_P = \sqrt{\frac{s_1^2(n_1 - 1) + s_2^2(n_2 - 1)}{n_1 + n_2 - 2}}$$

$$= \sqrt{\frac{(20)^2(99) + (15)^2(99)}{198}} = 17.7$$

The test statistic is:

$$t = \frac{\overline{Y}_1 - \overline{Y}_2}{s_P\sqrt{\frac{1}{n_1} + \frac{1}{n_2}}} = \frac{145 - 120}{17.7\sqrt{\frac{1}{100} + \frac{1}{100}}}$$

$$= 9.99$$

5. *Decision rule.* The hypothesis described in Step 3 requires a two-tailed rejection region since the investigator is interested in detecting a difference in the means in either a positive or negative direction. The level of significance was chosen to be $\alpha = .01$ and the $df = n_1 + n_2 - 2 = 198$. In Appendix Table A we must find the column headed $\alpha/2 = .005$ (a two-tailed test) and the row labeled $df = 198$. Since the value for $df = 198$ is not included in the table, we will select the value closest to it, i.e., $df = 200$. The value at the intersection of this row and this column is the cut-off point for the rejection region.

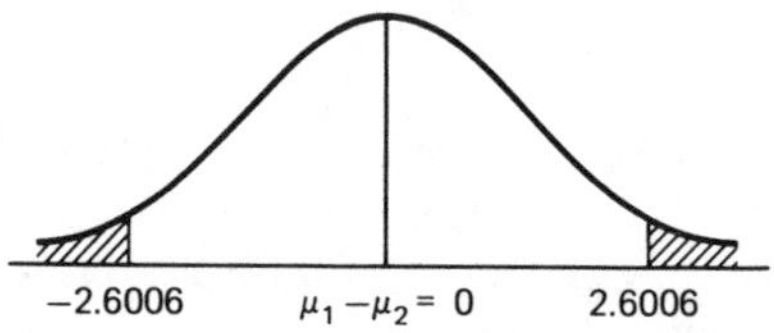

6. *Conclusion.* Since the calculated value of the test statistic falls in the rejection region, the null hypothesis is rejected and it may be concluded that there is a statistically significant difference (at the $\alpha = .01$ level) between true mean systolic blood pressure for

the exposed and not-exposed groups. The likelihood of rejecting the null hypothesis of no difference when in fact it is true is 1%.

t-TEST FOR PAIRED SAMPLES

As the name implies, for the paired case, *pairs* are randomly selected from the population. Each member of the pair is randomly assigned to one of two treatments. (The word "treatment" refers to the condition under study. A treatment may be a drug, a procedure, etc.)

The use of identical twins is a classic example of the paired situation. One twin receives Treatment A and one twin receives Treatment B. If one is interested in evaluating a particular teaching method and if pairings of identical twins are used, then the subjects are as nearly identical as possible (same age, sex, heredity, environment, etc.) except for the procedure under study. A significant difference in performance between the pairs can, therefore, for the most part, be attributed to the teaching method alone.

In addition to the use of naturally paired study subjects such as twins or animal litter mates, the investigator may artificially pair subjects on the basis of certain characteristics that have been deemed relevant to the variable being studied. For example, it is widely accepted that home environment, IQ, and age all play a role in learning ability. In a study to test the effect of a new teaching procedure, an investigator may wish to pair subjects on the basis of the above three characteristics. One member of the pair would then be randomly assigned to the treatment group (the group being taught by the new procedure) and the other member of the pair would be assigned to a control group (the group being taught by a traditional method).

The difficulty with the use of artificially matched study subjects is in selecting the relevant characteristics on which the pairings are to be based. Obviously, this is not a simple task. In most situations, it is impossible to match on all characteristics that could possibly affect subject response. Usually, subjects are matched on as many characteristics as can feasibly be carried out. The possibility, therefore, for unrecognized differences in matched study subjects is always present and must be kept in mind when results of such studies are evaluated.

The "before" and "after" measurements made on the same individual is another widely used method of pairing. In this type of paired situation, each individual serves as his own control.

Probably the most obvious advantage to using paired observations is the partial elimination of biological and other extraneous sources of variation. This results in a more precise comparison of sample means.

Schematically, the paired situation is shown below.

Sample I	*Sample II*	*Difference*
a_1	a_2	$a_1 - a_2 = d_1$
b_1	b_2	$b_1 - b_2 = d_2$
c_1	c_2	$c_1 - c_2 = d_3$
$\vdots$	$\vdots$	$\vdots$
		$\bar{d} = \frac{\Sigma d_i}{n}$

In the paired t-test, the individual differences between the pairs of observations are considered as a single sample of measurements and a single sample t-test is carried out using only these differences.

1. *Data*. The data consist of measurements made on pairs of individuals (or things), one member having received one treatment and another member having received another treatment.
2. *Assumptions*. The observed differences constitute a random sample from a population of differences that is normally distributed.
3. *Hypotheses*. The null hypothesis is that the population means of the two groups are equal, i.e., $\mu_{\text{I}} - \mu_{\text{II}} = 0$. In the paired situation, this hypothesis is equivalent to saying that the mean of the population of *differences* is equal to zero. Symbolically,

   $$H_0: \quad D = 0$$

   where D represents the true mean of the population of differences. The alternatives to be tested are analogous to the ones for the pooled t-case. The alternative may be:

   H_a: $D \neq 0$ (two-tailed)

   H_a: $D > 0$ (one-tailed)

   H_a: $D < 0$ (one-tailed)

4. *Test statistic*. The test statistic for the paired t-test is the same as the test statistic for the simple t-test.

   $$t = \frac{\bar{d} - D}{s_d / \sqrt{n}}$$

where $\bar{d}$ = mean of column of differences, s_d = standard deviation of column of differences, and n = number of *pairs* in the sample.

5. *Decision rule.* The rejection region is found in the same manner as for the simple *t*-test.
6. *Decision or conclusion.* If the value of the test statistic falls in the rejection region, it may be concluded that there is a statistically significant difference in the means of the two treatment groups (reject H_0) at the α level of significance.

EXAMPLE 8.6 A nurse researcher hypothesized that back rubs lower systolic blood pressure in geriatric patients. The systolic blood pressures of the nine randomly selected nursing home patients were taken prior to a back rub and then again immediately following a 10-minute back rub. The data are as follows:

Systolic Blood Pressure (in mmHg)

Patient	*Before Back Rub*	*After Back Rub*	*Differences*
1	140	130	−10
2	160	150	−10
3	180	180	0
4	140	150	+10
5	160	140	−20
6	180	160	−20
7	150	150	0
8	170	130	−40
9	180	180	0

$$\bar{d} = \frac{-90}{9} = -10$$

Can it be concluded that back rubs significantly lower systolic blood pressure in geriatric patients (at the .05 level of significance)?

SOLUTION

1. *Data.* The data consist of pairs of "before" and "after" blood pressure measurements on nine geriatric patients.
2. *Assumptions.* It is assumed that the column of differences constitutes a random sample from a population of differences that is normally distributed.

3. *Hypotheses*. The null hypothesis is that there is no difference in the "before" and "after" blood pressure measurements. This is equivalent to stating that the mean of the population of differences is zero.

$$H_0: \quad D = 0$$

The alternative of interest is that the "after" measurements are *less than* the "before" measurements. In terms of differences, this may be written:

$$H_a: \quad D < 0$$

4. *Test statistic*. To calculate the test statistic we must first calculate the mean of the column of differences ($\bar{d}$) and the standard deviation of the differences.

$$\bar{d} = \frac{\Sigma d_i}{n} = \frac{-90}{9} = -10$$

$$s_d = \sqrt{\frac{2700 - \frac{(90)^2}{9}}{8}} = 15$$

The test statistic is:

$$t = \frac{\bar{d} - D}{s_d/\sqrt{n}}$$

$$= \frac{-10 - 0}{15/\sqrt{9}} = \frac{-10}{5} = -2$$

5. *Decision rule*. To find the cut-off point for the rejection region, we locate the intersection of the column $\alpha = .05$ (one-tailed alternative) and the row $df = n - 1 = 8$. The value is -1.8595.

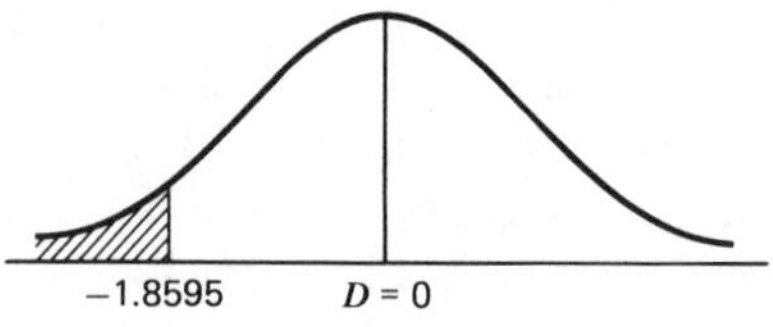

The rejection region consists of all values of the test statistic that fall *below* -1.8595.

6. *Decision or conclusion.* The calculated test statistic falls in the rejection region; the null hypothesis of no difference may therefore be rejected at the .05 level of significance. It may be concluded that back rubs significantly lower systolic blood pressure in geriatric patients (at the .05 level of significance). There is a 5% chance of erroneously rejecting H_0.

EXERCISE 8.6

Refer to Exercise 8.5.

a. Discuss how this study could be designed using a pooled *t*-test.
b. How could it be designed using a paired *t*-test?

EXERCISE 8.7

Refer to Exercise 2.6.

a. Discuss this study in terms of null hypothesis and method of analysis.
b. Suppose the investigator states that there is a significant increase in blood pressure after seven minutes of crying at $\alpha = .05$. Discuss what this conclusion means with respect to level of significance.

EXERCISE 8.8

Refer to Exercise 8.4.

a. How might this study be carried out using the pooled *t*-test?
b. The paired *t*-test? Discuss both of the above in terms of null hypothesis, sample selection, and method of analysis.

† EXERCISE 8.9

A researcher hypothesized that serum indirect bilirubin levels are different for premature and full-term infants. The serum indirect bilirubin levels were determined for five premature and five healthy full-term infants, as follows:

Premature	*Full-term*
1.0	2.0
2.0	4.0
3.0	6.0
2.0	4.0
2.0	4.0

If it is reasonable to assume that the two populations have equal standard deviations, i.e., $\sigma_1 = \sigma_2 = 1.0$ mg/100 cc, can it be concluded that the mean serum indirect bilirubin level is different for the two groups? (Use $\alpha = .05$.)

† EXERCISE 8.10

In order to determine whether a certain oral contraceptive results in weight gain, nine healthy females were weighed prior to the start of the medication and again at the end of a three-month period. The weights are listed below.

	Weight (in Pounds)	
Subject	*Initial*	*3 Months*
1	120	123
2	141	143
3	130	140
4	150	145
5	135	140
6	140	143
7	120	118
8	140	141
9	130	132

Is there sufficient evidence to conclude that females experience a weight gain following three months of oral contraceptive use? (Use $\alpha = 0.05$.)

† EXERCISE 8.11

A nurse midwife wished to determine if the mean age of females who retained IUD's was different from the mean age of females who expelled the devices. Fifteen subjects were randomly selected from among a group of females who had retained IUD's and 10 subjects from among those who had expelled IUD's. The data are shown below.

Expelled	*Retained*
$\overline{Y}_E = 28$	$\overline{Y}_R = 33$
$s_E = 10$	$s_R = 12$
$n_E = 10$	$n_R = 15$

Can it be concluded that the mean age of females who expel IUD's is different from the mean age of females who retain IUD's? (Use $\alpha = .05$.)

† EXERCISE 8.12

Nursing performance was evaluated in two hospitals using a standard nursing audit instrument. In each hospital 15 patients were selected at random and their nursing care was evaluated by an audit committee. The mean nursing performance score for Hospital 1 was found to be 85 with a standard deviation of 10 and the mean performance score for Hospital 2 was 70 with a standard deviation of 10. Can it be concluded that the nursing performance scores for the two hospitals are different? (Use $\alpha = .01$.)

† EXERCISE 8.13

A group of nine randomly selected senior nursing students were given a standardized pretest on medical-surgical nursing knowledge prior to their taking a special clinical elective. After the six-week course, the students were given a posttest covering the same material. The pre and posttest scores are given below.

Student	*Pretest*	*Posttest*
1	80	85
2	75	90
3	85	85
4	60	75
5	95	98
6	70	75
7	65	70
8	75	85
9	90	80

Can it be concluded that the scores are higher after taking the special elective? (Use $\alpha = .05$.)

†Type II Error (β), Power, and Determination of Sample Size

At this time, the reader should reread in Chapter 6 the sections entitled "Statistical Decisions and Their Outcomes" and "Power of the Test and Determination of Sample Size." The concepts in these sections are illustrated in terms of z statistics; however, the rationale is applicable to t statistics as well.

As stated in Chapter 6, one important use of the idea of power of a statistical test is that of determination of a sample size, in advance, that will ensure a specified power for the test. The calculation of sample size, based on power considerations, requires that the investigator specify the following:

1. The size of the effect that is clinically worthwhile to detect
2. The probability of falsely rejecting a true null hypothesis (α)
3. The probability of failing to reject a false null hypothesis (β)
4. The standard deviation (σ) of the population being studied

The first three items are under the control of the investigator. A value for the population standard deviation may be determined from previous work, work reported in the literature, a pilot study, or simply an educated guess.

For the one- and two-sample t-tests described in this chapter, Table L in Appendix 3 may be used to determine the sample size n for a given α, β, and Δ, where

$$\Delta = \frac{\mu - \mu_0}{\sigma}$$

Δ for the one-sample case is the difference in units of standard deviation between the hypothesized and the true population means.

EXAMPLE 8.7 A nurse researcher hypothesizes that the mean plasma potassium level for patients with a particular disease is lower than for nondiseased individuals. The mean plasma potassium level for nondiseased individuals is shown from the literature to be 4.5 mEq/liter. The mean plasma potassium levels will be determined for individuals with the disease. The investigator feels that a difference of 1 mEq/liter in the mean plasma potassium levels of nondiseased and diseased populations would be of clinical significance. The researcher wishes to be 80% sure that if a difference of at least 1 mEq/liter exists between the normal and diseased population, the statistical test will detect it (i.e., the null hypothesis of no difference will be rejected). Further, from previous knowledge, the researcher assumes that the true standard deviation for the population is $\sigma = 1$. How large a sample of diseased individuals must be taken if the test is to be carried out at an $\alpha = .01$ level of significance?

SOLUTION We are given the following:

$$\alpha = .01$$

$$\text{Power} = .80$$

$$\beta = 1 - .80 = .20$$

From Table L(a) in Appendix 3, we find that for the one-sided test (H_a: $\mu < 4.5$) for $\alpha = .01$, the sample size must be equal to 13. If the investigator wishes to detect a difference as small as 0.30 standard deviations with $\alpha = .01$ and power $= .80$, then the sample size must be increased to at least 115 observations. By increasing the sample size from 13 to 25 for $\alpha = .01$, the power of the test to detect a difference of one standard deviation is increased from .80 ($\beta = .2$) to .99 ($\beta = .01$).

EXAMPLE 8.8 A researcher hypothesized that there is a difference in serum sodium levels of normotensive and newly diagnosed hypertensive patients not yet on a sodium-controlled diet. In the study a sample of both normotensive and newly diagnosed hypertensive patients will be taken and serum sodium levels recorded. The investigator wishes to design the study such that if at least a five-unit difference exists between the means of the two populations (normotensive and hypertensive), then there will be a 90% chance that the statistical test will detect it (H_0 will be rejected). Based on information from a pilot study, the investigator estimates the common σ for the two populations to be 5. How large must the sample sizes be if the test is to be carried out at an $\alpha = .10$ level of significance?

SOLUTION The investigator wishes to detect, with 90% certainty, a difference of one standard deviation, if it exists, in the mean serum sodium levels of the normotensive and the newly diagnosed hypertensive populations. From Table L(b) for $\alpha = .10$ (two-sided test), power = .90, $\beta = 1 - .90 = .10$, and $\Delta = 1$, we find that $n_1 = n_2 = 18$. The sample size must be increased to $n_1 = n_2 = 70$ if we wish to detect a difference in population means as small as 0.5 standard deviations. Thus, as the difference in population means decreases, the required sample sizes increase sharply. Note also from the table that for fixed $\Delta = 1$, $\beta = .1$ (power = .9), as α decreases, sample size increases. Similarly, for fixed Δ and α, as β decreases, sample size increases.

For emphasis, we refer the reader back to the section in Chapter 6 entitled "Statistical Significance Versus Practical Significance."

SUMMARY OF CRITICAL CONCEPTS IN CHAPTER 8

The following concepts were presented in Chapter 8:

1. Hypothesis testing procedures usually involve six basic steps:
   a. Nature of the data
   b. Assumptions
   c. Hypothesis to be tested (null hypothesis)
   d. Decision-maker (test statistic)
   e. Decision rule (rejection region)
   f. Decision or conclusion
2. The simple t-test is used to test hypotheses about the mean of a single sample.
   a. Data—continuous with at least an interval scale.
   b. Assumptions—single-sample randomly selected from normally distributed population.
   c. Null hypothesis:

$$H_0: \quad \mu = \mu_0$$

   where μ_0 = any value
   d. Test statistic:

$$t = \frac{\overline{Y} - \mu_0}{s/\sqrt{n}}$$

where $\overline{Y}$ = sample mean, μ_0 = value hypothesized under H_0, s = sample standard deviation, and n = sample size

e. Decision rule

(1) Two-tailed rejection region (alternative hypothesis: $\mu \neq \mu_0$)

(a) $\alpha/2$ in each tail

(b) $df = n - 1$

(2) One-tailed rejection region (alternative hypothesis: $\mu < \mu_0$, $\mu > \mu_0$)

(a) α in either upper ($\mu > \mu_0$) or lower ($\mu < \mu_0$) tail

(b) $df = n - 1$

f. Conclusion—reject H_0 if test statistic falls in rejection region (outer tails of curve).

3. All tests of hypothesis refer to a null hypothesis that must be either explicitly stated or clearly impled.
4. The level of significance (p or α) must *always* be clearly stated.
5. Tests of hypothesis are based on the assumption of random sampling. Method of sample selection should be evaluated in terms of departure from strict random sampling, and how this departure may bias generalizability to the target population.
6. Caution should be exercised when comparing the results of a current study to historical controls.
7. The use of concurrent sampling (i.e., sampling from two populations simultaneously) overcomes many of the pitfalls of using historical controls.
8. In hypothesis testing about two means, two groups are treated alike except for the treatment or procedure under study. The hypothesis test answers the question "Is the observed difference between the two groups due to chance (sampling error) or to 'true' differences in the treatments being applied?"
9. The *pooled t-test* is used when two independent random samples are selected from populations that are normally distributed.
10. The *paired t-test* is used when subjects have been "paired" on the basis of certain characteristics such as age, IQ, family background, etc. The purpose of pairing is to eliminate the effects of extraneous sources of variability in the data, thus leading to a more precise comparison of sample means.

ANALYSIS OF VARIANCE

OVERVIEW

In Chapter 8 methods for testing hypotheses about single population means (simple *t*-test) and differences between two population means (pooled and paired *t*-tests) were discussed. Chapter 9 will present methods that may be used to test hypotheses about more than two means. The technique is called Analysis of Variance (ANOVA) and may be considered as an extension of the two sample *t*-tests presented in Chapter 8.

The simplest forms of the Analysis of Variance, the Completely Randomized Design, and the Randomized Complete Block Design will be presented. The focus will be on use and interpretation rather than computation. For the interested student, computational procedures will be presented in special daggered sections.

OBJECTIVES

Upon completion of this chapter, the student will be able to:

1. State when Analysis of Variance procedures are used.
2. Discuss the Completely Randomized Design in terms of the six steps of hypothesis testing.
3. Be familiar with common terms associated with Analysis of Variance procedures.
†4. Carry out the analysis for a Completely Randomized Design.
5. Discuss the Randomized Complete Block Design in terms of the six steps of hypothesis testing.

† Sections marked by a dagger (†) cover more difficult or theoretical topics.

†6. Carry out the analysis for a Randomized Complete Block Design.

†7. Determine which pairs of means are different using the Bonferroni t procedure.

DISCUSSION

Some Basics of Analysis of Variance

Consider the schematic listing in Table 9.1. This table represents four randomly selected groups of individuals who have received four different treatments or procedures. We are interested in testing whether there is a statistically significant difference in the true means of the four populations from which these four groups were selected.

A question that usually comes to mind is why not use the pooled t-test and test the equality of all possible combinations of sample means. This approach is not feasible for several reasons. First of all, when the number of groups becomes large, the amount of work required in computing all possible t values becomes restrictive. For four groups, there are six pairs that must be tested; for five groups, ten pairs, etc. Second, the results of all possible t-tests are difficult to interpret. With one t-test using an .05 level of significance, the chance of obtaining a significant t value (one that falls in the rejection region) is .05. This means that rejecting H_0 for any one test carries with it a 5% chance of error. When multiple t-tests are performed on the same data, the chance of having at least one t value be significant just by chance (i.e., a significant t when H_0 really is true) becomes greatly increased. For example, if five means are involved and multiple pooled t-tests are used, the probability of rejecting at least one hypothesis of equal mean values

TABLE 9.1 FOUR GROUPS RECEIVING FOUR DIFFERENT TREATMENTS

Group A	*Group B*	*Group C*	*Group D*
a_1	b_1	c_1	d_1
a_2	b_2	c_2	d_2
⋮	⋮	⋮	⋮
a_n	b_n	c_n	d_n
$\bar{Y}_a$	$\bar{Y}_b$	$\bar{Y}_c$	$\bar{Y}_d$

($\mu_1 = \mu_2$) just by chance is much higher than 5%. In order to avoid these pitfalls, a new approach—the Analysis of Variance—must be utilized.

Returning to the table, we know that differences in the four sample means could be due to chance (sampling error) or to a true difference in the four treatments being tested. The basic idea underlying all Analysis of Variance techniques is that the total variability in a set of measurements may be divided into two (or more) categories: in this case, the variability among the measurements within a group and the variability between the means of the different groups. If the four treatments are all alike in their effect (i.e., $\mu_A = \mu_B = \mu_C = \mu_D$) then we would expect the variability between mean values to be no greater than the variability of the measurements within a group. The within-group variability should only reflect the effects of random biological differences between individuals who have been treated alike within groups.

The test statistic used in the Analysis of Variance is called the F *statistic* and is defined as:

$$F = \frac{\text{Variance among groups}}{\text{Variance within groups}}$$

If there is no difference in the true means of all the treatment groups then the value of the F ratio should be close to 1 (within variation = among variation). As the differences between the sample mean values become larger and larger, the numerator in the F statistic becomes larger also, while the variation within groups of individuals treated alike remains unchanged. Thus for large F values we reject the hypothesis of equal means, while for values of F close to 1, we do not reject and must conclude that we cannot show a difference in the treatments under study.

The simplest Analysis of Variance method using the Completely Randomized Design will be presented in terms of the six-step hypothesis testing procedure used in Chapter 6 and 8. (The Analysis of Variance technique using the Completely Randomized Design is called a one-way analysis of variance.)

Completely Randomized Design

1. *Data*. The data consist of measurements made on samples of individuals from more than two populations. The data are continuous in nature and possess at least an interval scale of measurement.
2. *Assumptions*. It is assumed that each sample was randomly and

independently selected from populations whose underlying distributions of measurements are normal and have equal variance. The Completely Randomized Design is an extension of the pooled t-test to more than two groups.

3. *Hypotheses*. The null hypothesis to be tested is that all population (group) means are equal, i.e.,

$$H_0: \quad \mu_1 = \mu_2 = \mu_3 = \cdots = \mu_k$$

where k is the number of treatment groups. This hypothesis is equivalent to stating that there is no treatment effect; all treatments produce the same result. The alternate hypothesis is that *at least one* of the pairs of means is not equal. Note that the alternate hypothesis does not specify which pairs are not equal, only that one or more pairs are unequal.

4. *Test statistic*. The test statistic or decision-maker is the F statistic.

$$F = \frac{\text{Among-group variance}}{\text{Within-group variance}}$$

Usually the results of an Analysis of Variance are presented in the form of an ANOVA table, as in Table 9.2. The among-group variation is the variation due to difference in treatments; the within-group variation is called the *error* variation since it is due in part to the failure of individuals treated alike to react alike. In addition, k = number of treatment groups, N = total number of measurements in all groups combined, SST = sum of squares for treatments (among groups), SSE = sum of squares for error (within groups), MST = mean square for treatments, MSE = mean square for error, and $F = MST/MSE$. The computational formulas for sum of squares will be presented in a special section at the end of this chapter.

TABLE 9.2 ANALYSIS OF VARIANCE TABLE (ANOVA)

Source of Variation	*df*	*Sum of Squares*	*Mean Square*	*F*
Among groups	$k - 1$	SST	$MST = SST/k - 1$	MST/MSE
Within groups	$N - k$	SSE	$MSE = SSE/N - k$	
Total	$N - 1$	SS_{Total}		

5. *Decision rule*. The rejection region for the F-test is determined by locating the appropriate cut-off point in an F table (Appendix Table C). To find a value in the F table, degrees of freedom for the numerator in the F ratio and degrees of freedom for the denominator must be computed. These values are in the ANOVA table.

$$df_{\text{numerator}} = k - 1$$

$$df_{\text{denominator}} = N - k$$

The numerator df are located across the top of the table (columns) and the denominator df are located down the rows. A different table is given for each level of significance. For an analysis that has four groups and five measurements per groups, $k = 4$ and N, the total number of measurements, $= 20$. Thus df for numerator is equal to $k - 1 = 3$ and df for denominator is $N - k = 20 - 4 = 16$. For an $\alpha = .05$ level of significance, the value in Appendix Table C is 3.24. The rejection region is shown below.

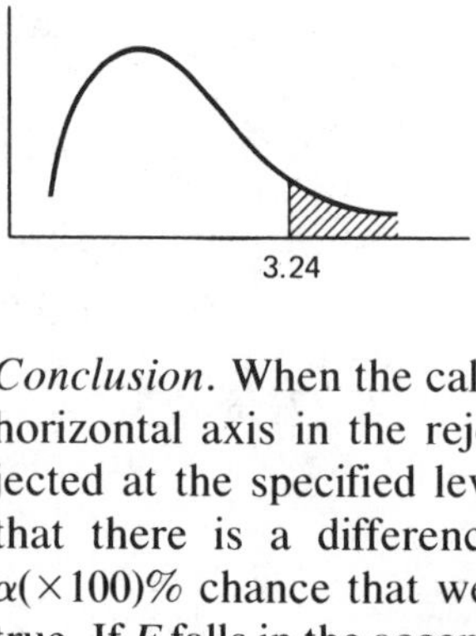

6. *Conclusion*. When the calculated value of F (Step 4) falls on the horizontal axis in the rejection region (shaded area), H_0 is rejected at the specified level of significance and it is concluded that there is a difference in treatment effects. There is an $\alpha(\times 100)\%$ chance that we have rejected H_0 when, in fact, it is true. If F falls in the acceptance region, it must be concluded that there is not enough evidence to show that there is a difference in the treatments at the α level of significance.

EXAMPLE 9.1 An investigator is interested in evaluating four different methods of instructing children in dental hygiene. Twenty children are randomly selected and assigned to four groups of five children each. Each group is instructed using one of the four methods. At the end of the instruction period each child is given a hygiene performance score based on both a written and practical exam. An Analysis of Variance is carried out on the mean scores for the four methods; an F value of 19.2 is reported ($\alpha = .01$). Discuss this study in terms of the six steps of hypothesis testing.

SOLUTION

1. *Data*. The data consist of four sets of five hygiene scores that are continuous in nature. The score data are treated as if they possess an interval scale. The data layout would be as shown below.

Teaching Method

Method 1	*Method 2*	*Method 3*	*Method 4*
a_1	b_1	c_1	d_1
a_2	b_2	c_2	d_2
a_3	b_3	c_3	d_3
a_4	b_4	c_4	d_4
a_5	b_5	c_5	d_5
$\overline{Y}_1$	$\overline{Y}_2$	$\overline{Y}_3$	$\overline{Y}_4$

2. *Assumptions*. It is assumed that the four samples were randomly selected from populations whose underlying distributions are normally distributed with equal variances.
3. *Null hypothesis*. The null hypothesis is that the mean hygiene scores of the children are the same for all four teaching methods; that is, the investigator is interested in testing whether the true means of the populations of scores for each of the four methods are equal. Symbolically,

$$H_0: \quad \mu_1 = \mu_2 = \mu_3 = \mu_4$$

The alternative of interest is that at least one of the pairs is not equal.
4. *Test statistic*. The test statistic for the Analysis of Variance is the F ratio.

$$F = \frac{\text{Among-method variation}}{\text{Within-method variation}} = \frac{MST}{MSE} = 19.2$$

The ANOVA table would be as shown below.

Source of Variation	*df*	*SS*	*MS*	*F*
Among methods	3	*SST*	*MST* = *SST*/3	*F* = 19.2
Within methods	16	*SSE*	*MSE* = *SSE*/16	
Total	19			

5. *Decision rule*. The cut-off point for the rejection region is located at the intersection of column 3 (numerator *df*) and row 16 (denominator *df*) in Appendix Table C headed α = .01. The value is 5.29. The rejection region consists of all points on the horizontal axis under the shaded area shown below.

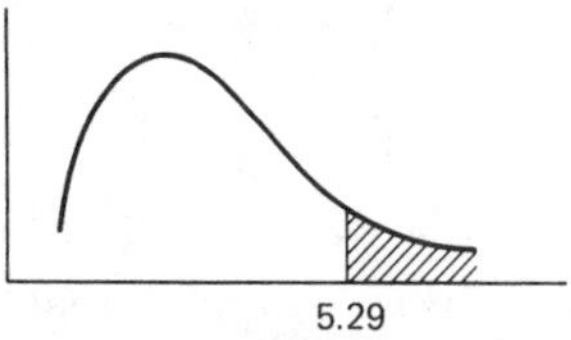

6. *Conclusion*. Since the calculated *F* value (Step 4) falls in the rejection region, the null hypothesis may be rejected at the α = .01 level of significance. It may be concluded that at least one of the treatment means is not equal (at the .01 level of significance). The investigator is willing to accept a 1% chance of rejecting H_0 if in fact it is true.

EXERCISE 9.1

A dietician was interested in comparing the effects of three different diets on reduction of serum cholesterol levels. Individuals were assigned at random to one of the three treatment (diet) groups. There were five individuals in the first group, six in the second, and six in the third. At the end of three months, serum cholesterol measurements were made on each participant. The investigator analyzed the data using an Analysis of Variance and reported an *F* value of 2.75. Interpret the results of the study (using α = .05) in terms of the six steps of hypothesis testing.

Often after an Analysis of Variance has been carried out and the F value is found to be significant (i.e., there is a difference between means), an investigator may wish to determine which of the pairs are different from each other. There are several techniques for doing such pairwise comparisons. The most popular of these are *Duncan's Multiple Range Test, Least Significant Difference (LSD), Tukey's Procedure*, and *Scheffe's Procedure*.* A procedure gaining in popularity is the Bonferroni t procedure. This method will be presented in a special section at the end of the chapter.

† Computational Methodology for Completely Randomized Design (CRD)

Formally stated, in a Completely Randomized Design, there are k treatments, each of which is assigned at random to a group of experimental units. (The objects being measured are called experimental units. In a clinical setting, the experimental units are usually the patients receiving the treatments or procedures being investigated.) The null hypothesis is that the treatment means for the k populations are all equal (i.e., H_0: $\mu_1 = \mu_2 = \cdots = \mu_k$).

In a Completely Randomized Design each experimental unit has an equal and independent chance of receiving any one of the treatments. The basic assumption underlying this design is that the observed values in any one group represent a random sample of all possible values of all experimental units under that particular treatment. Further, we assume that the responses are normally distributed about the treatment mean and that the variation among observations treated alike is identical for all k treatments.

Example 9.2 will be used to illustrate the computation procedure for the Completely Randomized Design.

EXAMPLE 9.2 A nursing audit was carried out for all services in a large community hospital. The total audit score for each patient reflects nursing performance for seven basic areas: (1) application and execution of physician's legal orders; (2) observations of symptoms and reactions; (3) supervision of the patient; (4) supervision of those participating in care (except the physician); (5) reporting procedures and techniques; (6) application and execution of nursing procedures and techniques; and (7) promotion of total health

* For a detailed discussion of these procedures see R. G. Steele and J. H. Torrie, *Principles and Procedures of Statistics* (New York: McGraw-Hill, 1960), pp. 106–114.

by direction and teaching. The following table presents the total audit scores for five randomly selected patients on services within the given hospital:

	Medical	*Surgical*	*Pedia-trics*	*Ob-Gyn*	
	75	75	70	50	
	65	80	70	65	
	50	85	60	70	
	60	70	75	70	
	70	90	80	75	
Total (T_i)	320	400	355	330	$\sum_{\text{all}} Y = 1405$
n_i	5	5	5	5	$\sum_{\text{all}} Y^2 = 100675$
$\bar{Y}_i$	64	80	71	66	

In order to complete the Analysis of Variance, we must first calculate a quantity called *sum of squares* for each source of variation specified in the ANOVA table (Table 9.2).

Total Sum of Squares (SS_{Total})

As stated previously, the rationale behind the analysis of variance technique is to partition the total variation in a set of data into recognizable sources such as treatment differences and experimental error. For ease of discussion, sum of squares rather than actual variances are partitioned.

The total sum of squares is the total of the squared deviations of the observations from the overall mean of the data. It is simply the numerator in the familiar formula for calculating the variance of all the observations considered as a single group. Symbolically,

$$SS_{\text{Total}} = \sum_{\text{all}} Y^2 - \frac{(\sum_{\text{all}} Y)^2}{N}$$

where $N = n_1 + n_2 + n_3 + \cdots + n_k$, k = number of treatments. For convenience of calculation, the term:

$$\frac{(\sum_{\text{all}} Y)^2}{N}$$

is given a special name. It is called the *correction factor* and is used in several calculations.

For the data in Example 7.2,

$$CF = \frac{(\sum_{\text{all}} Y)^2}{N} = \frac{(1405)^2}{20} = 98701.25$$

and

$$SS_{\text{Total}} = \sum_{\text{all}} Y^2 - CF = 100675. - 98701.25 = 1973.75$$

Among-groups Sum of Squares (*SST*)

The among-treatments variation (the failure of the k treatments to be alike) is calculated by the formula:

$$SS_{\text{Among}} = SST = \sum_{i=1}^{k} \frac{(T_i)^2}{n_i} - CF$$

where T_i represents the totals of each of the k treatment groups and n_i is the total number in each group.

For our example,

$$SST = \left[\frac{T_1^{\,2}}{n_1} + \frac{T_2^{\,2}}{n_2} + \frac{T_3^{\,2}}{n_3} + \frac{T_4^{\,2}}{n_4}\right] - CF$$

$$= \left[\frac{(320)^2}{5} + \frac{(400)^2}{5} + \frac{(355)^2}{5} + \frac{(330)^2}{5}\right] - 98701.25$$

$$= 99465 - 98701.25 = 763.75$$

Within-groups Sum of Squares (*SSE*)

Since the within-treatments variation is the variation associated with observations treated alike, it is the variation associated with experimental error. The computation formula for *SSE* is given by:

$$SS_{\text{within}} = SSE = \sum_{\text{all}} Y^2 - \sum_{i=1}^{k} \frac{(T_i)^2}{n_i}$$

However, since:

$$SS_{\text{Total}} = SS_{\text{Among}} + SS_{\text{within}}$$

$$= SST + SSE$$

we may obtain SSE by subtraction. That is,

$$SSE = SS_{\text{Total}} - SST$$

For our example,

$$SSE = 1973.75 - 763.75 = 1210$$

(You may confirm that the value obtained by subtraction is the same as that computed directly by the formula.)

The general procedure for computing the mean square column for the ANOVA table (see Table 9.3) is first to compute the sum of squares column and enter in the table; then to compute the degrees of freedom and enter in the table; and finally to compute the mean squares by dividing the degrees of freedom into the corresponding sum of squares.

For example,

$$MST = \frac{SST}{k - 1} = \frac{763.75}{3} = 254.583$$

and

$$MSE = \frac{SSE}{N - k} = \frac{1210}{16} = 75.625$$

The F statistic is computed by the formula:

$$F = \frac{MST}{MSE} = \frac{254.583}{75.625} = 3.37$$

† EXERCISE 9.2

Complete the above problem (Example 9.2) in terms of the six steps of hypothesis testing. (Use $\alpha = .01$.)

TABLE 9.3 ANOVA OF AUDIT SCORES FOR EXAMPLE 9.2

Source	*Degrees of Freedom*	*Sum of Squares*	*Mean Squares*	*F*
Among	3	763.75	254.583	3.37
Within	16	1210	75.625	
Total	19	1973.75		

†EXERCISE 9.3

A nurse investigator wished to compare the elapsed time of three different methods of collecting information on patients in an emergency room. Nine patients were selected and each was randomly assigned to one of the three methods. Response is time (in minutes) to collect the information.

Method 1	*Method 2*	*Method 3*
30	25	21
20	15	20
40	35	25

Test the hypothesis that the three methods of collecting information require equal amounts of time. (Use $\alpha = .05$.)

Randomized Complete Block Design*

For some experiments, it may be possible to "block" experimental units into homogeneous groups. A block can be considered an extension of the concept of a paired experiment as discussed in Chapters 7 and 8. In this case, the experimental units are subdivided into homogeneous groups called blocks, and the treatments are then randomly assigned to members of the block. Each block receives every treatment and each treatment appears in every block. This is called the randomized complete block design (RCB).

The randomized complete block is used when the investigator feels that the response of the experimental unit may be affected by a factor or factors other than the treatment under study. For example, patients in a drug study may be in different age groups. It is likely that response to drug would be affected by age of the patient. By blocking the experimental units (patients) into homogeneous age groups, the variation attributable to the extraneous factor(age) can be removed. Examples of a RCB are given below.

EXAMPLE 9.3 In a study of weight gain associated with three supplemental diets, it was felt that initial weight of the participants would be reflected in the final response to the diet. The patients were divided according to weight at the outset of

*The section is adapted from Duncan, Knapp, Miller, *Introductory Biostatistics for the Health Sciences*, 2nd Ed., 1983, pp. 141–155. Used by permission of John Wiley & Sons, Inc.

the experiment. At the end of a 3-month period the weight gain of each participant was recorded.

Block	*Treatments*		
Initial Weight	*Diet 1*	*Diet 2*	*Diet 3*
100–120	10	12	8
120–140	8	10	10
140–160	5	12	10
160–180	10	5	5

Note: Within each block, the participants were randomly assigned to the treatments (diet).

EXAMPLE 9.4 An investigator wished to compare the elapsed time of three different methods of collecting information on patients in an emergency room. Three different hospital emergency rooms were utilized in the study. Response is time (in minutes) to collect the information.

Block	*Treatments*		
Hospital	*Method 1*	*Method 2*	*Method 3*
1	30	25	20
2	20	15	20
3	40	35	25

As was the case with the Completely Randomized Design (CRD) presented in previous sections, the general method of analysis appropriate for the Randomized Complete Block Design (RCB) is the *analysis of variance*. We now outline the five steps for hypothesis testing appropriate for the analysis of variance procedure for an RCB design.

1. *Data*. The data consist of measurements made on samples of individuals from more than two populations. The data are continuous in nature and possess at least an interval scale of measurement. In the RCB design, the experimental units (e.g., patients) are subdivided into homogeneous groups called *blocks*. Within each block, we must have available experimental units (patients) equal in number to the number of treatments. The k treatments are then assigned at *random* to the k *individuals within each*

block. Thus, each block receives every treatment. The data layout is as follows:

	Treatments					
Blocks	*1*	*2*	*3*	...	*k*	*Totals*
1	Y_{11}	Y_{12}	Y_{13}		Y_{1k}	B_1
2	Y_{21}	Y_{22}	Y_{23}	...	Y_{2k}	B_2
3	Y_{31}	Y_{32}	Y_{33}	...	Y_{3k}	B_3
|	|					|
b	Y_{b1}	Y_{b2}	Y_{b3}		Y_{bk}	B_b
Totals	T_1	T_2	T_3	...	T_k	$G = \sum_{\text{all}} Y$

The first subscript on Y indicates the block and the second subscript the treatment. Thus, Y_{12} is the measurement of an individual in Block 1 receiving Treatment 2.

2. *Assumptions*. Each measurement Y_{ij} represents an independent random sample of size 1 from the population of patients in Block i who receive Treatment j. It is assumed that each of the kb populations represented is normally distributed and has the same variance σ^2. A further assumption of the RCB design is that treatment and block effects are additive; that is, the combined effect of a particular treatment block combination is not greater than the sum of the individual treatment or block effects. In statistical terminology, this last assumption is equivalent to stating that there is *no interaction* between treatments and blocks. Daniel (1978) states, "The consequences of a violation of this assumption (of no interaction) are misleading results." He further cites Anderson and Bancroft (1952), who suggest that one need not become concerned over the violation of the assumption of no interaction unless the largest mean is more than 50% greater than the smallest.
3. *Statistical hypotheses*. The null hypothesis to be tested is that there are no differences in treatment effects versus the alternate hypothesis that not all treatment effects are equal. A test of hypothesis concerning differences among block effects is, in general, of no interest and, thus, is usually not carried out. The variable on which the experimental units are blocked or "matched" is usually a nuisance variable whose effects we wish

to eliminate in order to get a "clean" comparison of treatment effects.

4. *Test statistic*. The rationale for calculating a test statistic for the analysis of variance using an RCB design is similar to the rationale described for the Completely Randomized Design (CRD). We compute the variation among treatment groups just as was done for the CRD. In addition, for the RCB design we also calculate the variation attributable to the blocking factor. The error term for the RCB equivalent to the *within groups* error term of the CRD, is the residual variation in observations, which is accounted for neither by variation in treatments nor by variation attributable to blocks. See Table 9.4. As was the case with the CRD (Table 9.2), the test statistic for testing difference in treatment effects is

$$F = \frac{\text{Among treatment group variance}}{\text{Residual error}} = \frac{MST}{MSE}$$

The computation formulas are presented in the next section. If there is variation in the response variable Y due to the extraneous variable used as a blocking factor, then by incorporating this extraneous variable into the design of the study by means of the RCB, we draw this variation out of the "unaccounted for" variation (residual error) category. The effect of this reduction in the error variation line is to *decrease* the denominator in the test statistic F and hence to increase its overall magnitude. As before with the CRD, the larger the F value, the more likely we are to reject H_0.

TABLE 9.4 ANOVA FOR THE RANDOMIZED COMPLETE BLOCK DESIGN

Source of Variation	*Degrees of Freedom*	*Sum of Squares*	*Mean Squares*	*F*
Among blocks	$b - 1$	SSB		
Among treatments	$k - 1$	SST	$MST = SST/(k - 1)$	MST/MSE
Error	$(b - 1)(k - 1)$	SSE	$MSE = SSE/(b - 1)(k - 1)$	
Total	$bk - 1$	SS		

5. *Decision rule*. The rejection region is determined exactly as described for the CRD. The degress of freedom in this case are

$$df_{\text{numerator}} = k - 1$$

$$df_{\text{denominator}} = (b - 1)(k - 1)$$

6. *Conclusion*. When the calculated value of F falls on the horizontal axis under the outer tail shaded area of the curve (rejection region), that is, when the calculated F is greater than the tabulated F, we reject H_0. When H_0 is rejected, we conclude that there is a difference in treatment effects at the α level of significance. When the calculated F is less than the tabulated F, we do not reject H_0 and conclude that there is insufficient evidence to demonstrate a difference in treatment effects.

EXAMPLE 9.5 Three large hospitals agreed to a cooperative study of laboratory costs by hospital service. In this study it was recognized that laboratory fees within a hospital would be uniform across all types of service. That is, a hospital with low laboratory fees should be low for all patients regardless of whether they are surgical, medical, or so on. Since the purpose of the study is to compare laboratory fees across services, the effect of the extraneous variable "hospital" must be accounted for. In other words, it is recognized that the response being measured, "laboratory fee" can be affected by "type of service" and by "hospital." Since the purpose of the study is to compare types of services, then "type of service" is considered the "treatment" variable and hospital is the "blocking" variable. The investigator hypothesized that laboratory fees differ among the different services. Test this hypothesis at the α = .05 level of significance. The data are shown in Table 9.5.

SOLUTION

1. *Data*. The data consist of cost in dollars for four different services (treatments) made in each of three hospitals (blocks). Hospitals is the "matching" factor whose effect is to be eliminated in order to achieve a more precise comparison of services.
2. *Assumptions*. It is assumed that the 12 populations from which the single observation samples have been obtained are normally distributed with the same stan-

TABLE 9.5 PATIENT LABORATORY COSTS (DOLLARS) BY HOSPITAL AND TYPE OF SERVICE

Blocks	*Treatments (Service)*			
(Hospital)	*Surgery*	*Medicine*	*Pediatrics*	*Ob-Gyn*
Hospital 1	44	59	60	41
Hospital 2	33	19	49	71
Hospital 3	44	40	45	31

dard deviation σ. Further, it is assumed that there is no *interaction* between services and hospitals.

3. *Statistical hypotheses.* The null hypothesis is that there is no difference in mean costs for the four *services* studied. The alternate hypothesis is that at least one pair of treatment (services) means is different.
4. *Test statistic.** The ANOVA table is shown in Table 9.6. The test statistic for testing differences among services is

$$F = \frac{\text{Variation among services}}{\text{Residual error}} = 0.375$$

5. *Decision rule.* The cutoff point for the rejection region is located at the intersection of column 3 (numerator *df*) and row 6 (denominator *df*) in Appendix Table C headed $\alpha = .05$. The value is 4.76. The rejection region consists of all points beyond 4.76 on the horizontal axis underneath the shaded area of the curve shown below.

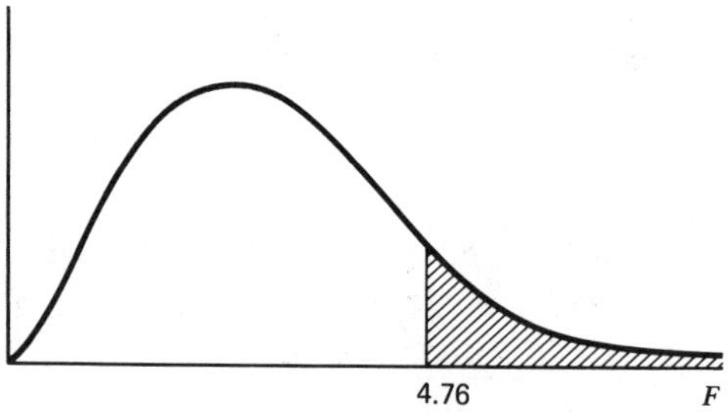

* Computational procedures for obtaining the table are given in the next special section. The purpose of this section is to emphasize interpretation rather than computation.

TABLE 9.6 ANOVA OF LABORATORY COST

Source of Variation	*Degrees of Freedom*	*Sum of Squares*	*Mean Squares*	*F*
Among hospitals	2	258.7		
Among services	3	302.0	100.7	0.375
Error	6	1610.0	268.3	
Total	11	2170.7		

6. *Conclusion*. Since the calculated $F = .375$ (step 4) does not fall in the rejection region ($F_{\text{cal}} < 4.76$), the null hypothesis cannot be rejected at the $\alpha = .05$ level of significance. There is insufficient evidence to demonstrate a difference in mean costs among the four services studied.

EXERCISE 9.4

In a study similar to that described in Exercise 9.1, the dietician recognized that the differences in final serum cholesterol measurements for patients in the study could be attributed to differences in initial (prediet) serum cholesterol levels as well as to the three diets being studied. In order to take into account prediet levels of serum cholesterol, the investigator grouped study participants into three levels of initial (prediet) serum cholesterol values (i.e., those with low, moderate, and elevated levels). Three subjects were available at each of the three initial levels of serum cholesterol. Within each initial level of serum cholesterol, the three patients were assigned at random to the three experimental diets, one patient to each diet. At the end of three months, serum cholesterol levels were again measured for each participant. The investigator analyzed the data using an Analysis of Variance for an RCB design and reported an F value of 5.88. Interpret the results of the study (using $\alpha = .05$) in terms of the six steps of hypothesis testing.

†Computational Methodology for the Randomized Complete Block Design*

The procedure for computing the needed sums of squares is summarized as follows:

* This section is reprinted from Duncan, Knapp, Miller, *Introductory Biostatistics for the Health Sciences*, 2nd Ed., 1983, pp. 149–151. Used by permission of John Wiley & Sons, Inc.

STEP 1. Compute the correction factor (CF) as for the completely randomized design.

STEP 2. Compute the block sum of squares (SSB) by summing the squares of block totals, dividing this sum by the number of treatments, and subtracting the correction factor.

STEP 3. Compute the treatment sum of squares, SST (as before), by summing the squares of the treatment totals, dividing by the number of blocks, and subtracting the correction factor.

STEP 4. Compute the error sum of squares (SSE) by subtracting the block sum of squares and the treatment sum of squares from the total sum of squares ($SSE = SS - SSB - SST$).

STEP 5. Fill in and complete the ANOVA as in Table 9.4.

The individual observations and the necessary totals are shown in Table 9.7. Note: The B_j are block totals and the T_i are treatment totals. There are b blocks and k treatments.

1. Correction factor

$$CF = \left(\sum_{\text{all}} Y\right)^2 \Big/ N = (536)^2/12$$

$$= 23941.3$$

where N = total number of observations = bk

2. Total sum of squares

$$SS_{\text{total}} = \sum_{\text{all}} Y^2 - CF$$

$$= (44)^2 + (59)^2 + \cdots + (31)^2 - CF$$

$$= 26112 - 23941.3$$

$$= 2170.7$$

3. Sum of squares for treatments

$$SST = \sum_{i=1}^{k} \frac{(T_i)^2}{n_i} - CF$$

$$= \frac{(121)^2}{3} + \frac{(118)^2}{3} + \frac{(154)^2}{3} + \frac{(143)^2}{3} - 23941.3$$

$$= 302.0$$

TABLE 9.7 PRELIMINARY CALCULATIONS FOR THE ANOVA OF LABORATORY COSTS

Blocks (*Hospitals*)	*Treatments (Service)* Surgery	*Medicine*	*Pediatrics*	*Ob-Gyn*	*Totals* (B_j)
1	44	59	60	41	204
2	33	19	49	71	172
3	44	40	45	31	160
Totals (T_i)	121	118	154	143	536

4. Sum of squares for blocks

$$SSB = \sum_{j=1}^{b} \frac{(B_j)^2}{n_j} - CF$$

$$= \frac{(204)^2}{4} + \frac{(172)^2}{4} + \frac{(160)^2}{4} - 23941.3$$

$$= 258.7$$

5. Sum of squares for error

$$SSE = SS_{\text{total}} - SST - SSB$$

$$= 2170.7 - 302.0 - 258.7$$

$$= 1610.0$$

The ANOVA can now be completed (Table 9.6). The tabulated F value for three (MST) and six (MSE) degrees of freedom is 4.76. Since the calculated value does not exceed the table value, it is concluded that there is insufficient evidence to show that patient laboratory costs differ among the various types of hospital services studied.

EXERCISE 9.5

In a study to evaluate hypoglycemic effectiveness, each of five maturity-onset diabetics was given a treatment that consisted solely of dieting, a treatment that consisted of chlorpropamide (100 mg/day), and one that consisted of chlorpropamide (250 mg/day). The order of administration of each treatment was assigned at random. At the end of a specified time period, Hb A_{1c}(percentage) levels were determined and the following data were obtained.

Patient	*Diet Alone*	*Chlorpropamide (100 mg/day)*	*Chlorpropamide (250 mg/day)*
1	8	5	5
2	7	6	5
3	9	8	7
4	7	5	5
5	8	6	7

Do these data provide sufficient evidence to indicate a difference in Hb A_{1c}(percentage) among the three different treatments? (Use $\alpha = .01$.)

Interpretation of the ANOVA*

The F-test for significance of differences among three or more mean values is a cumulative test in the sense that it takes into account the differences among all pairs of means but does not single out any particular pair of mean values as being different. All a significant F-test indicates is that at least some of the means differ among themselves. That is, at least one $\mu_i \neq \mu_j$, so the null hypothesis is rejected.

Usually, in addition to testing the global hypothesis that all population means are equal, we are interested in testing hypotheses about specific mean differences. That is, we wish to determine just which pairs of means are different.

†Testing Hypotheses About Specific Differences Following the Analysis of Variance*

One method that immediately comes to mind for testing differences between individual pairs of means is performing t-tests on all possible pairs of means. When, for example, a study consists of three treatments, we may test the hypotheses $\mu_1 = \mu_2$, $\mu_1 = \mu_3$, $\mu_2 = \mu_3$ by carrying out the two-sample t-test procedure as described in Chapter 8 for each hypothesis. There is, however, a pitfall associated with this "multiple t-test" procedure. When each individual comparison is tested at a specified level of significance, α, the corresponding level of significance for the set of C comparisons is at most $C\alpha$, where $C\alpha$ is called the *error rate per experiment*. Thus, if we do five comparisons, each at $\alpha = .05$, the per experiment error rate could be as much as $5 \times .05 = .25$, which far exceeds the overall 5 percent rate we usually want. To avoid this we make the individual comparisons at a smaller significance level as described below.

* These two sections are reprinted from Duncan, Knapp, Miller, *Introductory Biostatistics for the Health Sciences*, 2nd Ed., 1983, pp. 151–155. Used by permission of John Wiley & Sons, Inc.

The per experiment error rate may be conceptualized as the average number of comparisons that would be falsely declared significant per experiment if, theoretically, the experiment was repeated many times. Here, the term experiment refers not only to the physical conduct of the study but also to the repeated testing of the set of comparisons defined by the investigator to be of interest. If, as usual, we want our per experiment error rate to be α, then we should test each of the C individual comparisons at the significance level α/C. That is, we would compare the computed t statistic to the tabulated t value ot $t_{\alpha/C}$ for a one-tailed test or $t_{(\alpha/C)/2}$ for a two-tailed test. This procedure, called the *Bonferroni procedure,* ensures that the error rate for *all* the comparisons of interest considered as a group will be at most α per experiment.

EXAMPLE 9.6 A study was carried out to compare four different drugs with respect to changes in blood pressure. Summary statistics and the ANOVA table are given in Table 9.8.

	Drug			
	A	*B*	*C*	*D*
$\overline{Y}$:	9.0	13.0	11.0	15.0
n_i:	4	4	4	4
s:	1.826	1.826	1.826	1.826

Suppose the investigator has defined the following five hypotheses to be of interest.

$$\mu_A = \mu_B$$

$$\mu_A = \mu_C$$

$$\mu_A = \mu_D$$

$$\mu_B = \mu_C$$

$$\mu_C = \mu_D$$

TABLE 9.8. ANOVA

Source	*df*	*SS*	*MS*	*F*
Among treatments	3	80	26.7	8.1
Within treatments	12	40	3.3	
Total	15	120		

$F_{3,12,.05} = 3.49$

These individual hypotheses may be tested using the t-statistic.

$$t = \frac{\overline{Y}_i - \overline{Y}_j}{\sqrt{MSE\left(\frac{1}{n_i} + \frac{1}{n_j}\right)}}$$

where MSE is the pooled estimate of within group variation obtained from the ANOVA table. The degrees of freedom for this test are the degrees of freedom for the within treatments sums of squares in the ANOVA table. Returning to the example, the five t statistics are summarized below.

Hypothesis	*t statistic*
$\mu_A = \mu_B$	$\dfrac{\overline{Y}_A - \overline{Y}_B}{\sqrt{MSE\left(\frac{1}{n_A} + \frac{1}{n_B}\right)}} = \dfrac{9.0 - 13.0}{\sqrt{3.3\left(\frac{1}{4} + \frac{1}{4}\right)}} = -3.11$
$\mu_A = \mu_C$	$\dfrac{\overline{Y}_A - \overline{Y}_C}{\sqrt{MSE\left(\frac{1}{n_A} + \frac{1}{n_C}\right)}} = \dfrac{9.0 - 11.0}{\sqrt{3.3\left(\frac{1}{4} + \frac{1}{4}\right)}} = -1.56$
$\mu_A = \mu_D$	$\dfrac{\overline{Y}_A - \overline{Y}_D}{\sqrt{MSE\left(\frac{1}{n_A} + \frac{1}{n_D}\right)}} = \dfrac{9.0 - 15.0}{\sqrt{3.3\left(\frac{1}{4} + \frac{1}{4}\right)}} = -4.67$
$\mu_B = \mu_C$	$\dfrac{\overline{Y}_B - \overline{Y}_C}{\sqrt{MSE\left(\frac{1}{n_B} + \frac{1}{n_C}\right)}} = \dfrac{13.0 - 11.0}{\sqrt{3.3\left(\frac{1}{4} + \frac{1}{4}\right)}} = 1.56$
$\mu_C = \mu_D$	$\dfrac{\overline{Y}_C - \overline{Y}_D}{\sqrt{MSE\left(\frac{1}{n_C} + \frac{1}{n_D}\right)}} = \dfrac{11.0 - 15.0}{\sqrt{3.3\left(\frac{1}{4} + \frac{1}{4}\right)}} = -3.11$

To ensure that the overall error rate per experiment does not exceed $\alpha = .05$, we compare each calculated t statistic with the tabulated t value:

$$t_{.05/5} = t_{.01} = 2.681 \qquad \text{(12 degrees of freedom)}$$

for a one-tailed test or

$$t_{(.05/5)/2} = t_{.005} = 3.0545$$

for a two-tailed test. Assuming that we are interested in the two-tailed alternatives $\mu_i \neq \mu_j$, we may conclude simultaneously that drug A is different from drug B and drug D, and drug C is different from drug D.

Suppose, instead of testing only the five hypotheses we have just illustrated, we are interested in testing all six possible pairwise hypotheses:

$$\mu_A = \mu_B$$

$$\mu_A = \mu_C$$

$$\mu_A = \mu_D$$

$$\mu_B = \mu_C$$

$$\mu_B = \mu_D$$

$$\mu_C = \mu_D$$

For the two-tailed alternative $\mu_i \neq \mu_j$, we compare the calculated t statistics with the tabulated value $t_{(\alpha/6)/2} = t_{(.05/6)/2} = t_{.004}$. This procedure often requires tabulated t values not found in standard t tables. One may use linear interpolation to obtain values from standard t tables, or an approximate value for t that cuts off the upper $\alpha/2$ proportion with v degrees of freedom can be determined from the standardized normal distribution by

$$t_{\alpha/2,v} = z + \frac{z^3 + z}{4(v - 2)}$$

where z is the corresponding value in a normal distribution.* Using this approximation,

$$\begin{aligned} t_{.004,12} &= z_{.496} + \frac{z_{.496}^3 + z_{.496}}{4(10)} \\ &= 2.65 + \frac{(2.65)^3 + 2.65}{40} \\ &= 3.18 \end{aligned}$$

*Roger E. Kirk, *Experimental Design: Procedures for the Behavioral Sciences*. Belmont, California: Wadsworth Publishing Co., Inc., 1968.

The six calculated t statistics appropriate for each of the six possible pairwise hypotheses are compared with the approximate table value of 3.18 to determine which pairwise differences are significant.

When the number of comparisons is large, the Bonferroni procedure is very conservative. That is, as the number of comparisons increases, the tabulated value of t against which the calculated t statistics are compared increases, and it becomes increasingly difficult to declare a difference in means to be significant. Often it is more difficult than is really necessary. Under these circumstances, other multiple-comparison procedures may be preferred. These procedures are discussed in most intermediate-level statistics textbooks.

EXERCISE 9.6

Refer to Exercise 9.5. If there are differences in HbA_{1c} (percentage) among the three different treatments as indicated by the ANOVA, determine which treatments are different. (Use a per experiment error rate of $\alpha = .01$).

EXERCISE 9.7

Refer to Example 9.2 and Table 9.3. If there are differences among the four services as indicated by the ANOVA, which services are different? (Use a per experiment error rate of $\alpha = .01$).

Final Note on Analysis of Variance

Often in reading the literature, you will encounter terms that are closely associated with the Analysis of Variance technique. The terms listed in Table 9.9 are used for different "designs" or ways in which the study may be structured. Each of these designs suggests use of the ANOVA;

TABLE 9.9 COMMON TERMINOLOGY ASSOCIATED WITH ANALYSIS OF VARIANCE

Completely Randomized Design
Randomized Block Design
Latin Square Design
Cross Classification
Nested Classification
Hierarchical Classification
Factorial Experiment
One-way ANOVA
Two-way ANOVA

the calculational techniques for obtaining the F statistic are different for each design but the basic reasoning discussed for the Completely Randomized Design holds for all designs.

Summary of ANOVA Computational Procedures for the CRD and the RCB*

The steps for computing the ANOVA for the CRD and RCB are:

STEP 1. Compute the square of the total of all observations and divide by the total number of observations. This is called the correction factor.

STEP 2. Compute the total sum of squares (SS) by summing the squares of all of the individual observations and subtracting the correction factor.

STEP 3. Compute the treatment sum of squares (SST) by summing the squares of treatment totals, each square divided by its own number of observations, and subtracting the correction factor.

STEP 4. For RCB compute the block sum of squares (SSB) by summing the squares of block totals, each square divided by its own number of observations, and subtracting the correction factor.

STEP 5. Compute the error sum of squares:

for CRD:

$$SSE = SS_{\text{total}} - SST$$

for RCB:

$$SSE = SS_{\text{total}} - SST - SSB$$

STEP 6. Compute degrees of freedom.

STEP 7. Compute mean squares by dividing each sum of squares by its own degrees of freedom.

STEP 8. Compute test statistic.

$$F = \frac{MST}{MSE}$$

* This section is reprinted from Duncan, Knapp, Miller, *Introductory Biostatistics for the Health Sciences,* 2nd Ed., 1983, pp. 155–156. Used by permission of John Wiley & Sons, Inc.

SUMMARY OF CRITICAL CONCEPTS IN CHAPTER 9

The following concepts were presented in Chapter 9:

1. Carrying out multiple *t*-tests on the same set of data increases the chance of having a significant *t* value (rejecting H_0) when in fact H_0 is true.
2. The procedure used for testing differences in mean values for more than two groups is called the Analysis of Variance (ANOVA) procedure.
3. The basic idea underlying all Analysis of Variance techniques is that the total variability in a set of measurements may be divided into variability among the measurements within a group and the variability between the means of the different groups.
4. If the *k* treatments are all alike in their effect, i.e., all the means are equal, then the variability between the mean values should be no greater than the variability of the measurements within a group.
5. The test statistic used in the Analysis of Variance (CRD) is:

$$F = \frac{\text{Variance among treatment groups}}{\text{Variance within groups}}$$

6. When there is no difference in the true means of all the treatment groups, the value of *F* is close to 1.
7. The null hypothesis of equal treatment means is rejected for "large" *F* values.
8. Rejection of the null hypothesis of equal mean values says that *at least one* pair of means is not equal; the alternative hypothesis does not specify *which* pairs are not equal, only that one or more pairs are unequal.
9. To determine which pairs of means are not equal, such techniques as Duncan's Multiple Range Test, Least Significant Difference (LSD), Tukey's, Scheffe's procedure, or the Bonferroni *t* procedure must be used.
10. In a *Completely Random Design* (CRD) the *N* experimental units (patients) are randomly assigned to the *k* treatments being studied (or the treatments are randomly assigned to the experimental units).
11. In a CRD it is assumed that the *N* experimental units are a reasonably homogeneous group.

12. In a CRD there may be unequal numbers in the k samples receiving the k treatments.
13. In a CRD it is assumed that the randomization process equalizes treatment groups with respect to extraneous variables that could influence the response being measured. (This is not always assured, particularly with small samples).
14. In the RCB design an extraneous variable may be recognized and taken into account through the design of the study.
15. In the RCB design subjects (experimental units) are "blocked," or grouped in homogeneous groups with respect to the blocking variable. Within each of the b blocks, subjects are assigned at random to the k treatments (or treatments assigned randomly to subjects).
16. In general, with an RCB design we have k subjects available (k = number of treatments) within each level of the "blocking" variable.
17. In an RCB design the variation in the response variable attributable to the blocking (extraneous) variable is computed and arithmetically removed from the residual error term. The residual error term is the denominator in the F statistic. The more the residual error term can be reduced, the more likely it is that the F statistic will be significant, that is, the more likely we are to detect differences in treatment means.

BIBLIOGRAPHY

Anderson, R. L., and Bancroft, T. A., *Statistical Theory in Research*. New York: McGraw-Hill, 1952.

Daniel, W. W., *Biostatistics: A Foundation for Analysis in the Health Sciences*, 2nd Ed. New York: John Wiley & Sons, 1978.

CHAPTER 10

CORRELATION AND REGRESSION

OVERVIEW

One of the most frequently encountered problems in statistics is how to describe a relationship between two variables. For example, a nurse educator may wish to determine the relationship between final grade point average of nursing students and scores on a state board examination; an investigator may wish to determine the degree of agreement between blood pressure readings on the same patient made by two different nurses using the same instrument; or it may be desirable to determine the relationship between patient's age and survival time following a heart attack.

The above are examples of two widely used statistical techniques—*correlation analysis* and *regression analysis*. The first part of Chapter 10 will discuss correlation in terms of its meaning, interpretation, and limitations. The basic computational formulas will be given for the most often used correlation procedures. Each of the following correlation techniques will be discussed:

Pearson's Product Moment Correlation
Correlation Ratio
Spearman's Rank Order Correlation
Kendall's Tau
Partial Correlation
Coefficient of Concordance
Multiple Correlation

The remainder of the chapter will be devoted to a presentation of simple linear regression.

OBJECTIVES

Upon completion of this chapter, the student will be able to:

1. Recognize a positive linear, negative linear, and a curvilinear relationship.
2. Represent a set of data by means of a scatter diagram.
3. Define the term "correlation."
4. Discuss the interpretation of correlation, i.e., what is meant by range of values −1 to +1 and by the squared correlation.
5. Discuss precautions in the interpretation of correlation.
6. Determine the correlation between two variables using Pearson's Product Moment Correlation Coefficient.
7. Determine the correlation between two variables using Spearman's Rank Order Correlation Coefficient.
8. Describe in words the use of regression analysis.

†9. Carry out a simple linear regression analysis.

DISCUSSION

Correlation: Its Meaning and Interpretation

As you may have noticed from the Overview, there exists a variety of correlation procedures. All basically provide the same type of information: the *direction* (either positive or negative) and the *magnitude* of the relationship between variables. The need for several different procedures is a result of the fact that different investigations involve different types of variables (discrete or continuous) using different measuring scales. The type of variable measured is reflected in the choice of correlation procedure to be used.

While there is a variety of different techniques, the basic meaning and interpretation is the same for all. The remainder of this section will be a general discussion of the meaning, interpretation, and limitations of correlation; the final section will discuss several of the more widely used procedures.

THE DATA

The data in a correlation study are usually arranged as shown in Table 10.1 where the (X_i, Y_i) pairs may, for example, represent college

† Sections marked by a dagger (†) cover more difficult or theoretical topics.

TABLE 10.1 DATA LAYOUT FOR CORRELATION STUDY

Subject	*Measurement on Variable 1*	*Measurement on Variable 2*
1	X_1	Y_1
2	X_2	Y_2
⋮	⋮	⋮
n	X_n	Y_n

GPR and scores on state board examination; blood pressure readings for Nurse 1 and Nurse 2; or age in years and survival time following a heart attack. Thus X_1 may be the GPR and Y_1, the state board score for student No. 1, etc.

THE SCATTER DIAGRAM

Usually, the first step in the investigation of a relationship between variables is to display the data graphically. This graphical representation, called a *scatter diagram*, gives the investigator a visual image of the relationship being studied. In a scatter diagram, each of the n pairs (X_i, Y_i) is plotted as a single point on the graph. The X's are plotted on the horizontal axis and the Y's are plotted on the vertical axis.

EXAMPLE 10.1 Suppose the following data represent systolic blood pressure readings (in mmHg) on five patients read by two nurses using the same instrument:

Patient	*Nurse 1(X)*	*Nurse 2(Y)*
1	120	123
2	140	138
3	136	138
4	160	160
5	130	128

The scatter diagram showing the above pairs of measurements is given in Figure 10.1.

By looking at the arrangement of points on the scatter diagram, one may be able to discern a pattern that is indicative of the nature of the relationship between the variables. From Figure 10.1, we see that for large values of X we also have large values of Y and for small X values we have small Y values. In addition, the points seem to fall (or nearly so)

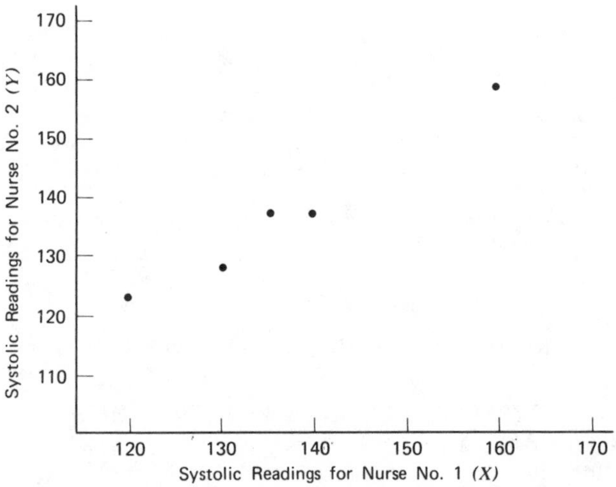

FIGURE 10.1 SCATTER DIAGRAM OF SYSTOLIC BLOOD PRESSURES (IN mmHg) READ BY TWO NURSES USING THE SAME INSTRUMENT

on a straight line. This type of relationship is called a *positive linear* relationship between X and Y.

Consider Figure 10.2. In this scatter diagram, as the X measurements become larger, the Y measurements become smaller. This type of relationship is a *negative linear* relationship between X and Y. In Figure 10.3, we see that the relationship is clearly not representable by a straight line. The plots (*a*) and (*b*) are representative of the class of relationships called *nonlinear,* or *curvilinear* relationships. It is apparent from the pattern of points in Figure 10.4 that values for X have no relationship to the values for Y. This is the case of no relationship between the variables X and Y.

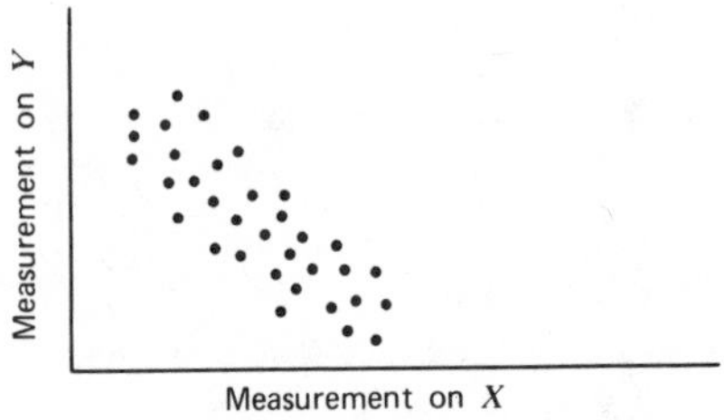

FIGURE 10.2 A NEGATIVE LINEAR RELATIONSHIP BETWEEN X AND Y

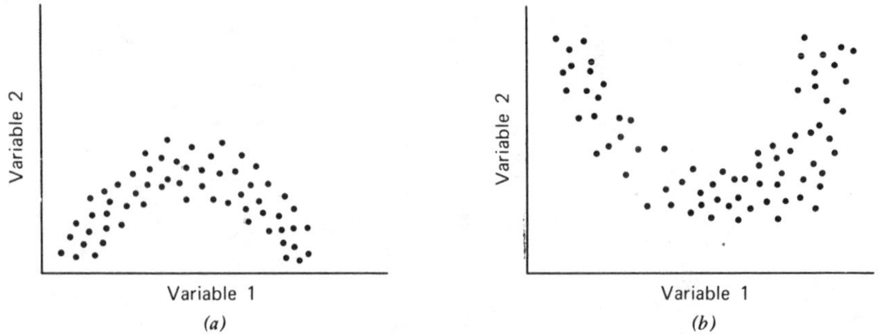

FIGURE 10.3 CURVILINEAR RELATIONSHIPS BETWEEN *X* AND *Y*

A NUMERICAL REPRESENTATION OF RELATIONSHIPS BETWEEN VARIABLES

While scatter diagrams provide a simple and useful means of studying relationships between variables, often it is necessary to describe a relationship by means of a more precise numerical value. This numerical value, usually called a correlation coefficient, or simply a *correlation*, is a measure of the magnitude or the strength of the relationship between variables. Among the most common of these are Pearson's Product Moment Correlation Coefficient (r) and Spearman's Rank Order Correlation Coefficient (r_s). These two measures will be discussed in detail, while some of the less common ones will be mentioned briefly along with references when appropriate.

Pearson's Product Moment Correlation Coefficient (r)

Pearson's Product Moment Correlation Coefficient, symbolized by r, measures the *linear* relationship between two *continuous* variables. (A linear relationship, either positive or negative, is one that may be adequately described by a straight line, as in Figures 10.1 and 10.2.) The correlation coefficient r is an index number whose value ranges from -1 to $+1$. A correlation of 0 indicates no linear relationship between

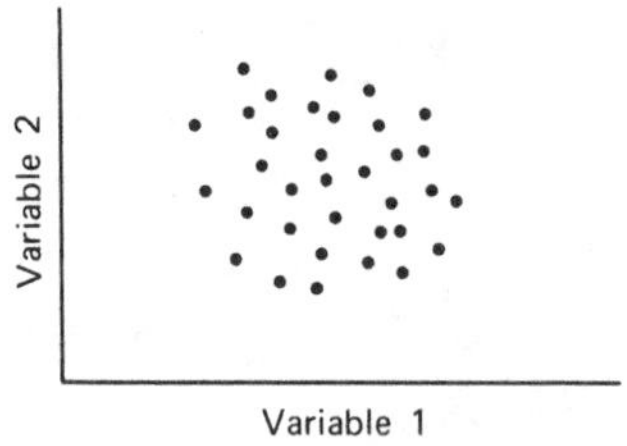

FIGURE 10.4 NO RELATIONSHIP BETWEEN *X* AND *Y*

variables, while a correlation of -1 or $+1$ indicates either a perfect negative linear relationship or a perfect positive linear relationship. In practice, perfect correlations are rarely encountered; more often correlations will lie somewhere in the range -1 to $+1$.

The following describe situations in which the product-moment correlation is applicable:

EXAMPLE 10.2 A nurse educator wishes to determine the degree of association between students' clinical performance scores and scores on the NLN.

EXAMPLE 10.3 A nurse midwife wishes to measure the degree of association between uterine size and age of patient.

EXAMPLE 10.4 To ascertain the observer reliability of a tool for evaluating student performance in Medical-Surgical nursing, an investigator determined the degree of agreement between student scores evaluated by two different observers.

COMPUTATION OF r

The product moment correlation coefficient is obtained using the formula:

$$r = \frac{S_{xy}}{\sqrt{S_{xx}S_{yy}}}$$

where

$$S_{yy} = \Sigma Y^2 - \frac{(\Sigma Y)^2}{n}$$

$$S_{xx} = \Sigma X^2 - \frac{(\Sigma X)^2}{n}$$

$$S_{xy} = \Sigma XY - \frac{(\Sigma X)(\Sigma Y)}{n}$$

While at first glance the above formula looks quite formidable, you should recognize the familiar sum of squares pattern found in the numerator of the standard deviation formula.

EXAMPLE 10.5 A nurse educator, investigating the degree of the relationship between student scores on a battery of personality tests and performance in nursing school, gave a random sample of applicants for nursing school a personality inventory evaluation. The scores on the battery of tests range from 0 to 10. At the end of the first year, college grade point average was recorded for each student in the study. The data are shown below.

Student	*Personality Score*	*GPA*
1	8.0	3.0
2	3.0	1.0
3	7.0	3.0
4	5.0	1.5
5	6.0	2.0

$$\Sigma X = 29.0 \qquad \Sigma Y = 10.5$$
$$\Sigma X^2 = 183.0 \qquad \Sigma Y^2 = 25.25$$
$$\Sigma XY = 67.5$$

(Note: $\Sigma XY = X_1 Y_1 + X_2 Y_2 + \cdots + X_n Y_n$

SOLUTION To calculate the correlation coefficient, we must first calculate S_{xx}, S_{yy}, S_{xy}.

$$S_{xx} = \Sigma X^2 - \frac{(\Sigma X)^2}{n} = 183 - \frac{(29)^2}{5} = 14.8$$

$$S_{yy} = \Sigma Y^2 - \frac{(\Sigma Y)^2}{n} = 25.25 - \frac{(10.5)^2}{5} = 3.2$$

$$S_{xy} = \Sigma XY - \frac{(\Sigma X)(\Sigma Y)}{n} = 67.5 - \frac{(29)(10.5)}{5} = 6.6$$

The strength of the association between scores on the personality inventory and grade point average at the end of the freshman year is given by:

$$r = \frac{S_{xy}}{\sqrt{S_{xx}S_{yy}}} = \frac{6.6}{\sqrt{(14.8)(3.2)}} = .96$$

EXERCISE 10.1

In order to evaluate the efficacy over time of a new tool for measuring anxiety among hospitalized children, a pediatric nurse investigator gave the exam to the same group of children on two different days. Determine the magnitude of the association between the two sets of scores using *r*.

	Scores	
Patient	*Day 1*	*Day 2*
1	5	4
2	3	3
3	4	5
4	6	6
5	2	3

$\Sigma X = 20$ $\Sigma Y = 21$

$\Sigma X^2 = 90$ $\Sigma Y^2 = 95$

$\Sigma XY = 91$

One assumption for the use of Pearson's Product Moment Correlation Coefficient is that the data possess at least an interval scale of measurement. The score data used in the above examples are treated as if the underlying scale is interval.

INTERPRETATION OF CORRELATION

Suppose the correlation between student GPR (grade point averages) and scores on state boards is found to be 0.5. This does not mean that the strength of the relationship is "halfway" between zero correlation (no relationship) and a perfect positive correlation. How then do we interpret such a value? One simple technique is to square the value of the correlation coefficient. This squared correlation indicates the percentage of variation in one variable that is explained by variation in the other. In the above example of the association between student GPR and state board scores, the square of the correlation is $(.5)^2 = .25$. This tells us that 25% of the variability in state board scores is attributable to variability in student GPR. Alternatively, given a student's GPR, we have 25% of the necessary information to predict the state board score. By looking at the square of the correlation, we realize that often the size of the actual correlation may be misleading. The following lists give values of the correlation and the percent of accounted for variation between the variables (variation in one variable attributable to variation in the other):

Correlation	*Squared Correlation*
.5	.25
.6	.36
.7	.49
.8	.64
.9	.81

From the above it may be seen that a correlation must be greater than .7 before 50% of the variation in one variable may be attributed to variation in the other. Using the above figures, we may generate a very crude guideline for interpreting correlations: from 0–.25, little or no relationship; .5–.75, a moderate to good relationship; .8 or greater, a very good relationship.

On the other end of the spectrum, one must be wary of extremely high correlations, especially in the biological or social sciences. Often correlations of .95 or larger may result when one quantity is correlated with another quantity of which the first is a component. For example, grades in clinical courses when correlated with overall GPR will, in all likelihood be high because grades in clinical courses are used in determining GPR. This is not to suggest that all high correlations are "flukes" but rather to point out that when extremely large correlations do occur, the critical evaluator will mentally eliminate the possibility of these spurious correlations before accepting the results of the study.

INFERENCE ON ρ, THE POPULATION CORRELATION

The value for r ranges from -1 to $+1$ with a zero correlation indicating no *linear* relationship between variables. It is important to note that r is a value calculated from a *sample* of data and is an estimate of the "true" correlation (ρ) between the two populations of interest. Therefore, when one wishes to determine whether a "significant" correlation exists between two variables, a test of hypothesis must be carried out. This test of hypothesis, which is analogous to tests of hypothesis about μ discussed earlier, determines whether a calculated value of r is significantly different from zero. It answers the question, is r different from zero only because of chance variation (sampling error) or because the true population correlation (ρ) is not zero. Hence, a value of r reported in the literature should be accompanied by a level of significance.

To determine whether a correlation is significantly different from zero, the calculated value of r is compared with a value found in Table F in the Appendix. The appropriate table value is located at the intersection of the column α and the row $df = n - 2$, where n is the number of *pairs* in the sample. To be "significant," a calculated correlation coefficient (r) must exceed the value obtained from the table. For example, for a calculated correlation coefficient to be significant at the $\alpha = .05$ level of significance when there are 12 pairs of measurements in the sample, it must exceed the value .5760 located in Table F at the intersection of the column .05 and the row $df = n - 2 = 10$. For negative values, the sign of r is ignored when comparing it with the appropriate table value.

EXAMPLE 10.6 For 20 patients on an antihypertensive drug, measurements of dose level and blood pressure were recorded. The correlation between these two sets of measurements was found to be $-.68$. Is the value of r significant (at $\alpha = .01$)? What does this mean in relation to the two variables being measured?

SOLUTION To determine whether the correlation coefficient is significantly different from zero, the calculated value of r must be compared with a value found in Table F. For $\alpha = .01$ and $df = n - 2 = 18$, the appropriate table value is .5614. Since .68 (sign ignored) is greater than .5614, it is said that a significant *negative* correlation exists between the dose of antihypertensive drug and blood pressure. This means that as dose level *increases*, blood pressure *decreases*.

A word of caution: an r that is *not* significantly different from zero indicates that there appears to be no *linear* relationship between variables. This does not eliminate the possibility of another type of relationship between variables, i.e., curvilinear, exponential, logarithmic, etc.

EXERCISE 10.2

A test was devised to measure the aptitude of applicants for admission to a nursing school. To estimate the *reliability* over time of the tool, it was given to 10 applicants twice within a two-week period. The scores for the 10 applicants were:

	Scores	
Applicant	*Week 1*	*Week 2*
1	90	85
2	50	60
3	85	80
4	60	60
5	40	45
6	60	55
7	85	90
8	70	75
9	75	75
10	90	95

a. Plot the data on a scatter diagram.
b. Based on the scatter diagram, what type of relationship does there appear to be between the two measuring periods (e.g., negative linear, positive linear, curvilinear, etc.)?
c. What does this tell us about the relationship between the two sets of scores?

EXERCISE 10.3

In Exercise 10.2, the magnitude of the correlation coefficient was found to be .95. In your own words, how would you interpret the meaning of this value? What does it tell you about the reliability of the tool?

EXERCISE 10.4

In a study to determine the relationship between vocabulary proficiency and grades in a senior level research course, it was found that there is a significant positive linear correlation of 0.90 between these variables. The researcher concluded that based on the highly significant correlation, it may be said that having a good vocabulary will cause a student to do well in a research course. Discuss the validity of this statement.

EXERCISE 10.5

A correlation of .98 was reported between grades in Ob-Peds and overall clinical grade point average. What statistical factors might explain this high correlation?

EXERCISE 10.6

The following table gives grade point average at the end of four years of nursing school and scores on state boards for 12 nursing students:

Student	*GPR(X)*	*Score on State Board (Y)*
1	1.95	435
2	2.23	430
3	2.55	515
4	2.98	560
5	1.96	430
6	2.85	415
7	3.00	570
8	1.94	295
9	3.10	560
10	3.13	445
11	2.37	445
12	3.35	620

a. Plot the scatter diagram. What type of relationship does the diagram suggest?
b. The correlation between GPR and scores on state boards is found to be 0.75. Is this correlation significant? (Use α = .05 level of significance.) Explain what this means.

EXERCISE 10.7

In a study of the relationship between amphetamine metabolism and amphetamine psychosis, six chronic amphetamine users were given a psychosis intensity score. Plasma amphetamine levels (ng/ml) were also determined for these patients:

Patient	*Psychosis Intensity Score*	*Plasma Amphetamine (ng/ml)*
1	30	500
2	55	250
3	30	250
4	45	200
5	40	100
6	15	150

a. Plot the scatter diagram. What type of relationship is suggested?
b. The correlation coefficient is found to be −0.047. Is this value significant? (Use $\alpha = .05$.) What may be concluded about the relationship between psychosis intensity score and plasma amphetamine level?

EXERCISE 10.8

An investigator was interested in determining the interobserver reliability for a tool that had been developed to measure depression among women using oral contraceptives. The scores for five patients were recorded for two observers.

Patient	*Observer 1*	*Observer 2*
1	9	8
2	7	7
3	6	6
4	4	5
5	10	10
	$\Sigma X = 36$	$\Sigma Y = 36$
	$\Sigma X^2 = 282$	$\Sigma Y^2 = 274$

$$\Sigma XY = 277$$

a. Plot the scatter diagram. What type of relationship does it suggest?
b. Determine the product-moment correlation coefficient for the two groups of scores.
c. How does the correlation coefficient relate to the interobserver reliability for the tool being used? (Use $\alpha = .01$ to test the significance of r.)

EXERCISE 10.9

An investigator wished to develop a new tool for measuring nursing proficiency. A traditional tool exists and is known to be a valid device for measuring proficiency. How might this traditional tool be used in establishing the validity of the new instrument?

Precautions in the Interpretation of Correlation

One of the most important rules of interpretation is "correlation does not mean causation." Since correlations are reversible, i.e., X correlated with Y implies Y correlated with X, then one cannot state that X "causes" Y. It would be as easily said that Y "causes" X. In most cases, correlation studies do not *prove* that one variable causes another. A high correlation only implies that the two variables tend to vary together; a third variable (or a multitude of other variables) may be at work to produce the observed phenomenon.

A second precaution that must be observed when interpreting correlations is that the correlation only applies to the range of values observed for the two variables. For example, when correlating height and age, an investigator observed individuals younger than 20 years old and found a strong positive association between height and age. This does not necessarily mean that the same correlation holds for individuals outside the age range observed in the sample. In fact, for people older than 20 years the linear correlation between height and age becomes negligible.

A final word of caution: correlation is a mathematical technique that may be applied to any set of data regardless of whether it is meaningful to do so. Often a nonsensically high correlation will be reported between two logically unrelated events, such as number of stork nests and number of births in a given country. These nonsense correlations often result when data has been collected over a long period of time where many factors may operate to affect both variables simultaneously.

In the final analysis, the interpretation of correlations is dependent on the particular investigation and the professional judgment of the investigator and the consumer. Ultimately only those with some knowledge of the field can determine whether a particular correlation is noteworthy.

Correlation Ratio (η)

A measure of association that is similar to r in most respects is the correlation ratio. The correlation ratio may be used when the relationship between two *continuous* variables is *curvilinear* (Figure 10.3). The computational formula for the correlation ratio is more complex than that for r and therefore will not be presented. The values for η ("eta") range from -1 to $+1$ and its magnitude is an indication of the degree of association between the two variables. The "form" of the relationship should be indicated by means of an accompanying scatter diagram.*

* For a more detailed discussion see Frank J. Kohout, *Statistics for Social Scientists* (New York: John Wiley & Sons, 1974), pp. 193–209.

Spearman's Rank Order Correlation Coefficient

Spearman's Rank Order Correlation Coefficient belongs to a class of statistical techniques known as *nonparametric* or distribution-free methods. These techniques are most commonly employed when the assumptions underlying the traditional techniques are not met (e.g., for a t-test we assumed the values are normally distributed). Nonparametric methods, while to some extent less capable of detecting relationships, are appealing because they usually involve less calculation and fewer assumptions. (See Chapter 12.)

When data are reported in terms of ranks (ordinal data) or when it seems feasible to assign ranks to the data values, the correlation between ranks rather than between raw scores may be determined. The correlation between two sets of ranks is called *Spearman's Rank Order Correlation Coefficient*. Consider the following examples.

EXAMPLE 10.7 The quality of clinical performance for five student nurses was evaluated by two different observers. Student performance was ranked from 1 to 5 by each observer.

	Rank	
Nurse No.	*Observer 1*	*Observer 2*
1	3	2
2	2	3
3	5	4
4	1	1
5	4	5

Spearman's Rank Order Correlation Coefficient would be used to measure the degree of association between the scoring of the two observers.

EXAMPLE 10.8 In a public health survey 20 large cities were ranked by fluoride concentration (ppm) in the water and number of cavities per 100 children. The degree of association may be determined using Spearman's Rank Order Correlation.

EXAMPLE 10.9 An investigator wishes to determine the association between class rank of senior nursing students and rank on a standardized nursing proficiency exam. The procedure is illustrated below.

Nurse No.	Class Standing	Rank on Exam	Difference (d) Between Ranks
1	4	2	2
2	1	5	−4
3	2	1	1
4	3	3	0
5	5	4	1

COMPUTATIONAL FORMULA

The formula for computing Spearman's Rank Order Correlation Coefficient is:

$$r_s = 1 - \frac{6(\Sigma d^2)}{n(n^2 - 1)}$$

where d = difference between ranks for each individual and n = number of pairs of ranks.

For the above example, the Rank Order Correlation between class standing and rank on a nursing proficiency exam is:

$$\begin{aligned} r_s &= 1 - \frac{6(\Sigma d^2)}{n(n^2 - 1)} \\ &= 1 - \frac{6(22)}{5(25 - 1)} = 1 - \frac{132}{120} \\ &= -0.1 \end{aligned}$$

Spearman's Rank Order Correlation may also be applied to data that were not originally in the form of ranks. In this case, the raw data must first be ranked in order from smallest to largest. For ties, the observations are assigned their average rank. For example, if the first and second smallest observation is the value 20, both values are assigned a rank of 1.5. After ranks have been assigned, the computational procedure is as described for Example 10.9.

INTERPRETING THE RANK ORDER CORRELATION COEFFICIENT

Like the Product Moment Correlation Coefficient, Spearman's Rank Order Correlation Coefficient may take any value from −1 to +1 with 0 indicating no linear association between ranks. Also like r, a test of hypothesis to determine if r_s is significantly different from zero may be carried out.

The calculated value of r_s must be compared with a value from Appendix Table G to determine if the calculated value is significantly different from zero. (A zero value for the true population correlation indicates no linear relationship between variables.) Table G is applicable for r_s values when n is between 4 and 30. The cut-off points are located at the intersection of the column $\alpha/2$ and the row equal to n. The lower tail value is obtained by affixing a negative sign to the positive value obtained from the table.

To determine whether the correlation between class standing and rank on the nursing proficiency exam ($r_s = -0.1$) is significantly different from 0, we must compare it with a value from Table G. For the $\alpha = .05$ level of significance, the value at the intersection of the column $\alpha/2 = .025$ and the row $n = 5$ is 0.90. Thus the computed value of r_s (ignoring the sign) must be larger than .90 to be significantly different from zero at the 5% level of significance. Clearly, $r_s = -0.1$ does not qualify as a significant correlation coefficient ($\alpha = .05$). Thus, we cannot say that there is a significant *linear* relationship between class standing of student and rank on the nursing proficiency exam.

EXERCISE 10.10

A study was carried out to determine the degree of association between the creativity of children and the creativity of their parents. Six children and their parents were interviewed and ranked according to creativity. The ranks are:

Child	*Parent*
4	5
2	2
1	3
3	1
5	6
6	4

a. Calculate the degree of association between the creativity ranking of the children and their parents.
b. Interpret this value. (Use $\alpha = 0.05$.)

EXERCISE 10.11

Compute Spearman's Rank Order Correlation for Example 10.7. Interpret the results. (Use $\alpha = .05$.)

Kendall's Tau

Kendall's Tau is another measure used to determine degree of association between two variables whose values have been ranked. It is similar to the Rank Order Correlation in computation and interpretation.*

Multiple Variables

Often, especially in nursing and the social sciences, one may wish to determine the relationship among several variables acting together. There are several correlation techniques for evaluating multiple variable relationships. A brief summary of several of the most used of these techniques will be presented.

1. *Partial Correlation.* This technique allows the researcher to evaluate the relationship between two variables when the effects of the remaining variables in the study have been removed. For example, in a study of the relationship between three variables, weight, height, and daily caloric intake of obese female patients, the partial correlation of weight (X) and caloric intake (Y) on height (Z) gives a measure of the association between weight and caloric intake after the effect of height has been removed.
2. *Kendall's Coefficient of Concordance.* This technique is used to measure the overall association between several variables that are in the form of ranks. For example, if three evaluators rank order five student nurses, the coefficient of concordance may be used to measure the agreement among the evaluators.
3. *Multiple Correlation.* This technique is used more in association with regression (next section) than with correlation. The multiple correlation coefficient measures the association between the observed values of the dependent variable and the values predicted by the regression equation.

Predicting One Variable from Another

The final topic to be covered in this chapter is that of *prediction*. The statistical technique for predicting the value of one variable given the

* A more detailed discussion of correlation may be found in: Kohout, *Statistics for Social Scientists,* especially pp. 225–236; Quinn McNemar, *Psychological Statistics* (New York: John Wiley & Sons, 1969); and Sidney Siegal, *Nonparametric Statistics for the Behavioral Sciences* (New York: McGraw-Hill, 1956).

value of a second variable is called *regression analysis* and is closely associated with correlation analysis.

In regression, the relationship between two variables (or multiple variables) is expressed by means of "fitting" a line or a curve to the pairs of data points. The variable that is to be predicted is called the *dependent* variable and the variable that serves as the "predictor" is called the *independent* variable. The following examples are situations in which regression analysis would be used:

EXAMPLE 10.10 An institution wishes to predict a student's score on state boards based on college GPR.

EXAMPLE 10.11 A nurse wishes to predict a patient's blood pressure decrease for a given dose of an antihypertensive drug.

EXAMPLE 10.12 A physician wishes to predict a patient's survival time based on age following a heart attack.

† Simple Linear Regression

As an introduction to the problem of regression analysis, data relating systolic blood pressure to dose level of an antihypertensive drug will be used.

EXAMPLE 10.13 The following list gives the systolic blood pressures for a group of hypertensive patients along with the dose level of a certain antihypertensive drug.

Dose Level of Drug (mg)	*Mean Systolic Blood Pressure (mmHg)*
2	278
3	240
4	198
5	132
6	111

Looking at these data, we can see that some relationship does exist. As the dose level increases, the mean systolic blood pressure decreases. The data are plotted in the scatter diagram in Figure 10.5. Observe that all of the points do not fall exactly on a straight line but do seem to follow a uniform downward trend. This is an indication that the relationship could be *linear*, and can be described by a straight line.

We now wish to determine the "equation" for the line shown in

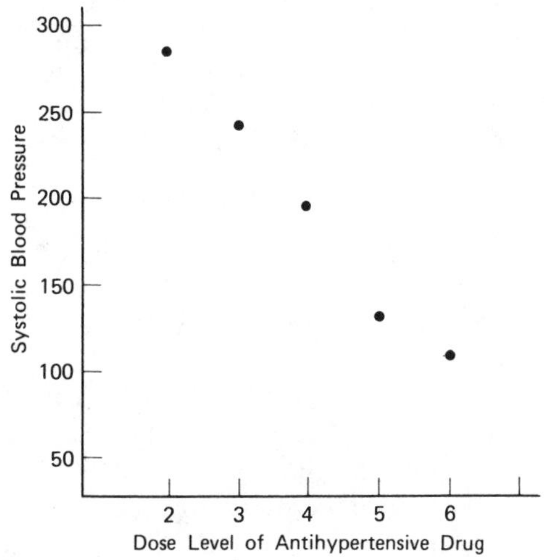

FIGURE 10.5 SCATTER DIAGRAM FOR DATA IN EXAMPLE 10.13

Figure 10.5. Recall that any straight line has the general form:

$$Y = a + bX$$

where b is the slope of the line and a is the point where the line intercepts the Y axis. For any two points, it is a simple matter to determine the equation for the line connecting the two points. However, in statistical problems involving three or more sample points, it is generally impossible to find a line that goes through all the points simultaneously. In this case, we attempt to find the line that "best" fits all the points. A stylized illustration of data and such a prediction line drawn by eye are shown in Figure 10.6. The "hat" over Y, read "Y hat," in Figure 10.6 is used to denote that this line is just an estimate of some theoretically true (population) line.

The distances of the observations from the line are given by:

$$d_i = Y_i - \hat{Y}_i$$

where

$$\hat{Y}_i = a + bX_i$$

It seems to be a function of the human eye that such lines seem naturally to try to minimize the distances (or deviations) d_i of the observed data points Y_i from the line.

In Figure 10.6, d_1 is positive, d_2 is negative, and d_3 is positive. As before, deviations add to zero so it is more convenient to compute the sum of squared deviations and use the square root to measure the overall "fitness" of the line. That is, we will deal with:

$$\Sigma d_i^2 = \Sigma(Y_i - \hat{Y}_i)^2$$

Further, since different people would probably draw different lines, a standard statistical procedure is needed to find the line such that the sum of squared deviations is a minimum.

The statistical procedure for finding this best-fitting line is called the *method of least squares* and the line is called the *regression line*. The formal derivation of this procedure, which requires differential calculus, is presented in advanced statistical texts. The results from the method of least squares will be applied to the data in Example 10.13 to illustrate the principles involved.

First, it is necessary to introduce some useful new notation:

$$(X_i, Y_i) = i^{th} \text{ pair of observations}$$

$$S_{xy} = \Sigma XY - \frac{(\Sigma X)(\Sigma Y)}{n}$$

$$S_{yy} = \Sigma Y^2 - \frac{(\Sigma Y)^2}{n}$$

$$S_{xx} = \Sigma X^2 - \frac{(\Sigma X)^2}{n}$$

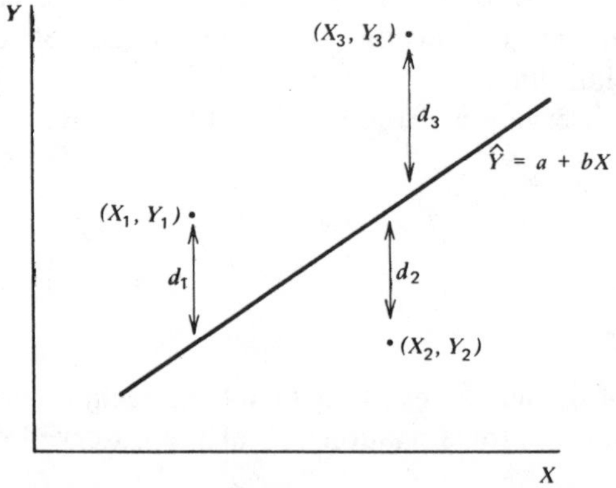

FIGURE 10.6 STYLIZED ILLUSTRATION OF PREDICTION LINE

The sample regression line is written:

$$\hat{Y} = \hat{\beta}_0 + \hat{\beta}_1 X$$

where the least squares estimates $\hat{\beta}_0$ and $\hat{\beta}_1$ are (without proof):

$$\hat{\beta}_1 = \frac{S_{xy}}{S_{xx}}$$

and

$$\hat{\beta}_0 = \overline{Y} - \hat{\beta}_1 \overline{X}$$

The values $\hat{\beta}_0$ and $\hat{\beta}_1$ are calculated from a *sample* of observations from the entire population of interest and are estimates of the "true" population values β_0 and β_1. As was the case with $\overline{Y}$ and s, the values $\hat{\beta}_0$ and $\hat{\beta}_1$ are subject to sampling variation and therefore may vary from sample to sample.

The data in Example 10.13 are rewritten in Table 10.2 together with the required computations and regression data. From the totals in Table 10.2, we can now compute:

$$\overline{Y} = \frac{\Sigma Y}{n} = \frac{959}{5} = 191.8 \qquad \overline{X} = \frac{\Sigma X}{n} = \frac{20}{5} = 4.0$$

$$S_{yy} = \Sigma Y^2 - \frac{(\Sigma Y)^2}{n} \qquad S_{xx} = \Sigma X^2 - \frac{(\Sigma X)^2}{n}$$

$$= 203{,}833 - \frac{(959)^2}{5} \qquad = 90 - \frac{(20)^2}{5}$$

$$= 19{,}896.8 \qquad = 10$$

$$S_{xy} = \Sigma XY - \frac{(\Sigma X)(\Sigma Y)}{n}$$

$$= 3{,}394 - \frac{(20)(959)}{5}$$

$$= -442$$

Thus, for Example 10.13:

$$\hat{\beta}_1 = \frac{-442}{10} = -44.2$$

$$\beta_0 = 191.8 - (-44.2)4$$

$$= 368.6$$

and

$$\hat{Y} = 368.6 - 44.2X$$

The estimated regression line is plotted in Figure 10.7. Using the regression equation, one may now *predict* the systolic blood pressure for a given dose of the drug for any patient on the medication. Suppose we wish to predict systolic blood pressure for a patient who is given 4 mg of the drug. Substituting in the equation, we have:

$$\begin{aligned}\hat{Y} &= 368.6 - 44.2X \\ &= 368.6 - 44.2(4) \\ &= 368.6 - 176.8 \\ &= 191.8\end{aligned}$$

Thus for a 4-mg dose of the drug, we would expect the patient to have a systolic blood pressure equal to 191.8 mmHg.

The values of $\hat{Y}$ for each X and the differences between the predicted values ($\hat{Y}$) and the observed values (Y) are shown in Table 10.3. The value $\hat{Y}$ obtained for a given X is the predicted *mean* of the population of all possible Y values that could occur at the given value X.

VARIATION ABOUT THE REGRESSION LINE

A measure of the scatter or variation of the observed points (Y) about the regression line is given by:

$$s^2_{y.x} = \frac{\Sigma(Y_i - \hat{Y}_i)^2}{n - 2}$$

TABLE 10.2 RELATIONSHIP BETWEEN DOSE LEVEL OF ANTIHYPERTENSIVE DRUG AND SYSTOLIC BLOOD PRESSURE

Dose Level (X)	*Systolic Blood Pressure (Y)*	X^2	Y^2	XY
2	278	4	77284	556
3	240	9	57600	720
4	198	16	39204	792
5	132	25	17424	660
6	111	36	12321	666
Total 20	959	90	203,833	3,394

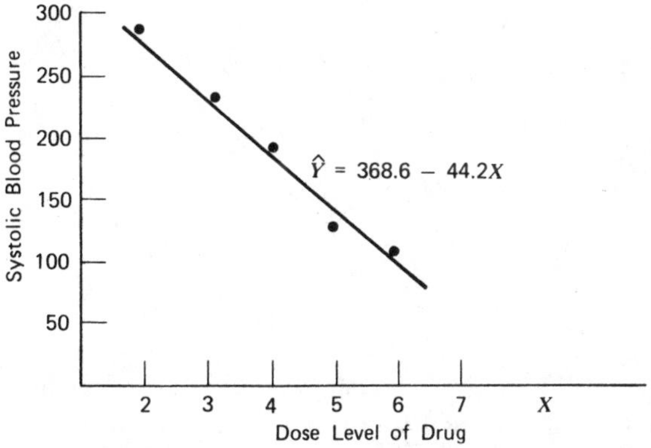

FIGURE 10.7 REGRESSION OF SYSTOLIC BLOOD PRESSURE ON DRUG DOSE LEVEL

where n = number of pairs (X_i, Y_i), Y_i = observed values of Y, and $\hat{Y}_i$ = value of Y predicted by regression line.

An alternate method for calculating $s^2_{y.x}$ is:

$$s^2_{y.x} = \frac{S_{yy} - \hat{\beta}_1 S_{xy}}{n - 2}$$

The positive square root of $s^2_{y.x}$, called the *standard error of the estimate*, is a measure of the "average" deviation of the observed values (Y) from the values ($\hat{Y}$) predicted by the regression line.

The standard error of the estimate $s_{y.x}$ for the regression equation relating systolic blood pressure and dose level of an antihypertensive

TABLE 10.3 RELATIONSHIP BETWEEN DOSE LEVEL OF ANTIHYPERTENSIVE DRUG AND SYSTOLIC BLOOD PRESSURE

Dose Level X	*Values of Systolic Blood Pressure* ($\hat{Y}$) *for Each X*	*Difference Between Predicted* ($\hat{Y}$) *and Observed* (Y) $(Y - \hat{Y})$
2	280.2	−2.2
3	236.0	4.0
4	191.8	6.2
5	147.6	−15.6
6	103.4	7.6

drug (Example 10.13) is given by:

$$s_{y.x} = \sqrt{\frac{S_{yy} - \hat{\beta}_1 S_{xy}}{n-2}}$$

$$= \sqrt{\frac{19{,}896.8 - (-44.2)(-442)}{3}} = \sqrt{\frac{360.4}{3}}$$

$$= 11.0$$

TESTS OF HYPOTHESES CONCERNING β_1

The procedure for obtaining the least squares regression line is a mathematical one and can be applied to any set of data regardless of the underlying functional form; that is, the statistical procedure can be applied to data that resembles that shown in Figure 10.4, although there is clearly no relationship between X and Y. It is for this reason that the sample regression line must be evaluated to determine if it adequately describes the relationship between the variables X and Y. This may be accomplished by testing the null hypothesis that the true slope β_1 of the population regression line is equal to zero, i.e., H_0: $\beta_1 = 0$.

As stated previously, $\hat{\beta}_0$ and $\hat{\beta}_1$ are estimates of the true population parameters β_0 and β_1 and are subject to variation from sample to sample. It is important, therefore, to be able to make inferences concerning the true population regression line based on the information obtained from the sample regression line.

The most important inference to be made concerns the "true" value of the slope β_1 of the population line. If the true population β_1 is zero, then the value of Y in no way depends on the value of X. In other words, a value of $\beta_1 = 0$ indicates that no *linear* relationship exists between X and Y.

The principles involved in tests of hypotheses about β_1 are exactly the same as described in Chapter 6. The steps for hypothesis tests on β_1 are outlined below.

1. *Set up hypotheses.*

   H_0: β_1 = Some specified value (usually 0)

   H_a: $\beta_1 \neq$ Some specified value (usually 0)

2. *Calculate sample values.*

$$\hat{\beta}_1 = \frac{S_{xy}}{S_{xx}}$$

$$s_{y.x} = \sqrt{\frac{S_{yy} - \hat{\beta}_1 S_{xy}}{n-2}}$$

3. *Calculate test statistic.*

$$t = \frac{\hat{\beta}_1 - \beta_1}{s_{y.x}/\sqrt{S_{xx}}}$$

4. *Determine the rejection region.* The cut-off points for the rejection region are located in Appendix Table A at the intersection of the column $\alpha/2$ and the row $df = n - 2$. The rejection region is located under the outer tails of the curve.

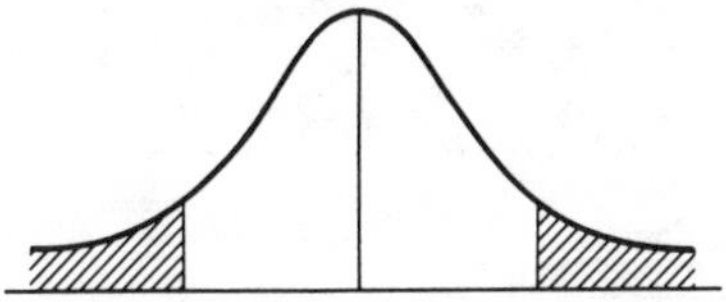

5. *Compare the calculated test statistic with the value found in Appendix Table A.* If the test statistic falls in the rejection region, the null hypothesis is rejected at the α level of significance.

The most common form of the null hypothesis is $\beta_1 = 0$ (i.e., no linear relationship between X and Y). If this null hypothesis is rejected, we conclude that there is a significant *linear* relationship between X and Y. If it is not rejected, we must say that there is not enough evidence to show that β_1 differs significantly from zero and therefore must conclude that X does not contribute information for the prediction of Y (X is not *linearly* related to Y).

Returning to Example 10.13, we wish to test whether a significant linear relationship exists between dose level of an antihypertensive drug (X) and systolic blood pressure (Y). This is equivalent to testing the hypothesis that $\beta_1 = 0$.

1. *Hypothesis.*

$$H_0: \quad \beta_1 = 0$$
$$H_a: \quad \beta_1 \neq 0$$

2. *Sample data.*

$$\hat{\beta}_1 = \frac{S_{xy}}{S_{xx}} = -44.2$$

$$s_{y.x} = 11.0$$

3. *Test statistic.*

$$t = \frac{\hat{\beta}_1 - \beta_1}{s_{y.x}/\sqrt{S_{xx}}} = \frac{-44.2 - 0}{11.0/\sqrt{10}}$$

$$= -12.6$$

4. *Rejection region.* For a 5% level of significance, the cut-off point for the rejection region is located in Appendix Table A at the intersection of the column $\alpha/2 = .025$ and the row $df = n - 2 = 3$. The value is 3.1825.

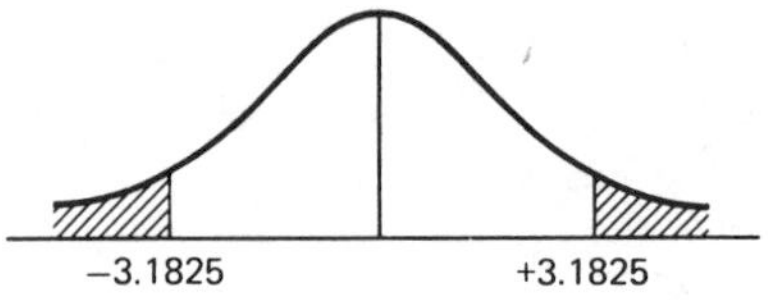

5. *Conclusion.* Since the calculated test statistic falls in the rejection region, we may conclude that there is a significant negative linear relationship between systolic blood pressure and dose level of the antihypertensive drug at an .05 level of significance.

Other Types of Regression

The preceding discussion has dealt primarily with the simplest type of regression: linear regression between a dependent variable Y and a single independent variable X. Often, in practice, we wish to determine the relationship between a dependent variable Y and 2 or more independent variables. For example, we may wish to predict uterine size (Y) based on age (X_1), parity (X_2), and gravidity (X_3), or we may wish to predict student performance in nursing school (Y) based on high school average (X_1), entrance exam score (X_2), and score on psychological evaluation (X_3). These are examples of *multiple linear regression*.

In addition, an investigator may be faced with finding the equation for predicting the variable (Y) from one or more X variables when it is known that the relationship is not described by a straight line. For example, the relationship may be exponential, logarithmic, or the familiar sigmoid-shaped curves which represent dose-response relationships in pharmacology. In such cases, it is often possible to transform or manipulate the data in such a way as to make it more amenable to linear regression analysis.

For the reader interested in a more detailed discussion of regression, the following works are listed:

Daniel, Wayne W. *Biostatistics: A Foundation for Analysis in the Health Sciences.* New York: John Wiley & Sons, 1974.

Duncan, Robert C., Rebecca G. Knapp, and M. Clinton Miller III. *Introductory Biostatistics for the Health Sciences.* New York: John Wiley & Sons, 1977.

Dunn, Olive Jean. *Basic Statistics: A Primer for the Biomedical Sciences.* New York: John Wiley & Sons, 1964.

Steel, Robert and James Torrie. *Principles and Procedures of Statistics.* New York: McGraw-Hill, 1960.

†EXERCISE 10.12

The admissions committee of a particular college of nursing wished to develop an equation for predicting a student's clinical performance based on a battery of personality-intelligence tests given to the students on application for admission to college. Scores on the personality-IQ test along with grade point average for clinical performance were obtained for a random sample of 10 nursing students. The data are shown below.

Student	*Personality-IQ Test Score*	*Clinical Grade Point Average*
1	46	2.7
2	41	2.2
3	22	1.3
4	19	1.2
5	32	1.6
6	60	3.4
7	48	2.7
8	30	1.7
9	61	3.5
10	24	1.5

$$\Sigma X = 383 \qquad \Sigma Y = 21.8$$
$$\Sigma X^2 = 16767 \qquad \Sigma Y^2 = 54.06$$
$$\Sigma XY = 951.1$$

a. Plot the scatter diagram.
b. Determine the equation of the line relating clinical grade point average to scores on the personality-IQ test.
c. Test the hypothesis that there is a significant linear rela-

tionship between clinical grade point average and scores on the personality-IQ test (use the 1% level of significance).

†EXERCISE 10.13

An investigator was interested in predicting weights of premature infants based on their age in weeks. The following data were collected on eight premature infants.

Infant	*Age (in Weeks)*	*Weight (in Pounds)*
1	4	5
2	4	4
3	5	5
4	1	2
5	1	2.5
6	2	2.5
7	3	3
8	6	6.5

$\Sigma X = 26$ $\quad \Sigma Y = 30.5$

$\Sigma X^2 = 108$ $\quad \Sigma Y^2 = 133.75$

$\Sigma XY = 118.5$

a. Plot the scatter diagram.
b. Determine the equation of the line relating weights of premature infants to age (in weeks).
c. Test the hypothesis that there is a significant linear relationship between weights of infants and age in weeks. (Use the 5% level of significance.)
d. What is the predicted weight of an infant who is three weeks old based on the above equation?

SUMMARY OF CRITICAL CONCEPTS IN CHAPTER 10

The following concepts were presented in Chapter 10:

1. The first step in a correlational study is to display the data graphically by means of a scatter diagram.
2. In a scatter diagram each pair of observations (X_i, Y_i) is plotted as a single point on a graph with the X values plotted

on the horizontal axis and the Y values plotted on the vertical axis.

3. When the points seem to fall approximately on a straight line and the X and Y values increase or decrease concurrently (as X increases Y increases), the relationship is said to be a *positive linear* relationship. If as the X measurements become larger, the Y measurements become smaller, then the relationship is said to be a *negative linear* relationship.
4. When the pattern of points in a scatter diagram appears to be representable by a curved rather than a straight line, then the relationship between variables is said to be *curvilinear*.
5. There is no apparent relationship between the variables X and Y when as X increases the value for Y remains essentially unchanged.
6. A correlation coefficient is a measure of the magnitude or strength of the relationship between X and Y.
7. Most correlation coefficients are index numbers that range from -1 (perfect negative correlation) to $+1$ (perfect positive correlation). A correlation of 0 indicates no linear relationship between variables.
8. In practice, perfect correlations are rarely encountered.
9. The correlation scale does not represent an interval decimal scale with equal increments from -1 to $+1$. For this reason, a correlation of 0.5 does not represent a relationship whose strength is "halfway" between no relationship and a perfect positive relationship.
10. The square of the correlation coefficient indicates how much knowledge of one variable will tell us about the second variable. It gives the percentage of variation in one variable that is explained by variation in the other.
11. Often very large correlations ($\geq .95$) may result when one quantity is correlated with another quantity of which the first is a component.
12. *Correlation does not necessarily mean causation*!
13. Correlation only applies to the range of values observed for the two variables.
14. *Pearson's Product Moment Correlation Coefficient* (r) measures the linear relationship between two continuous variables. A test of hypothesis that the true population correlation (ρ) equals zero must be carried out to determine if r is "significant." A value of r that is not significantly different from zero only indicates that there is no *linear* relationship between

variables; the possibility of curvilinear or other relationships still exists.

15. The *correlation ratio* is used to determine the degree of association between two continuous variables when there is a curvilinear relationship between the variables.
16. *Spearman's Rank Order Correlation Coefficient* determines the degree of association between two variables that have been rank ordered.
17. *Kendall's Tau* also measures degree of association between two variables whose values have been ranked.
18. The *partial correlation* measures the degree of association between two variables given that the effects of other variables have been removed.
19. The *coefficient of concordance* measures the overall association between multiple variables that have been rank ordered.
20. *Regression analysis* is a statistical technique for predicting the value of one variable, given the value of a second variable.

CHAPTER 11

THE CHI-SQUARED TESTS OF PROPORTIONS

OVERVIEW

Most of the statistical techniques presented previously are applied to measurement data that are continuous or nearly so. In this chapter, techniques for analyzing discrete (count) data will be presented. Count data result when the variable(s) being studied has a *nominal* level of measurement. In addition, data obtained on *ordinal* level variables or *equal-interval* level variables whose values have been categorized may be in the form of counts.

OBJECTIVES

Upon completion of this chapter, the student will be able to:

1. Describe what is meant by a contingency table.
2. Describe the general situation in which X^2 analysis is appropriate.
3. Describe what is meant by a "significant" X^2.
†4. Calculate the X^2 statistic and carry out an X^2 test of hypothesis.

DISCUSSION

Counts Versus Measurements

All of the procedures discussed previously (except for some of the correlation techniques) involved data that resulted from "measuring" the

† Sections marked by a dagger (†) cover more difficult or theoretical topics.

"amount" of a specific characteristic, such as height, weight, or blood pressure, on each member of a specified group. When a characteristic such as weight is measured, *any* value within a certain range is possible; we are limited only by the accuracy of the measuring device.

In contrast to measurement data, some data may be expressed in terms of the number of individuals (or objects) who fall into two or more discrete categories. Data of this type are called count or *frequency* data. By their nature, measurements on a nominal level variable are in the form of counts of individuals falling in the different categories of the variable. Measurements on an ordinal level variable or an equal-interval level variable may be either in form of "amount" of the variable being studied or in terms of counts of the number of individuals having specified values of the variable. (With equal-interval data, particularly, the use of counts rather than "amounts" results in a loss of information and hence is not advocated.) The examples given below illustrate.

EXAMPLE 11.1 In a study of a teaching method, 100 nursing students were asked whether they believe all courses should be taught on the semester rather than the block system. Each student responded either "yes" or "no." The data consisted of the number of "yes" and number of "no" responses.

EXAMPLE 11.2 A particular nursing procedure was used on a group of 10 patients. Another group of 10 patients did not receive the procedure. The number of "improved" patients for both groups was recorded. The data layout for the study would be:

	Improved	*Not Improved*	
Procedure	A	B	10
No procedure	C	D	10

where A, B, C, and D are the number of patients falling into each category. The investigator is interested in determining if the proportion of "improved" is the same for both groups.

EXAMPLE 11.3 In a study of anxiety level among nursing students, 60 freshmen, 50 sophomores, 50 juniors, and 40 seniors were classified according to anxiety level. The results are shown below:

	Anxiety Level			
	Low	*Moderate*	*High*	*Total*
Freshmen	15	15	30	60
Sophomores	15	20	15	50
Juniors	20	15	15	50
Seniors	10	10	20	40
				200

The investigator wishes to determine if the proportion of students falling into each anxiety level is constant from class to class.

The data from each of the above studies may be analyzed by means of chi-squared (X^2) analysis.

The Use of Chi-Squared (X^2)

The basic question answered by X^2 analysis is whether the counts or frequencies observed in a sample deviate significantly from some theorized population frequencies. In Example 11.3, we may hypothesize that there is no difference in anxiety levels across classes. If this is the case, then we would "expect" the same proportion of low, moderate, and high responses for all four classes. The question that must be answered is whether the differences between the frequencies *observed* from the sample and the frequencies *expected* in each category if there is no difference in proportion among the categories are due to chance variation (sampling error) or whether there is a true difference in population proportions. To answer this question, we calculate for each category the frequencies we would expect if the null hypothesis is true, that is, if there is *no difference* in proportion among the categories. We then compare these expected frequencies with the frequencies for each category actually observed in the sample. The X^2 statistic is a measure of the disparity between the observed and expected frequencies. For "large" disparities between observed and expected frequencies, the null hypothesis is rejected. The decision about whether the discrepancy between observed and expected frequencies is "large" or "small" is made by comparing the size of the computed test statistic (X^2) with a critical value obtained from a table of X^2 values (Appendix Table D). The X^2 analysis will now be described in terms of the six steps of hypothesis testing.

EXAMPLE 11.4 In a study on the effect of dental caries in children, 100 children were randomly selected: fifty received instruction and 50 did not. At the end of a 6-month period, the children were classified according to the following: no cavities, one to three cavities, more than three cavities. The data are given below.

	Number of New Cavities			
	None	*1–3 Cavities*	*>3 Cavities*	*Total*
Instruction	30	15	5	50
No instruction	20	15	15	50

SOLUTION

1,2. *Data and assumptions*. In this study we are comparing two groups, those who received instruction versus those who did not receive instruction. The response of interest, number of cavities for each study participant, is a discrete equal-interval level variable. We have recorded the *number* (count) of individuals whose values for this variable fall into the three categories: "none," "1–3 cavities," ">3 cavities."

The data for a chi-square analysis are displayed in a *contingency table*, which gives observed frequency of individuals falling into each category within the variables of interest. The categories within each variable should be independent of one another; that is, the response of any individual should not be dependent on the response of any other individual.

3. *Hypotheses*. In Example 11.4 we wish to determine whether the two populations (those who receive instruction and those who do not) are homogeneous with respect to number of cavities. The research question is, "Is the proportion of individuals having no cavities (or 1–3 cavities or >3 cavities) the same for the instruction and the no instruction groups?" The null hypothesis may be written symbolically as

H_0: $p_{11} = p_{21}$ (proportion with no cavities equal for the two groups)

$p_{12} = p_{22}$ (proportion with 1–3 cavities equal for the two groups)

$p_{13} = p_{23}$ (proportion with >3 cavities equal for the two groups)

The alternate hypothesis is obtained by a "not equal to" ($\neq$) in H_0 above.

In this example, we have assumed that the row totals are *fixed*; that is, the number of individuals in the instruction and the no instruction groups is under the control of the investigator, and the number of individuals falling in the categories of the other variable (number of cavities) is random. However, in some situations we may select the total sample and then classify the participants according to the two criteria of classification. For example, instead of assigning 50 individuals to the instruction group and 50 to the no instruction group and following them to determine number of new cavities, the investigator may have selected 100 individuals and asked them, "How many new cavities have you had within the last six months?" and "Have you received instruction on dental hygiene procedures during this time period?" In this case, neither row totals nor column totals are under the investigator's control, and hence both row and column totals are chance quantities. The null and alternate hypotheses, in this latter case, are often stated in terms of associations between the two criteria of classification. Stated another way, we wish to determine if there is an "association" between number of cavities and whether or not instruction was received. If there is no association between these two variables, then we would expect the proportion of cavities to be the same for both the instruction and no instruction groups. The null hypothesis is usually stated as "There is no association between number of cavities and type of instruction received" while the alternate hypothesis states "There is an association between number of cavities and type of instruction received."

4. *Test statistic*. The X^2 statistic is a measure of the

disparity between the "observed" frequencies and the frequencies we would "expect" if the null hypothesis is true. To calculate X^2, we must first find the expected frequencies. Using an easy rule of thumb, expected frequencies for each cell in the contingency table are found by multiplying the total of the row in which the given cell falls by the total of the column in which it falls and dividing the product by the total number of observations in the table below.

	Number of New Cavities			
	None	*1–3 Cavities*	*>3 Cavities*	*Total*
Instruction	30 (25)	15 (15)	5 (10)	50
No instruction	20 (25)	15 (15)	15 (10)	50
Total	50	30	20	100

For example, the expected frequency for Cell 1 (circled in the table) is found by multiplying 50 (its row total) by 50 (its column total) and dividing the product by 100.

$$E_{11} = \frac{50 \times 50}{100} = 25$$

where E_{11} represents the expected frequency for the cell in row 1, column 1. The expected frequencies are given in parentheses in the table.

$$E_{12} = \frac{50 \times 30}{100} = 15$$

$$E_{13} = \frac{50 \times 20}{100} = 10$$

$$E_{21} = \frac{50 \times 50}{100} = 25$$

$$E_{22} = \frac{50 \times 30}{100} = 15$$

$$E_{23} = \frac{50 \times 20}{100} = 10$$

Note that the sum of the expected and observed frequencies must be the same. This serves as a handy check on arithmetic. Once expected frequencies have been determined, the test statistic X^2 may be calculated by the formula:

$$X^2 = \Sigma \frac{(O - E)^2}{E}$$

This formula says "For each cell, square the difference between the observed and expected frequencies and divide by the expected frequency for that cell. Then add the results for each cell." For Example 11.4,

$$X^2 = \frac{(30-25)^2}{25} + \frac{(15-15)^2}{15} + \frac{(5-10)^2}{10} + \frac{(20-25)^2}{25} + \frac{(15-15)^2}{15} + \frac{(15-10)^2}{10}$$

$$= 1 + 0 + 2.5 + 1 + 0 + 2.5$$

$$= 7.0$$

5. *Decision rule.* As for all other tests of hypothesis, a cut-off point for the rejection region must be determined. This value is found in Appendix Table D. To obtain a value from Table D, we must first determine degrees of freedom. For X^2:

$$df = (r - 1)(c - 1)$$

where r = number of rows in the contingency table and c = number of columns in the contingency table. In our example,

$$df = (2 - 1)(3 - 1) = 2$$

For a level of significance of $\alpha = .05$ we find the column labeled $(1 - \alpha)$, or $X^2_{.95}$ and the row $df = 2$. The cut-off value for the rejection region is 5.991 shown below. The rejection region consists of all values of the test statistic falling on the horizontal axis under the shaded area of the curve.

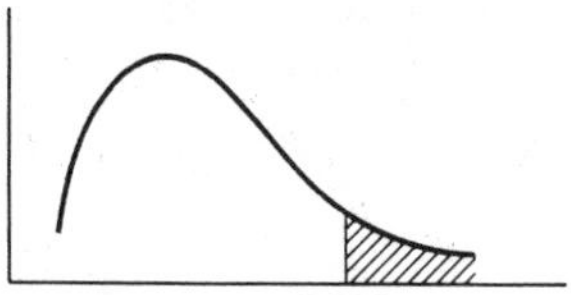

6. *Conclusion.* Since the calculated X^2 (Step 4) falls in the rejection region (horizontal axis under shaded area), the null hypothesis may be rejected at the $\alpha = .05$ level of significance. We conclude that there is a difference in proportion of cavities for the instruction and no instruction groups. Note: A significant X^2 tells the researcher that there are some statistically significant differences in the data but does not specify the direction of the differences.

EXERCISE 11.1

A public health nurse wished to determine if there is an association between parents' educational level and whether or not children are vaccinated against polio. One hundred children were selected at random and classified according to vaccination status and parents' educational level. The data are given below. Describe in terms of the six steps of hypothesis testing how the data would be analyzed using X^2. (Let $\alpha = .05$.)

	Educational Level Completed by Parents			
	Elementary	*Secondary*	*College*	*Total*
Children vaccinated	15	30	25	70
Children not vaccinated	15	10	5	30

EXERCISE 11.2

In a study of several leading remedies for relief of headache pain, it was reported that a significant X^2 ($\alpha = .01$) was found. Discuss this study in terms of what the data would look like, the null hypothesis tested, and what a significant X^2 means in terms of this hypothesis.

EXERCISE 11.3

Senior nursing students were asked to rate a particular course according to level of difficulty of the material presented in lectures. The following table shows this response cross-tabulated with grade received in the course.

	Perceived Difficulty of Course			
Class Rank (Final Grade)	*Not Difficult*	*Intermediate Difficulty*	*Very Difficult*	*Total*
Upper third	5	10	5	20
Middle third	10	5	5	20
Lower third	5	5	10	20
				60

Is there an association between perceived difficulty of the course and class rank? (Use $\alpha = .05$.)

a. What is the null hypothesis for this study?
b. Calculate the X^2 value.
c. What conclusions may be drawn?

EXERCISE 11.4

In order to examine the effects of tranquilizers and stimulants on driving skills, 150 people selected at random were given either a stimulant, a tranquilizer, or an identical looking placebo. After receiving the medication, the participants were given a battery of coordination and reaction time tests and the number of "mistakes" for each was recorded. The total number of mistakes for the entire battery of tests for the three groups is given below.

	0–5 Mistakes	*6–10 Mistakes*	*11–15 Mistakes*	*Total*
Stimulant	10	20	20	50
Depressant	5	15	30	50
Placebo	25	15	10	50

Based on these data, can it be concluded that the proportion of mistakes is different for the three groups? (Use $\alpha = .01$.)

a. What is the null hypothesis?
b. Calculate the value for the X^2 statistic. Based on this value, what conclusions may be drawn?

The 2 × 2 Contingency Table

Often the situation arises in which two variables of classification in a contingency table are broken down into only two levels of response. The data layout would be:

Variable 1

Variable 2	A	B
	C	D

where A, B, C, and D are the observed frequencies in the four cells. The above table is called a *2 × 2 contingency table* since it has only two rows and two columns. Applying the general formula for degrees of freedom, we find for the 2 × 2 table that:

$$df = (r - 1)(c - 1)$$
$$= (2 - 1)(2 - 1) = 1$$

When degrees of freedom is equal to 1, as in this case, a special formula is used for computing the X^2 statistic. The formula that contains a "correction factor" is:

$$X^2 = \Sigma \frac{(|O - E| - .5)^2}{E}$$

where the straight lines bracketing the difference "observed−expected" ($|O - E|$) mean "disregard the sign and consider the difference to be positive" *before* subtracting the 0.5.

EXAMPLE 11.5 In a study on the relationship of sleeping habits and performance in clinical work, 100 student nurses were cross-classified according to clinical performance and sleeping habits. The investigator wished to determine if there is an association between clinical performance and sleeping habits of student nurses at the .01 level of significance. The data are given below.

	Sleeping Habits	
Clinical Performance	*Poor*	*Good*
Poor	20	10
Satisfactory	20	50

SOLUTION

1, 2. *Data and assumptions.* The data consist of two variables of classification (sleeping habits and clinical performance), which are each broken down into two levels of response. The response of any one individual does not affect the response of any other individual; that is, the fact that Student Nurse No. 1 is classified as having poor sleeping habits and good clinical performance should not affect the classification of Student Nurse No. 2. The data consist of counts or frequencies.

3. *Hypotheses.* The investigator wishes to determine if there is an association between level of clinical performance and whether or not a student has good sleeping habits. Stated formally, the null hypothesis is that "there is no association between level of clinical performance and sleeping habits of student nurses." The alternate hypothesis is that there *is* an association between level of clinical performance and sleeping habits of student nurses.

4. *Test statistic.* To determine a value for X^2, the test statistic, we must first obtain the expected frequencies if H_0 is true (i.e., if there is no association between the categories). The expected frequencies, obtained as before, are shown in parentheses in the following table.

	Sleeping Habits		
Clinical Performance	*Poor*	*Good*	*Total*
Poor	20 (12)	10 (18)	30
Satisfactory	20 (28)	50 (42)	70
Total	40	60	100

The expected frequency for the cell in row 1, column 1 is:

$$E_{11} = \frac{(30)(40)}{100} = \frac{1200}{100} = 12$$

and row 1, column 2 is:

$$E_{12} = \frac{(30)(60)}{100} = 18$$

and row 2, column 2 is:

$$E_{22} = \frac{(70)(60)}{100} = 42$$

The test statistic X^2 for the 2×2 contingency table is calculated using the formula:

$$\begin{aligned} X^2 &= \Sigma \frac{(|O - E| - .5)^2}{E} \\ &= \frac{(|20 - 12| - .5)^2}{12} + \frac{(|10 - 18| - .5)^2}{18} \\ &\quad + \frac{(|20 - 28| - .5)^2}{28} + \frac{(|50 - 42| - .5)^2}{42} \\ &= \frac{(7.5)^2}{12} + \frac{(7.5)^2}{18} + \frac{(7.5)^2}{28} + \frac{(7.5)^2}{42} \\ &= 11.2 \end{aligned}$$

Note: $|10 - 18|$ equals +8, not −8.

5. *Decision rule*. The X^2 statistic computed in Step 4 above is a measure of the discrepancy between the expected frequencies if H_0 is true and the frequencies observed in the sample. "Large" values of X^2 lead to rejection of H_0 while for "small" values of X^2, H_0 cannot be rejected. To determine what constitutes "large," a critical value must be obtained from Appendix Table D. In this case,

$$df = (r - 1)(c - 1) = 1$$

Let $\alpha = .01$. The value in Table D, located at the intersection of the column $(1 - \alpha)$ or $X^2_{.99}$ and the row $df = 1$, is 6.635. The rejection region consists of values of the test statistic on the horizontal axis under the shaded area shown below.

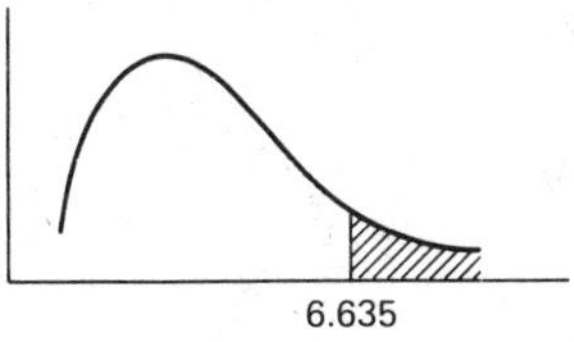

6. *Conclusion*. Since the calculated test statistic falls in the rejection region, the null hypothesis is rejected and it may be concluded that there is an association between the level of clinical performance and sleeping habits at the $\alpha = .01$ level of significance.

A Test of Hypothesis Concerning Specified Cell Probabilities

A common occurrence in the analysis of frequency data is a test of the hypothesis that proportions of observations falling into the different levels of classifications have specified values.

EXAMPLE 11.6 In a clinical trial to determine the preference for a new analgesic in headache relief, 100 patients were given the preparation. After a specified time interval each patient was asked whether he preferred the new preparation over the standard preparation he had been using. Of the 100 patients, 60 claimed to prefer the new analgesic. Based on this response, can it be said that the new is preferred over the old?

SOLUTION

1, 2. *Data*. The data may be displayed as follows:

Preferred	*Not Preferred*
60	40
	Total = 100

3. *Hypotheses*. The null hypothesis of interest is that the proportion of patients preferring the new treatment is the same as the proportion of patients preferring the old. That is, we are testing whether $P_{new} = P_{old} = 0.5$ where P refers to the proportions in the populations. If the null hypothesis is true and the proportion of patients in the population who prefer the new treatment is 0.5 (50%) then we would "expect" 50 out of the sample of 100 patients to prefer the new analgesic. The number of patients in the sample who actually preferred the new treatment is 60. As was the case in previous tests of hypothesis, this difference may have arisen because of sampling error or the true population proportion may not in fact be equal to 0.5. The following gives the observed and expected frequencies of patients preferring the new analgesic (the expected values are given in parentheses):

Preferred	*Not Preferred*
60 (50)	40 (50)
	Total = 100

4. *Test statistic*. The X^2 statistic is:

$$X^2 = \Sigma \frac{(|O - E| - .5)^2}{E}$$

$$= \frac{(|60 - 50| - .5)^2}{50} + \frac{(|40 - 50| - .5)^2}{50}$$

$$= 3.61$$

Note: The degrees of freedom for data from a single sample that has been classified according to the presence or absence of only one condition is $c - 1$. The formula containing the "correction factor" is used when $df = 1$, as in this example.

5. *Decision rule*. As in the previous sections, the calculated X^2 must be compared with a value from Appendix Table D to determine whether H_0 can be rejected. The value from Table D for $\alpha = .05$ and $df = 1$ is 3.841. The rejection region consists of all values of the test statistic greater than 3.841.

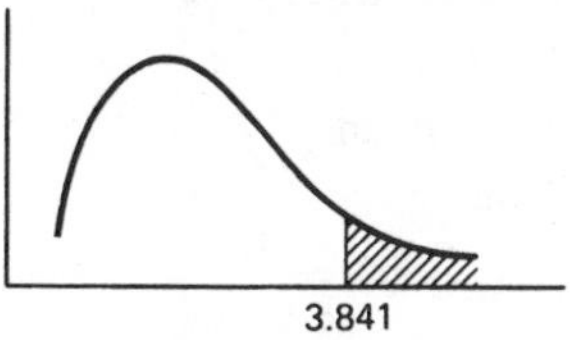

6. *Conclusion*. Since the calculated X^2 does not fall in the rejection region, the null hypothesis of equal preference cannot be rejected. There is not enough evidence to support a claim that the new preparation is preferred for relieving chronic headaches.

In the preceding example, the null hypothesis was that $P = .5$. This is not always the case. For example, a particular nursing technique may have associated with it a success rate of 80%. Suppose a nurse wishing to test whether a new technique improves success rate, treats a sample of 100 patients using the new procedure. She observes that 85 respond successfully. The data would be:

Successful	*Not Successful*
85	15
	Total = 100

The null hypothesis is that the true proportion of successes for the new procedure is equal to the true proportion for the old procedure; that is, the proportion of successes is equal to .8. What would be the expected frequencies in this case?

EXAMPLE 11.7 In a large medical-school-affiliated hospital, all women from ages 15–45 who were discharged with a diagnosis of idiopathic thromboembolism were identified and then classified as to type of contraceptive used. The data are given below. Is there a difference in contraceptive usage among women suffering from thromboembolic disease? (Use $\alpha = .05$.)

SOLUTION 1, 2. *Data.* Contraceptive usage among thromboembolic women was as follows:

	Observed
Oral contraceptive	30
IUD	25
Diaphragm	20
None	25
	100

3. *Hypotheses.* The null hypothesis is that there is no difference in proportion of thromboembolic disease among the four categories of contraceptives. Symbolically,

$$H_0: \quad P_1 = P_2 = P_3 = P_4 = .25$$

The alternate hypothesis is that the proportion among the categories is not equal.

4. *Test statistic.* As before, expected values for each category must be obtained. If the true population proportion of thromboembolic women using oral contraception is $P_1 = .25$, then we would expect $(.25)(100) = 25$ women in the sample to fall in this category. This is compared to 30 women actually observed in the category. The remaining expected values are shown below.

	Observed	*Expected*
Oral contraceptive	30	(25)
IUD	25	(25)
Diaphragm	20	(25)
None	25	(25)
Total	100	100

The calculated X^2 is:

$$X^2 = \Sigma \frac{(O - E)^2}{E}$$

$$= \frac{(30 - 25)^2}{25} + \frac{(25 - 25)^2}{25}$$

$$+ \frac{(20 - 25)^2}{25} + \frac{(25 - 25)^2}{25}$$

$$= 1 + 0 + 1 + 0 = 2.0$$

5. *Decision rule*. For $\alpha = .05$ and $df = 3$, the value from Appendix Table D is 7.815. The rejection region consists of all values of the test statistic under the shaded area below.

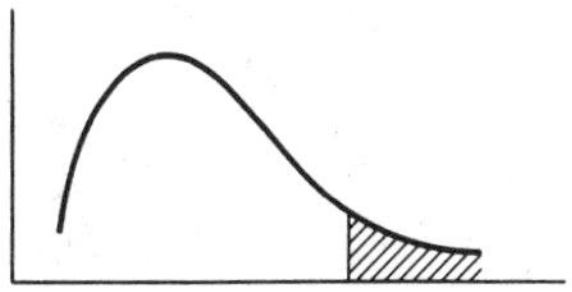

6. *Conclusion*. The test statistic does not fall in the rejection region. The null hypothesis of equal population proportions, therefore, cannot be rejected. Contraceptive usage cannot be shown to be different among thromboembolic women.

EXERCISE 11.5

Thirty patients receiving a tricyclic antidepressant for management of their depressive states were given a psychomotor skill test. Their overall response was rated either satisfactory or not satisfactory. Another group of 30 patients receiving the same medication was given alçohol in addition to the antidepressant and their psychomotor skills were evaluated. Twelve of the patients receiving only the antidepressant performed satisfactorily while eight of the patients receiving both the antidepressant and alcohol performed satisfactorily. Is there a difference in psychomotor performance for the two groups? (Use $\alpha = 0.05$.)

a. Lay out the data in a contingency table.
b. State the hypotheses for this study.

c. Calculate the X^2 statistic. State the conclusion for this study.

EXERCISE 11.6

Fifty inpatient insomniacs were treated with equipotent doses of three currently used hypnotics and a placebo. Of the 50 patients in each of the treatment groups, those who felt that they still did not get a "good night's sleep" were classified according to their perceived reasons for not sleeping well. The data are shown below.

Reason Given for Not Sleeping Well	*Hypnotic A*	*Hypnotic B*	*Hypnotic C*	*Placebo*
Restlessness	7	12	9	16
Trouble getting to sleep	8	4	6	14
Bad dreams	5	9	5	5
Total	20	25	20	35

Is there an association between reasons given for not sleeping well and type of medication received by the patient? (Use $\alpha = .01$.)

a. State the hypotheses.
b. Calculate the X^2 statistic. State the conclusions for this study.

EXERCISE 11.7

One hundred institutionalized schizophrenic patients were classified according to the season of the year in which they were born, as follows:

	Observed
Fall	20
Winter	35
Spring	20
Summer	25

Based on the above data, can it be said that there is a difference in season of birth among institutionalized schizophrenic patients? (Use $\alpha = .05$.)

Additional Points Regarding X^2

Two important points must be kept in mind when working with X^2.

First, the levels of classification in a contingency table must have some logical basis for categorization. For example, when using a low, medium, and high scale for classifying a variable, as in Example 11.3, the values that delineate each classification should not be arbitrary but should preferably be supported by previous work.

Second, when expected values are small, the validity of the X^2 test is questionable. While much controversy exists about what constitutes "small," a general rule of thumb is that no expected frequency should be less than five. The problem of small expected frequencies may sometimes be resolved by "collapsing" adjacent rows or columns. For example, student performance that has been classified according to five levels of performance ranging from "very good" to "very poor" may be reduced to two levels, "satisfactory" and "not satisfactory." Note, as stated in the paragraph above, levels of classification should only be collapsed when there is a logical basis for doing so.

SUMMARY OF CRITICAL CONCEPTS IN CHAPTER 11

The following concepts were presented in Chapter 11:

1. Data that may be expressed in terms of the number of individuals (or objects) who fall into two or more discrete categories are called *count* or *frequency* data.
2. The basic question answered by X^2 analysis is whether the counts or frequencies observed in a sample deviate significantly from some theorized or hypothesized population frequencies.
3. The null hypothesis in most X^2 analyses is that there is no association between two variables that have been classified into discrete categories.
4. "Large" values of the X^2 statistic imply large differences in the observed frequencies and the frequencies expected if H_0 is true; a "large" X^2 thus leads to rejection of H_0.
5. A "significant" X^2 (one for which H_0 is rejected) leads to the conclusion that there is an association between the two variables.
6. To find the expected frequency for Cell 1 in a contingency table, its row total is multiplied by its column total and the product is divided by the total table frequency.

7. The X^2 statistic is computed by the formula:

$$X^2 = \Sigma \frac{(O - E)^2}{E} \text{ or } X^2 = \Sigma \frac{(|O - E| - .5)^2}{E} \text{ when } df = 1$$

8. A significant X^2 tells the researcher that there are some statistically significant differences in the data but does not specify which categories are different.
9. When expected frequencies are small, the validity of the X^2 test is questionable.

CHAPTER 12

NONPARAMETRIC METHODS

OVERVIEW

In the material presented previously, with some exceptions, we have been dealing with classical *parametric* statistical techniques. These include *t*-tests, ANOVA, correlation,* and regression. All these procedures require knowledge of the nature of the distribution of the population from which the sample values were drawn. In most of these tests we assume that the sample measurements were drawn from a normally distributed population of measurements. The validity of this assumption of the normality of the population directly affects the meaningfulness of the results of the statistical tests. When the assumptions are not justifiable, the validity of the test results are questionable. In addition to the normality of the population, the observations or scores being analyzed by most parametric techniques must be measured on at least a continuous *interval* scale. An interval scale requires that the distances between any two numbers on the scale be of known size. This is in contrast to the nominal and ordinal scales. The *nominal* scale is used when scores or observations are simply classified according to discrete categories such as race, sex, or blood type. An *ordinal* scale of measurement exists when there is a predetermined ordering among the response classifications. For example, patients may be classified according to low, moderate, and high stress levels. This scale differs from the interval scale in that while we know that patients in the moderate stress category are more stressed than those in the low category, we do not know *how much* more stressed they are. For the ordinal scale, the relationships that exist between categories are "greater than" or "less than." For the equal interval scales (interval and ratio), the magnitude of the differences between any two numbers on the scale is known.

* Some correlation procedures, such as Spearman's Rank Order Correlation, are nonparametric.

In contrast to parametric techniques, nonparametric tests require few or no assumptions about the nature of the population from which the sample values were drawn. In addition, most parametric techniques require at least an interval scale of measurement, while most nonparametric tests require only a nominal or ordinal scale of measurement. Nonparametric techniques are viable alternatives when the researcher finds that the necessary assumptions for parametric tests cannot be made or when the data are inherently in the form of ranks or enumeration data. Because of their lack of dependence on assumptions about the nature of the population, nonparametric techniques provide greater generality in the conclusions to be drawn.

In this chapter, some of the most common nonparametric techniques will be briefly described. There will be no attempt to provide an in-depth discussion of the calculational procedures except for the most widely used of these techniques. The tests will be presented in three sections: the one-sample tests, the related-sample tests, and the independent-sample tests.

OBJECTIVES

Upon completion of this chapter, the student will be able to:

1. Discuss the general differences in parametric and nonparametric techniques.
2. State briefly the situations in which each test may be applicable.
3. Carry out a test of hypothesis on related samples using the Sign Test and the Wilcoxon Signed Rank Test.
4. Carry out a test of hypothesis on independent samples using the Median Test and the Wilcoxon Rank Sum Test.

DISCUSSION

One-sample Tests

The tests to be described in this section are analogous to the simple *t*-test in that they involve drawing a single sample of data and comparing a statistic calculated from the sample with some specified population parameter. In many of the one-sample tests we are concerned with testing the hypothesis that a given sample was drawn from a population with a specified distribution.

THE BINOMIAL TEST

The Binomial Test may be used when the study population consists of only two classes, i.e., male-female, Democrat-Republican, sick-well, married-single, ambulatory-institutionalized. All measurements drawn from the population fall into one or the other of two discrete categories.

The proportion of observations in the population falling into one of the classes is P while the proportion falling into the other category is $1 - P$. Due to sampling error, a sample drawn from this population will not necessarily have proportions exactly equal to P and $1 - P$. The null hypothesis is that the population proportion P is equal to a specified value. For example, a researcher may wish to test whether the proportion of males with a given disease is equal to the proportion of females with the disease. Here:

$$H_0: \quad P = \text{proportion of males} = .50$$

In a sample of 100 patients with the disease, it may be found that there are 44 males and 56 females. If the null hypothesis is true, we would expect 50 males and 50 females with the disease. The question to be answered then is whether the observed difference in proportions is due to chance (sampling error) or whether H_0 is in fact not true.

The Binomial Test determines whether the proportions or frequencies obtained in a sample could have been drawn from a population with a specified value of P.*

THE X^2 ONE-SAMPLE TEST

The data consist of measurements that may be classified into two or more discrete categories. Each observation is independent of every other observation.

The X^2 one-sample test determines whether the actual observed frequencies or proportions in the various categories equal some specified population proportion for the categories. For example, a nurse may classify a group of 12 pediatric patients according to four modes of play:

	Mode of Play		
A	*B*	*C*	*D*
2	4	3	3

The data consist of the number of children falling into each of the four

* The computational procedure is described in Sidney Siegal, *Nonparametric Statistics for the Behavioral Sciences* (New York: McGraw-Hill, 1956).

categories. The object of the study is to determine if the children have a preference for mode of play. In null form, the hypothesis is that there is no difference in proportion of children in each of the four categories; that is,

$$H_0: \quad P_1 = P_2 = P_3 = P_4 = .25$$

The X^2 test tells us whether the observed frequencies in each category differ from those expected if H_0 is true. The computational method is as described in Chapter 11.

Tests Used with Related Samples

Just as the one-sample nonparametric tests are used for evaluating a single sample of data, there are nonparametric procedures for evaluating two samples when the samples are related. For these procedures the sample subjects are paired on the basis of extraneous factors whose effects are to be eliminated from the results of the study. The pairings are as described for the paired *t*-test discussed in Chapter 8. Individuals or objects are matched on the basis of factors such as sex, educational background, heredity, environment, age, disease state, etc. Also, an individual may serve as his own control in the familiar "before and after" situation. In this case, the same individual may be exposed to different treatments at different points in time.

THE SIGN TEST

The Sign Test is analogous to the paired *t*-test in that it tests whether a real (statistically significant) difference exists between two populations of measurements that have been paired on the basis of some extraneous factor(s).

The data consist of "+" and "−" rather than numerical values. The difference between each individual pair in the two samples ($Y_A - Y_B$) is expressed in terms of "+" ($Y_A > Y_B$), "0" ($Y_A = Y_B$), or "−" ($Y_A < Y_B$). The Sign Test focuses on the median difference between groups rather than the mean difference, as used by the paired *t*-test. If there is no difference between the groups *A* and *B*, then we would expect the number of "+" signs to be equal to the number of "−" signs. Stated another way, the Sign Test tests the null hypothesis that the median difference for the matched pairs is zero.

EXAMPLE 12.1 Nine student nurses were asked to instruct pediatric patients in proper dental hygiene techniques. The performance of each student was rated by an observer. Subsequently, the students participated in a lecture-slide demonstration on teaching proper dental hygiene techniques. They were again rated on teaching techniques. Students were evaluated in terms of whether their performance was better or worse following the demonstration (see below). A plus "+" indicates better performance after the demonstration. If there was no change in performance, the student was dropped from the analysis. The investigator wishes to determine if the demonstration has an effect on student performance.

	Rating		
Student	*Before Demonstration*	*After Demonstration*	*Sign*
1	6	7	+
2	5	7	+
3	3	4	+
4	4	5	+
5	6	7	+
6	3	5	+
7	1	3	+
8	5	4	−
9	6	7	+

SOLUTION In null form, the hypothesis to be tested is that the median difference for the matched pairs is zero (i.e., the number of "+" signs is equal to the number of "−" signs). To test this hypothesis, a critical value must be obtained from Appendix Table J. The value from this table specifies the maximum number of differences in the minority direction (whichever sign is smallest) that can occur if H_0 is to be rejected. In our example, the smallest sign, "−", occurred once. From Table J for $N = 9$ and $\alpha = .05$, we see that H_0 can be rejected only when one *or fewer* "−" signs occur. Therefore, the null hypothesis can be rejected and we may conclude that there is a difference in before and after performance scores (at the

.05 level of significance). If the scores on a pair of observations are equal (i.e., no sign can be assigned to the pair), then the pair is dropped from the analysis.

THE WILCOXON SIGNED RANK TEST

The Wilcoxon Signed Rank test is another analog of the paired *t*-test. This test differs from the Sign Test discussed previously in that it takes into account not only the *direction* of the differences but also the *magnitude* of the differences and thus is a more powerful test than the Sign Test.

The data for the Signed Rank Test consist of differences between individual pairs of observations. The differences between pairs are assigned ranks from smallest to largest without regard to the sign of the differences. A difference of zero is ignored in the ranking and the pair is dropped from the analysis. If two or more pairs have the same absolute differences, then the ties are assigned their average rank. After the ranking has been carried out, each rank is assigned the sign of the original difference between pairs.

The null hypothesis for the Wilcoxon Signed Rank Test is that the median difference among the entire population of pairs is zero; that is, if H_0 is true and there is no difference between groups A and B, we would expect the sum of the positive ranks to be approximately equal to the sum of the negative ranks. To determine whether there is a significant difference between the groups, the sum of either the positive or negative ranks (whichever is smallest) is compared to a table of critical values (Appendix Table H) for the Signed Rank Test. For example, in Table H for the Signed Rank Test, it may be seen for a sample with $n = 10$ pairs, a sum of signed ranks of 8 or less would indicate a statistically significant difference at the $\alpha = .05$ level of significance.

EXAMPLE 12.2 A researcher hypothesized that patients would feel less anxious after an orientation talk than before the talk. Ten newly hospitalized patients were evaluated upon admittance and assigned an anxiety score from 1 to 20 with 20-most anxious and 1-least anxious. The same 10 patients were evaluated again after they had been given a 20-minute orientation talk by a nurse.

SOLUTION The measurements are made on the ordinal scale and may be ranked in order of absolute size. The data are given on page 271.

	Anxiety Scores				
Patient	*Before Talk*	*After Talk*	*Differences*	*Rank Disregarding Sign*	*Signed Rank*
1	8	6	−2	2.5	−2.5
2	15	10	−5	6.5	−6.5
3	10	10	0	—	—
4	5	8	+3	4	+4
5	18	19	+1	1	+1
6	10	6	−4	5	−5
7	12	7	−5	6.5	−6.5
8	6	8	+2	2.5	+2.5
9	7	7	0	—	—
10	9	3	−6	8	−8

Sum of positive signed ranks = 7.5
Sum of negative signed ranks = 28.5
Sample size $n = 8$ (2 zeros discarded)

Referring to Appendix Table H for the critical values for the Signed Rank Test, we see that for $n = 8$ and $\alpha = .05$, the value is 4. Since the smaller of the sum of signed ranks (7.5) is *not less than* the table value, we *cannot* say that there is a difference in the two groups. To be significant, the smallest sum must be equal to or less than the critical value from Table H.

EXERCISE 12.1

A study was carried out to determine the effectiveness of a particular procedure for treating decubitus ulcers. Patients were assigned a score on the basis of number of ulcers and their degree of severity. The patients were evaluated before the treatment and again at the end of a four-week treatment period. Is there a difference in the before and after treatment scores? (Use $\alpha = .05$.) Test the above hypothesis using both the Sign Test and the Signed Rank Test. The data are given on page 272.

	Ulcer Severity Score	
Patient	*Before*	*After*
1	15	10
2	20	14
3	10	13
4	7	5
5	12	9
6	8	6
7	6	10

THE McNEMAR TEST FOR THE SIGNIFICANCE OF CHANGES

The McNemar Test is useful for detecting whether a change has occurred between a set of "before and after" type measurements.

The data are measured on the *nominal* scale and may be displayed in a 2 × 2 table as shown below. The 0 and 1 represent the two categories of response.

		After 0	1
Before	1	A	B
	0	C	D

The symbols A and D represent the frequencies of individuals who showed change in response for the two measuring periods, while B and C are the frequencies of individuals showing no change. The test statistic is X^2 with 1 *df* (degree of freedom; see Chapter 9).

$$X^2 = \frac{(|A - D| - 1)^2}{A + D}$$

The calculated X^2 value is compared to a tabulated critical value in a X^2 table to determine whether a significant change in response has occurred from the first measuring period to the second. If the calculated X^2 is larger than the value from the table, then it may be concluded that a significant change has occurred from one measuring period to the next.

Tests Used with Two Independent Samples

The tests discussed in the previous section are nonparametric analogs of the paired t-test. Similarly, the tests discussed in this section are analogous to the pooled t-test for testing the significance of the difference between two independent samples. The two samples are assumed to be randomly drawn from the two populations of interest or it is assumed that within a given study group the two treatments are randomly allocated to members of the group.

THE MEDIAN TEST

The Median Test determines the likelihood that two independent samples have been drawn from populations having the same median. The data consist of measurements on at least the ordinal scale made on a variable whose underlying distribution is continuous. The two samples do not necessarily have to have an equal number of measurements.

The procedure for performing the Median Test may be summarized as follows:

1. For the two groups combined, arrange all the observations in ascending order and determine the common median.
2. Determine the number of observations that fall above and below the common median for each of the two groups.
3. Arrange the resulting frequencies in a 2 × 2 contingency table as shown below.

	Group A	*Group B*	*Total*
No. of Scores Above Median	A	B	$A + B$
No. of Scores Below Median	C	D	$C + D$
Total	$A + C$	$B + D$	$N = n_1 + n_2$

$$n_1 = \text{number of observations in Group } A$$

$$n_2 = \text{number of observations in Group } B$$

If the null hypothesis that the two samples are from populations whose medians are the same is true, then we would expect about half the scores in each group to be above the common median and half to be below. In other words, if H_0 is true, the frequencies A and C should be about equal and likewise the frequencies B and D.

The test statistic for testing the equality of population medians is the X^2 statistic with degrees of freedom equal to 1.

$$X^2 = \frac{N\left(|AD - BC| - \frac{N}{2}\right)^2}{(A + B)(C + D)(A + C)(B + D)} \qquad df = 1$$

(The above formula is equivalent to the X^2 formula with the "correction factor" given in Chapter 11.) The calculated X^2 is compared with a critical value found in Appendix Table D to determine whether the null hypothesis can be rejected.

EXAMPLE 12.3 In a nursing audit carried out on two different services within a given hospital, patients were selected at random from each of the two services. Each patient was observed and assigned a "quality of care" score on a scale from 0–50 that reflected several areas of nursing practice. The results are shown below.

Service A	*Service B*
26	43
38	30
25	50
35	50
46	48
33	39
26	47
34	49
26	
27	

The investigator wishes to determine if there is a difference in median audit scores for the two services. In null form, the hypothesis is that the median scores for the two services are equal.

SOLUTION To test this hypothesis, we first arrange all the scores (both groups combined) in ascending order and determine the common median. For our example, the median score is 36.5. Next, we determine the number of scores in each group that fall above and below 36.5 and arrange these frequencies in a 2 × 2 contingency table.

	Service A	*Service B*	*Total*
No. of Scores Above Median	2	7	9
No. of Scores Below Median	8	1	9
Total	*10*	*8*	*18* = N

The value of X^2 is given by:

$$X^2 = \frac{N\left(|AD - BC| - \frac{N}{2}\right)^2}{(A + B)(C + D)(A + C)(B + D)}$$

$$= \frac{18\left(|(2)(1) - (7)(8)| - \frac{18}{2}\right)^2}{(9)(9)(10)(8)}$$

$$= 5.625$$

The value from Appendix Table D for $\alpha = .05$ and $df = 1$ is 3.841. Since the calculated value of X^2 is greater than the value obtained from Table D, the null hypothesis is rejected at the .05 level of significance. It may be concluded that the two samples have been drawn from populations whose medians are not equal (at the .05 level of significance).

The Median Test may be extended to test the hypothesis of equality of medians for k groups. The logical extension is to construct a $2 \times k$ contingency table consisting of the frequencies above and below the combined median for each of the k groups. The X^2 statistic would be calculated as described in Chapter 11.

WILCOXON RANK SUM TEST

The Wilcoxon Rank Sum Test is used to test whether two independent samples have been drawn from the same population. The null hypothesis is that the two populations from which the samples were drawn have identical underlying distributions. A summary of the procedure for the Rank Sum Test is given.

1. Combine the observations from both groups.
2. Arrange the combined observations in ascending order.
3. Rank each observation, assigning the average rank to tied observations.
4. Sum the ranks that correspond to the group having the smallest sample size. (For groups with equal numbers, either group of ranks may be summed.)
5. The rank sum calculated in (4) is compared with a critical value from Appendix Table I for the Rank Sum Test to determine

whether the difference between the groups is significant (i.e., H_0 rejected).

EXAMPLE 12.4 A researcher wished to compare student performance using two different teaching methods. Twenty student nurses were selected at random to participate in the special educational research project. A group of eight students was given a traditional lecture and demonstration on how to carry out a particular nursing technique. The remaining twelve students were asked to master the procedure using only slides and a self-instructional manual. At the end of a two-week period, each nurse was evaluated with regard to number of trials required to perform the task with 100% mastery. The data are shown below. An asterisk denotes a lecture score.

Number of Trials

*Lecture**	*Self-instruction*
1	4
4	5
2	2
8	6
3	10
6	9
12	11
15	1
	5
	4
	13
	3

Order and rank for the 20 combined scores:

Order	1*	1	2*	2	3*	3	4*	4	4	5
Rank	1.5	1.5	3.5	3.5	5.5	5.5	8	8	8	10.5
Order	5	6*	6	8*	9	10	11	12*	13	15*
Rank	10.5	12.5	12.5	14	15	16	17	18	19	20

SOLUTION The tied ranks are assigned the average rank for the tied group. For example, the first two observations in the ordered list are 1's. The rank assigned to these two values is 1.5, the average of ranks 1 and 2. Similarly, the next two values are tied for rank 3 and 4 and receive an average rank of 3.5. The Rank Sum Test is performed by summing the ranks for the group having the smaller sample size. In the example, the lecture group has $n_1 = 8$ scores. The sum of the ranks corresponding to the scores in the lecture group is 83. From Appendix Table I, for $n_1 = 8$, $n_2 = 12$, and $\alpha = .05$, the critical values are 58,110. Since the calculated rank sum value of 83 falls between the two critical values, it cannot be concluded that there is a difference between the scores of the two groups of students. To be statistically significant the rank sum for the smaller group must be less than 58 or greater than 110 for $\alpha = .05$.

EXERCISE 12.2

In a study to determine the effect of family support on length of hospitalization of critically ill patients, a researcher recorded the length of hospitalization in days for 10 patients whose families visited frequently and for 10 patients whose families did not visit. The data are shown below.

Length of Hospitalization (Days)

Family Visits	*No Family Visits*
54	54
10	84
21	60
14	8
17	79
74	64
10	27
52	10
14	9
20	19

Is there a statistically significant difference between length of hospitalization for the two groups? (Use $\alpha = .01$.) (Test the hypothesis using both the Median Test and the Rank Sum Test.)

MANN-WHITNEY U-TEST

The Mann-Whitney U-test is an alternative version of the Wilcoxon Rank Sum Test described above.*

Additional Nonparametric Tests: More-Than-Two Group Designs

The tests discussed in this unit have been the nonparametric analogs of the simple t, the paired t, and the pooled t. These nonparametric procedures are used when the assumptions necessary for the t-test are not valid for the data being analyzed, when measurements are made on the nominal or ordinal scale and do not meet the requirements of the t-test, or when it is preferred to avoid the assumptions of the t-test, thus giving greater generality to the conclusions.

The analogs of the Analysis of Variance are used when it is of interest to compare $k \geq 3$ groups. Among the procedures for k samples are the Cochran Q-test, the Friedman Two-way Analysis of Variance, and the Kruskal-Wallis One-way Analysis of Variance.**

KRUSKAL-WALLIS ONE-WAY ANALYSIS OF VARIANCE BY RANKS

The nonparametric analog of the one-way Analysis of Variance–Completely Randomized Design described in Chapter 9 is the Kruskal-Wallis One-way Analysis of Variance by ranks. Recall that for the one-way Analysis of Variance–Completely Randomized Design, we assumed that the k samples were randomly selected from normally distributed populations having equal standard deviations. When these assumptions are not met or when the data consist only of ranks (ordinal level measurements), then the nonparametric Kruskal-Wallis One-way Analysis of Variance is appropriate. As for the parametric case, we require that the k samples be independently selected; that is, the design does not involve matching of observations.

EXAMPLE 12.5 A nurse researcher wished to compare three methods for preparing mothers to greet their hospitalized children. Mothers participating in the study were randomly assigned to one of three teaching methods. For Method 1, a nurse spent approximately 30 minutes with the mother explaining the child's illness and advising her on proper greeting behavior. For Method 2, mothers were given literature discussing the importance of and procedures for

* For a description of the computational procedure, see Siegal, *Nonparametric Statistics*.

** See Siegal, *Nonparametric Statistics*.

proper greeting behavior. The third group (Method 3) acted as a control. Mothers were then evaluated on a five-point scale with respect to greeting behavior when they saw their hospitalized children for the first time. The researcher wished to test the hypothesis, "There is a difference in greeting behavior of mothers for the three teaching methods." What statistical analysis is appropriate?

SOLUTION Since we have three independent (nonmatched) samples and the data is measured on an ordinal scale, the appropriate test is the Kruskal-Wallis One-way Analysis of Variance.

FRIEDMAN TWO-WAY ANALYSIS OF VARIANCE BY RANKS

The nonparametric analog of the two-way Analysis of Variance–Randomized Complete Block Design is the Friedman Two-way Analysis of Variance by ranks. This test is employed when the assumptions of the parametric test (described in Chapter 9) are not met or when the data are measured on an ordinal scale.

EXAMPLE 12.6 Eight nurses were asked to rank three methods for giving an injection of heparin with respect to patient discomfort. The investigator hypothesized that there is a difference in rankings for the three methods. The data were as follows:

	Method		
Nurse	*A*	*B*	*C*
1	1	2	3
2	2	1	3
3	1	3	2
4	2	1	3
5	3	1	2
6	1	2	3
7	2	1	3
8	1	2	3

What statistical test is appropriate?

SOLUTION The nurses act as "blocks" and the level of measurement is ordinal; thus, the Friedman Two-way Analysis of Variance by ranks is appropriate.

Nonparametric Correlation Procedures

In addition to the nonparametric analogs of the one- and two-sample t-tests and the Analysis of Variance, there are also nonparametric correlation procedures. With the exception of Pearson's Product Moment Correlation Coefficient, all of the correlation procedures presented in Chapter 10 are nonparametric in nature. Among these are Spearman's Rank Order Correlation Coefficient, the Contingency Coefficient, the Coefficient of Concordance, and Kendall's Tau.

EXERCISE 12.3

Two groups of mothers were selected to participate in a study of greeting behavior between mothers and their hospitalized children. In one group, each mother met with a staff psychologist and the attending nurse. The nature of the child's illness was explained and the mother was counseled concerning her reactions to the illness. The second group of mothers received no such counseling and served as a control group. A nurse then observed the initial encounter between each mother and child and assigned a greeting behavior score. The scores for the two groups are given below.

Greeting Behavior Scores

Mothers Who Received Counseling	*Mothers Who Received No Counseling*
5	2
3	4
8	3
7	6
10	8

Is there a difference in the scores for the two groups? (Use $\alpha = .05$.)

EXERCISE 12.4

A group of eight hospitalized pediatric patients was rated according to dental hygiene practices. During their stay in the hospital, the children were instructed in proper dental hygiene techniques. Two weeks after dismissal from the hospital, the same eight children were visited and rated according to dental hygiene techniques. The scores are given on page 281.

Child	*Before Instruction*	*After Instruction*
1	4	2
2	2	3
3	3	5
4	6	4
5	5	6
6	1	1
7	1	4
8	4	7

Is there a difference in scores before and after instruction? (Use $\alpha =$ .05.)

EXERCISE 12.5

A random sample of 12 student nurses was given a test designed to measure feelings toward working with alcoholic patients. The students then attended a three-day workshop on alcoholism and were again given the test. The scores are given below.

	Scores	
Student	*Before Workshop*	*After Workshop*
1	8	10
2	5	10
3	14	15
4	22	20
5	17	20
6	8	16
7	16	24
8	10	23
9	20	25
10	14	24
11	10	23
12	20	18

Using the Sign Test, test the hypothesis that the workshop did not affect student attitudes toward working with alcoholic patients. (Use $\alpha = .05$.)

EXERCISE 12.6

In a study to determine the effect of counseling on anxiety of presurgery patients, 20 patients were randomly selected and randomly assigned either to the counseling group or to the control group. The counseling group was given a 15-minute counseling session with a nurse prior to being prepared for surgery. Afterwards, both the treatment group and the control group were observed and assigned an anxiety score. The scores for both groups are given below.

Anxiety Scores	
Counseled Group	*Control Group*
25	45
30	40
30	32
15	45
40	50
45	48
35	25
25	65
29	60
30	50

Using the Median Test, can it be said that there is a difference in anxiety scores for the two groups? (Use $\alpha = .05$.)

SUMMARY OF CRITICAL CONCEPTS IN CHAPTER 12

The following concepts were presented in Chapter 12:

1. Classical parametric statistical techniques require some knowledge of the population from which the sample values are drawn.
2. When the assumptions of the parametric tests are not justifiable, the validity of the tests is questionable.
3. Most parametric techniques require at least an *interval* scale of measurement.
4. Nonparametric techniques require few or no assumptions about the nature of the population from which the sample values are drawn.

5. Most nonparametric tests require only the nominal or ordinal scale of measurement.
6. Nonparametric procedures are used when the assumptions necessary for parametric tests are not valid, when measurements are made on the nominal or ordinal scale and do not meet the requirements for parametric tests, or when it is preferred to avoid the assumptions of parametric tests, thus giving greater generality to conclusions.
7. The *Sign Test* and the *Wilcoxon Signed Rank Test* are analogs of the paired t-test and are used when the two samples consist of matched pairs of observations.
8. The *Median Test* and the *Wilcoxon Rank Sum Test* are analogs of the pooled t-test and are used when the two samples are independent random samples from the population of interest.
9. Nonparametric analogs of the Analysis of Variance (compare $k \geq 3$ groups) are Cochran Q-test, Kruskal-Wallis One-way Analysis of Variance, and Friedman Two-way Analysis of Variance.
10. Nonparametric correlation techniques include Spearman's Rank Order Correlation Coefficient, Contingency Coefficient, Coefficient of Concordance, and Kendall's Tau.

CHAPTER **13**

METHODS OF SAMPLE SELECTION

OVERVIEW

As stated in Chapter 4, the relationship between population and samples is a cornerstone of statistics. The numerical quantities computed from a sample both *describe* the sample itself (descriptive statistics) and provide for inferences about the characteristics of the population from which the sample was collected (inferential statistics). The degree to which sample statistics provide valid and reliable information on population parameters is a reflection of the degree to which a sample is "representative" of the population. Only when a sample is judged to be representative can the researcher legitimately generalize his conclusions to the population of interest. This ability to generalize from the sample findings to the study population is an essential element of most meaningful research.

Sampling techniques are usually grouped into two main categories: (1) probability or random sampling, and (2) nonprobability sampling. Chapter 11 will present the most commonly used techniques under each grouping.

OBJECTIVES

Upon completion of this chapter, the student will be able to:

1. Distinguish between probability and nonprobability sampling schemes by discussing the meaning, advantages, and disadvantages of each.
2. Describe and recognize common probability sampling techniques.
3. Describe and recognize common nonprobability sampling techniques.

4. Define and recognize cross-sectional, retrospective, and prospective studies.
5. Define single-blind and double-blind trials.
6. Define the Hawthorne Effect.

DISCUSSION

Before beginning the presentation of sampling techniques, we will first review the concepts of target population and population sampled, as discussed in Chapter 4.

The first step in the conduct of any research is to specify the common characteristics that define the population of interest. Often a distinction is made between the population about which an investigator wishes to draw conclusions, called the *target population*, and the population about which a conclusion *can* be made. The latter population is called the *population sampled*.

The distinction between the two populations is a statistical one. The methods of statistical inference presented in the preceding chapters enable an investigator to generalize from the sample to the population sampled. Generalization from the population sampled to the target population is much more subjective and open to controversy. To make the final step in generalizing to the target population, the investigator must be assured that the characteristics of the population sampled and the target population are identical. The possibility for the existence of subtle or hidden differences between the two populations makes it mandatory that the investigator proceed with extreme caution when generalizing from the population sampled to the target population. In the discussion that follows the term "population" will refer to the population sampled.

Probability Sampling

Probability or random samples are selected according to some chance mechanism. Neither the investigator nor the study subject has any influence over the inclusion of a particular unit in the sample. It is important to note that randomness in the statistical sense does not mean haphazardness. To prevent conscious or unconscious bias on the part of the investigator, the selection of a truly random sample can be accomplished only by mechanical means similar to drawing names out of a hat. More scientifically, sampling units may be selected by means of a table of random numbers (Appendix Table K). Each possible sampling unit is assigned a number, a list of random numbers is obtained from a random number table, and those units whose numbers correspond to the list from the table are selected for inclusion in the sample.

The basic advantages of probability or random sampling are:

1. The influence of either conscious or unconscious bias on sample selection is minimized.
2. For techniques of statistical inference to be applicable a random selection scheme must be employed. With this type of sampling, the researcher can evaluate the precision with which sample statistics estimate or supply information about population parameters.

One point of caution with regard to probability or random samples: The use of a random sampling scheme does not necessarily insure that the sample is representative of the population, especially when the number of units in the sample is small. In some instances, a nonprobability sample selection scheme may provide a sample that is more typical of the parent population. However, this gain in representativeness is at the expense of being able to statistically evaluate the precision of sample statistics. In general, the generalizability of conclusions based on nonprobability samples is questionable.

Also, it should be pointed out that while tests of significance assume random sampling from the population under investigation, this type of sampling in clinical situations is often not feasible. This applies particularly to clinical trials in which two treatment groups or a treatment and control group are being compared simultaneously. Usually, the investigator must take the patients who are available and who satisfy the conditions of the study. In such cases, an alternate method called random allocation or random assignment is used rather than true random selection of units for inclusion in the study. Under the method of random assignment, a group of study units are assigned at random to the treatment groups. For example, an investigator may have available a group of 20 patients with decubitus ulcers. Ten patients may be assigned *at random* to the treatment group and 10 patients to the control (no treatment) group. Random assignment is similar to random selection in that it eliminates the chances of investigator-instilled biases and personal judgment in the allocation of patients to treatment groups. Random assignment of patients to treatment and control groups prevents the investigator from selecting those for the treatment group who have the best chances of responding or being cured.

Random assignment, like random selection, does not assure a

representative sample, especially with small groups of study units; that is, with random assignment we are not necessarily assured that the treatment groups (or treatment and control) are alike in all respects before application of the treatment(s) under study. Usually, it is wise to present as part of the research findings a table comparing the treatment groups with respect to relevant characteristics such as sex, age, etc.

With the use of random assignment of study units to treatment groups, tests of significance may still be carried out even though the groups do not constitute true random samples selected from some theoretical treatment and control populations. However, conclusions or inferences apply only to the particular situation at hand and may not statistically be generalized to some larger theoretical populations. Generalization to some larger target population in this situation is highly subjective and depends on the investigator's ability to convince others that the sample being studied is truly a representative, nonbiased replica of the population to which the inferences will be referred.

In general, all tests of significance assume random sampling or random selection of study units from the study population. In the absence of true random sampling, the generalizability of sample results to a larger target or study population may be seriously questioned. In reviewing any research finding, it is the duty of the research consumer to determine how the sample was selected, the degree of departure from true random selection, and how this departure may affect the representativeness of the sample studied with respect to the target population.

Types of Random Samples

SIMPLE RANDOM SAMPLE

The simple random sample is the most basic type of probability sample. With simple random sampling, each unit in the population has an equal chance of being included in the sample. In other words, no one unit in the population has a greater likelihood of being selected in the sample than any other unit in the population. This equal probability of inclusion in the sample among all sampling units is assured only through the use of mechanical devices or random number tables.

The popularity of simple random sampling is due to the fact that most statistical techniques assume this mode of sampling. With other probability samples, the estimates of certain quantities such as standard error of the estimates becomes much more involved and time-consuming.

While statistically the most expedient, simple random sampling is often not feasible. To obtain a simple random sample, an investigator must first have a complete listing of every element in the population. Obviously, for large populations, this task is impossible. However, when the population of interest is limited in nature, such as all patients, records, students, or personnel within a given institution or a limited geographic area, then the task of enumerating the population becomes more manageable.

SYSTEMATIC RANDOM SAMPLE

Another widely used probability sampling technique is the selection of a sample according to some predefined system, hence the name *systematic sample*. Examples would include selecting every third patient on the daily hospital admissions list, every fifth baby born in a given hospital, and the patients in even numbered rooms. The element of randomness is introduced into this scheme because the investigator selects the first unit in the sample at random and then proceeds from that point using the predefined selection scheme. Even with the random start, the possibility of obtaining a biased or nonrepresentative sample exists. For example, as the result of hospital construction or numbering scheme, the even-numbered rooms in a hospital may be higher priced, have easier access to nursing stations, have windows, etc., which may possibly affect patient response and bias the results of the study.

STRATIFIED RANDOM SAMPLE

In many studies the population of interest may be divided into homogeneous subpopulations such as males-females, different age groups, income levels, occupations, etc. Random samples of units within each of the sections or subpopulations may be taken.

With a stratified random sample the chance of obtaining a nontypical sample is reduced. For example, in an attitude study involving all types of nurses in a large hospital, attitudes may differ significantly according to educational level or area of service. With simple random sampling, it is possible just by chance to obtain a sample that contains a disproportionate number of LPN's or supervisors having a masters degree. To prevent this possible bias, a sample of each subgroup may be taken proportionate to the numbers in each nursing category within the institution.

With stratified random sampling, an equal number may be taken from each subsection, or each subsection may be sampled according to predetermined proportions existing in the population. Also, in some instances, disproportionate subsampling may be employed in order to insure adequate representation for a subsection that is very small in number but that plays a significant role in the variable being studied.

CLUSTER SAMPLE

The researcher is often faced with the problem of sampling from a large population in which the sampling units are widely dispersed. For example, an investigator may be interested in children hospitalized for lead poisoning in the southeastern United States. In this study, the basic sampling unit is the child hospitalized for lead poisoning. However, in such cases, a list of the population from which to select units usually does not exist. Even if such a list were available, the time and cost of visiting the patients scattered over such a large geographical area would probably be prohibitive.

The dilemma may be solved by selecting at random a sample of hospitals within the designated geographic area. The investigator may then visit each hospital in the sample and interview all children with lead poisoning at those institutions, or a subsample of children may be selected within each hospital in the sample. In this example, the hospital (the *cluster* of patients) is the primary sampling unit and the patients within each hospital are the secondary sampling units.

Other examples of cluster sampling include randomly selecting blocks within a given city (the cluster) and then within the block randomly selecting houses to participate in the study; randomly selecting counties (the cluster) within a state and then within the county randomly selecting nurses to be interviewed; or within a given hospital randomly selecting wards (the cluster) and then interviewing all nurses on these selected wards.

One advantage of cluster sampling is in the saving of time and cost relative to many other sampling schemes. However, while large amounts of data can usually be obtained at relatively lower costs, this type of sampling procedure often results in less precise data than the other probability sampling designs.

SEQUENTIAL SAMPLE

In sequential sampling schemes no fixed number of sampling units is decided on before the project begins. Instead, a study unit is selected and the appropriate measurement is made, another unit is selected and measured, and so on. As each measurement is collected, an evaluation with respect to some criterion is made and as soon as a predetermined level of this criterion is reached, sampling is discontinued.

EXAMPLE 13.1 A public health nurse wished to determine extent of knowledge of personal hygiene among all first graders in a particular city. A list of all currently enrolled first graders was obtained from the school superintendent's office and the names were numbered consecutively. From a table of random numbers, a set of 100 random numbers was obtained. Students on the enrollment list whose numbers corresponded to the random numbers obtained from the table were selected for inclusion in the study. What type of sample selection scheme was employed? Why? To what group may the findings be generalized?

SOLUTION A simple random sampling technique was employed. The use of this method was possible since a listing of the entire population (all first graders in the city) was available. The names of the children in the study sample were selected mechanically through the use of a random number table, thus eliminating the possibility of investigator-induced bias. In this study, statistical inferences may be made to all first graders in the city. To generalize to the larger group of all first graders in the state or nation, etc., is not statistically justifiable and involves subjective arguments about the representativeness of the population sampled.

EXAMPLE 13.2 Discuss how cluster sampling may be employed in Example 13.1.

SOLUTION The investigator would obtain a list of all elementary schools in the city. From this group, a simple random sample of schools would be selected. (The selection of schools would be as described in Example 13.1.) The investigator would then visit only the schools included in the study sample, rather than visit all schools in the city. At each school in the sample, either all first graders would be interviewed, or a simple random sample of first graders would be interviewed. In this example, the school is the cluster or primary sampling unit and the child is the secondary sampling unit.

EXAMPLE 13.3 In Example 13.1, suppose the investigator believes that there may be a difference in response according to whether the child attends a public or private school. From the literature, it is learned that for this city, 30% of the children attend private school and 70% attend public school. Discuss what type of sampling scheme would be used to insure proportionate representation for both groups of children.

SOLUTION The sampling scheme that would be employed in this situation would be stratified random sampling. If the investigator wishes to have 100 children in the study sample, 70 children would be randomly selected from public schools and 30 from private schools. The sampling that is carried out on the subpopulations (types of schools) could be simple random sampling of children from enrollment lists of public and private schools, as described in Example 13.1, or it could be cluster sampling from the two groups, as described in Example 13.2. The investigator is free to choose the type of sampling scheme to be used for the subpopulations. Often in stratified random sampling a disproportionate number may be selected from the subpopulations. For example, suppose in Example 13.1, the investigator was interested in *all* children of first-grade age regardless of whether they were enrolled in school. The number of children not enrolled in school may be very small relative to the other groups and therefore the investigator may wish to include all (100%) or as many as possible of this group in the sample.

EXAMPLE 13.4 A city is laid out in 20 major sections, each section divided into 100 blocks. A public health investigator is interested in surveying the attitudes of the city residents toward venereal disease. Discuss possible sampling schemes for carrying out this survey.

SOLUTION

1. A list of all city residents may be obtained and a simple random sample of names from among those on the list may be selected. Usually this approach is highly impractical since such lists either do not exist or are not up-to-date.
2. From among the 20 sections, a simple random sample of sections may be taken. Within each of the randomly selected sections, a simple random sample of blocks may be chosen. The investigator may then visit within each of these randomly selected blocks, either a random selection of houses or even num-

bered houses, or every third house, etc. This sample selection scheme involves techniques of simple random sampling, cluster sampling, and systematic sampling.

3. If the investigator feels that it is important to include groups from low, middle, and high socioeconomic income brackets, it may be determined which of the sections of the city fall into each of the three categories. A random selection of blocks from among the three income groupings may be taken. The investigator may choose equal proportions of blocks from among the income groups, or a proportionate number equal to some population proportion for low, middle, and high family income. Within the selected blocks the investigator may employ simple random or systematic random sampling to determine which houses to visit.

The above are three possible ways of approaching the sampling problem described in Exercise 13.4. Other combinations of sampling techniques are also possible.

EXAMPLE 13.5 In a clinical trial to compare the efficacy of two different drugs for treatment of depression, one group of patients receive A while another group receive B. When a patient enters the study, he is randomly assigned either A or B. He continues taking the medication until he either achieves an "improved" status or remains in a "not-improved" state for a specified time period. At this point his treatment is discontinued and he is categorized as "I" or "NI." The study is continued for different patients until one drug shows a significantly higher proportion of "improveds" than the other drug. At this point the study is discontinued.

SOLUTION The above is an example of sequential sampling in a clinical trial. No fixed number of patients was decided on before the study began. Patients were added to the study until a statistical difference in proportions was achieved. For ethical reasons, these trials are usually terminated after a specified period of time even if significance has not been achieved.

Nonprobability Sampling

The compliment of probability or random sampling is nonprobability or nonrandom sampling. As its name implies, a nonprobability sample is one in which the study subjects (sampling units) are not selected according to a random selection scheme. Lower costs and convenience are two of the major reasons for the use of nonprobability sampling. The primary limitation is that generalization from a nonrandom sample to the population of interest is statistically not justifiable. Among the widely used nonprobability samples are the convenience, the purposive, the quota, and the volunteer sample.

CONVENIENCE SAMPLING

A convenience or incidental sample is one whose units are selected because they happen to be in a particular place at a particular time. A study on outpatient visiting patterns for a particular outpatient clinic may involve interviewing the patients who arrive between 2 P.M. and 4 P.M. on a given afternoon; or all nurses who happen to be in the nurses' lounge at a particular time may be interviewed as part of a study on nurses' attitudes toward the hospital's administrative policies.

PURPOSIVE SAMPLING

Purposive sampling is also called judgment sampling because the researcher deliberately selects the units to be included in the study sample. The sampling units are selected because, in the investigator's judgment, they are representative of the population being studied. If a researcher wishes to study the effect on patients of "bad" versus "good" nursing practices, based on either the investigator's professional judgment or the professional judgment of others, "bad" and "good" nurses are selected for inclusion in the study sample. This sample would be a purposive or judgment sample.

QUOTA SAMPLING

Quota sampling is the nonrandom analog of stratified random sampling. The population is divided into homogeneous subpopulations and samples from these subgroups are selected on the basis of predefined quotas. The subsampling scheme is one of convenience or judgment rather than random selection.

VOLUNTEER SAMPLING

Volunteer samples may arise from two different sampling schemes. In one, the investigator simply asks for participation in a study. In the other, an investigator may interview or send questionnaires to a randomly selected group of study units. Those individuals who choose to respond represent a voluntary rather than a truly random sample of individuals.

For Exercises 13.1 to 13.10, discuss the type of sampling scheme used and the generalizability of results.

EXERCISE 13.1

To determine nursing students' attitudes toward the school honor code, an investigator interviewed all nursing students in the nursing student lounge on a particular afternoon.

EXERCISE 13.2

A new nursing care procedure for hospitalized diabetic patients was instituted using diabetic patients from hospitals randomly selected from among all hospitals in a given city.

EXERCISE 13.3

To determine career satisfaction among the graduates of a particular nursing program, an investigator sent questionnaires to all graduates of the program. A 60% response was received.

EXERCISE 13.4

From the hospital admissions list, an investigator selected every third name, choosing the first name at random. The patients were interviewed to determine their preference for private, semiprivate, or multiple bed rooms.

EXERCISE 13.5

A nurse was interested in determining the effect of different arm positions on blood pressure determinations. Fifteen patients were available on a general clinical research unit. The patients were randomly assigned to two groups. One group had their blood pressures taken using the standard arm position; the second group held their arms to their sides while blood pressure readings were made.

EXERCISE 13.6

In a study of nursing students' attitudes toward abortion, an investigator determined that 60% of all nursing students are single, 30% married, and 10% divorced or widowed. The investigator visited several classes during the day and interviewed 60 single students, 30 married, and 10 divorced or widowed students.

EXERCISE 13.7

An investigator wished to determine the differences in patient waiting times for well-run emergency rooms versus poorly run, inefficient ones. Based on judgment, two emergency rooms from each of the two categories were selected.

EXERCISE 13.8

An nurse educator wished to investigate home life factors that contribute to student motivation. A group of nursing students felt to be highly motivated and another group felt to be inadequately motivated were selected.

EXERCISE 13.9

In a study concerning student perception of the fairness of admissions policies of colleges of nursing in the United States, an investigator randomly selected 20 colleges. Within the colleges selected, questionnaires were sent to a random selection of students. The return on the questionnaires was 100%.

EXERCISE 13.10

In the above study, the investigator wished also to ascertain the colleges' perceptions of the fairness of their admissions policies. Questionnaires were sent to the admissions committees of the 20 colleges selected in the sample. A 100% return was received.

Some Further Notes on Sampling

Much of nursing research involves the conduct of surveys dealing with observations on humans. This type of research is nonexperimental in nature since manipulation of the study environment is usually impossible. (Experimental research designs require control over all extraneous factors except the variable being studied.)

Surveys may be broadly classified into three categories; the *cross-sectional study*, the *retrospective study*, and the *prospective study*. These studies are often carried out for the purpose of establishing associations between two disease states or conditions, or the association between disease states and presence of a risk factor such as lung cancer and smoking, or thromboembolic disease and use of oral contraceptives.

CROSS-SECTIONAL STUDIES

A cross-sectional survey is one that is carried out at one point in time. For example, in order to study the association between dental caries among children and the use of a fluoride toothpaste, an investigator visited a sample of houses in a particular community. The children were given a dental exam and asked whether they used a fluoride toothpaste. In this type of study, it cannot be said that using fluoride decreases the chances of developing cavities since the study was not carried out over time. In other words, we do not know whether people who had fewer cavities to start with tend to use fluoride toothpaste or vice versa. The only conclusion that could be made is that an association exists between use of fluoride toothpaste and number of cavities in children.

Another use of the cross-sectional study is to describe what is happening over time by taking a survey at only one point in time. An investigator may survey all freshman, sophomore, junior, and senior nursing students at a college of nursing on a given day. The data may be used to represent the feelings of students as they progress through nursing school. It should be pointed out that while the data were used to present a chronological picture of the nursing school experience, that is, freshman versus sophomore etc., the students interviewed were different for each year. In other words, all data were collected on freshmen, sophomores, juniors, and seniors at a single point in time rather than one class of students being followed through all four years of nursing school.

RETROSPECTIVE STUDIES

Retrospective studies, as the name implies, involve looking back in time. This type of study begins with a study group known to have the disease or condition in question and with a comparison group who do not have the disease or condition being studied. Both groups are traced back in time by means of case histories or personal recollections and compared with respect to the presence or absence of different factors that could have produced the disease or condition. For example, in a study on lung cancer, the histories of a group of patients with lung cancer and a matched control group may be traced in order to compare the two groups with respect to their smoking habits. Another example of a retrospective study may be in the investigation of thromboembolic disease and the use of oral contraceptives. A group of women with diagnosed thromboembolic disease and a control group that have been matched as closely as possible may be traced back to determine differences in contraceptive usage. The question the investigator attempts to answer is whether the risk factor (oral contraceptive usage) occurs more frequently among women with thromboembolic disease than among women in the control group.

One of the major limitations of a retrospective study is its reliance on records or personal recollections. Often records may contain incorrect or incomplete information. A retrospective study is no better than the records from which the data were obtained.

Another area of concern in the conduct of retrospective studies is the degree of matching between the study and the control groups. The investigator attempts to make the two groups as alike as possible with respect to all factors except presence or absence of the condition being studied. However, it is usually impossible to match subjects with respect to all variables and therefore the possibility exists that a variable or variables not taken into account may play a role in the presence or absence of the condition being investigated.

In reviewing records as part of a retrospective study, several facts must be kept in mind.

1. No information about presence or absence of the risk factor does not necessarily mean that it does not exist. For example, in reviewing the records of the control group in the lung cancer study, the failure of the record to contain information about whether the patient smoked or not does not mean that the patient did not smoke. It may mean that the information was not elicited. Records of patients with lung cancer are much more likely to contain information about smoking than the records of the control patients.
2. Criteria for designating a particular condition or disease state may differ. In a retrospective study on depression, the criteria for classifying a person as suffering from depression may differ from investigator to investigator.
3. The risk factor may be present in varying degrees. In the lung cancer study, patients may differ greatly with respect to kinds and amount of smoking.
4. The diagnosis of the disease under study may be incorrect. Thus, patients who have been included in the disease group may not have the disease.

PROSPECTIVE STUDIES

In contrast to a retrospective study, a prospective study goes forward in time. Two groups of study subjects are selected; one group is known to have the factor under study (the risk factor) and the other group (control) does not have the factor. The two groups are followed forward in time to determine if differences exist between the groups with respect to the development of new cases of the disease. For example, in a prospective study of oral contraceptives and thromboembolic disease, a group of women known to be users of oral contraceptives are selected along with a control group who use other means of contraception. The control group is matched with the study group on as many factors as possible except for the risk factor being investigated (contraceptive usage), i.e., age, number of pregnancies, etc. Both groups are followed in time to determine if there is a difference in occurrence of thromboembolic disease for the two groups.

Advantages of the prospective study include the following:

1. The investigator does not have to rely solely on information contained in records or personal recollection. There is much more freedom in determining what is measured and how it is measured.
2. Recognition of the disease or condition is more uniform than in the retrospective study. Also, the criteria for determining the presence of the risk factor can be more uniformly controlled.
3. The degree of matching for the control and study groups is under the control of the investigator and therefore may be more closely determined. However, the possibility of the existence of another unidentified variable that could explain the association between the factors being studied does exist in the prospective as well as in the retrospective study.

Two of the major disadvantages of the prospective study are:

1. The achievement of good followup procedures is not always possible. The studies depend in large part on the willingness of the study subjects to participate. Also, the enthusiasm of the investigators as well as the participants may decrease over time.
2. Prospective studies are generally very large, expensive, and time-consuming to carry out.

For Exercises 13.11 to 13.15, classify the studies as cross-sectional, retrospective, or prospective.

EXERCISE 13.11

From the records of a large university hospital, a large group of patients with diagnosed hepatitis was selected. Another group of patients whose characteristics matched those of the disease group except for the presence of the disease was also selected. The records of both groups were examined to determine the amount of drug usage for both groups during the 12 months prior to the study.

EXERCISE 13.12

In 1950 in Framingham, Massachusetts, approximately 5,000 men and women between 30 and 59 years old were recruited to participate in a study of the association of various risk factors and heart disease. At the beginning of the study all participants were given a thorough cardiovascular evaluation. The group has been examined biennially since the onset of the study.

EXERCISE 13.13

In a study of the association between length of hospitalization and stress, groups of patients falling into each of five time categories were interviewed on a particular day. Group I consisted of patients who had been hospitalized less than one week; Group II, from one to two weeks; Group III, from two to four weeks; Group IV, one to three months; and Group V, greater than three months.

EXERCISE 13.14

In a study of community attitudes toward repeal of marijuana laws, a telephone survey of households was conducted on a given day.

EXERCISE 13.15

In a study of the association of smoking and lung cancer, a group of smokers and a matched group of nonsmokers were selected. Both groups were followed for a 15-year period to determine the occurrence of new cases of the disease.

Two additional terms that must be defined are *single-blind* and *double-blind trials*. These terms are usually associated with controlled clinical trials and are means for eliminating from the study the biases and prejudices of the investigator and the study subject. Often in clinical investigations, a patient's knowledge that he is in the treatment group may affect his response to the treatment itself. To remove this possibility of bias, the patient in a single-blind trial does not know whether he is in the treatment or the control group.

To prevent investigator prejudice from affecting subject response, a double-blind trial may be carried out. Using this procedure, neither the investigator nor the study subject knows who is assigned to which group.

In both kinds of experimental situations, highly controlled clinical trials and the nonexperimental survey method of collecting information, the possibility exists that there mere act of participating in an investigation may and probably will affect subject response. The general term for this phenomenon is the "Hawthorne Effect" and its implication should be borne in mind when interpreting the results of research studies.

SUMMARY OF CRITICAL CONCEPTS IN CHAPTER 13

In Chapter 13, an overview of basic sampling techniques has been presented. The critical concepts presented in this chapter are:

1. The purpose of sampling is to obtain study units that are representative of the population of interest.
2. The two basic types of sampling techniques are probability or random sampling and nonprobability sampling.
3. Probability or random sampling *must* be carried out through the use of mechanical or other devices such as random number tables.
4. The use of a random sample eliminates investigator-induced bias in the sample selection process.
5. Techniques of statistical inference assume random sampling.
6. Conclusions based on randomly selected samples of study units may be generalized to the population from which the sampling units were taken.
7. Random sampling schemes are often costly, time-consuming, difficult, or impossible to carry out.
8. A random sample does not necessarily insure representativeness.
9. Nonprobability sampling schemes are those that are carried out in a nonrandom manner (i.e., without the aid of mechanical or other random selection devices).
10. Generalization of conclusions from a nonrandom sample to the study population is statistically not justifiable.
11. Cross-sectional studies involve surveys that are carried out at one point in time.
12. Retrospective studies involve identification of a study group and a matched control group and the following of these two groups back in time to determine if they differ with regard to some factor under study. Data is obtained by reviewing records or through the personal recollections of the study subjects.
13. In a prospective study, a group of subjects who have present a particular factor and a matched control group are identified and followed forward in time to determine new cases of the

condition or disease being studied. The prospective study is also called a *cohort study*.

14. A single-blind trial is a study in which the study subject does not know whether he is in the treatment or control group.
15. In a double-blind trial, neither the study subject nor the investigator knows who is in the control and the treatment groups.
16. Hawthorne Effect is the response brought about by the study subject's knowledge that he is a participant in a research investigation.

CHAPTER 14

COMPUTERS: A BRIEF DISCUSSION

Daniel R. Knapp, Ph.D.

INTRODUCTION

This chapter is intended to give the student a brief overview of what a computer is, how it works, and what it can and cannot do. Why should this book include a chapter on computers? Computers are intimately connected with most areas of research today, especially research involving statistical analysis of data. In this text, the problems were designed to be workable using manual arithmetic. In most real research, the data are much less easily manageable, and some type of calculation assistance is needed. This assistance could range from a small calculator to a multimillion dollar computer. While it is not necessary for every researcher to be a computer programmer, it is essential that almost every researcher have some understanding of what a computer is and what it can and cannot do.

A computer is basically a "dumb" machine but, on the other hand, it is a very fast and a very accurate machine. The computer can do only very simple tasks like *add, subtract*, and, very important, *compare*. But since it is a very fast machine, it can do many of these simple tasks in a very short time. By breaking a complex task into very simple steps, the computer can perform the complex task rapidly by performing each simple step with great speed.

If a computer is just a machine that can rapidly perform simple steps like adding and subtracting, how is it different from a ten-dollar calculator? The basic feature that distinguishes a computer is the *ability to perform a series of tasks from a stored list of instructions* known as a program. The list of instructions, or program, is known as *software*. We shall discuss software further, but first we will look more closely at the machine itself, the computer *hardware*.

HARDWARE

The first true electronic computer was built using vacuum tubes. The computer filled a very large room, required an enormous amount of power and air conditioning, and was not very reliable. Thanks to the transistor, and more recently the microscopic integrated circuit, computers have steadily decreased in size and cost and have reached the point where the basic electronic components of a small computer can be held in one hand. Whether we are talking about a large computer system like an IBM 370-145, which still fills a large room, or a three-inch long microprocessor integrated circuit, the basic components of the computer are the same. Figure 14.1 shows a diagram of the basic computer hardware.

The heart of the computer is the *central processing unit* or *CPU*.* This unit receives data, performs tasks on the data (according to its list of instructions or program), and passes data on, as shown by the arrows in Figure 14.1. The *input device* receives data from the outside world and converts it to a form that is understandable by the central processing unit. It should be stated that this data can of course be either numbers (numeric data), letters (alphabetic data), or a mixture of both (alphanumeric data). The input device might be a Teletype®, a punched card reader, or an electronic keyboard. The *output device* converts data from computer-understandable form to human readable form. A typewriter-style printer would be an example of an output device. (The computereze term I/O refers to input/output.) So far, with an input device, a CPU, and an output device, we could be describing a printing calculator. The memory, which can be used to store either data or instructions (usually both), distinguishes our hardware as a computer. More accurately, it is the ability to store a list of instructions and then perform tasks as directed by the instructions that distinguishes our machine as a computer.

The memory may be made from magnetic "cores" or from semiconductor devices. When a computer person speaks of "core," he is talking about memory. The memory area is organized in units of "*words*," and each word has an "*address*." The address is obviously the location of the word and is used in the instructions to tell the CPU where to get data or where to put it. A computer "word" is different from a normal word in that usually all the words in a computer are the same length. The actual length depends upon the design of the computer and is usually of

* Computer people like to speak in letter abbreviations. Since knowing some of these abbreviations helps in communicating, we will introduce a little computereze vocabulary.

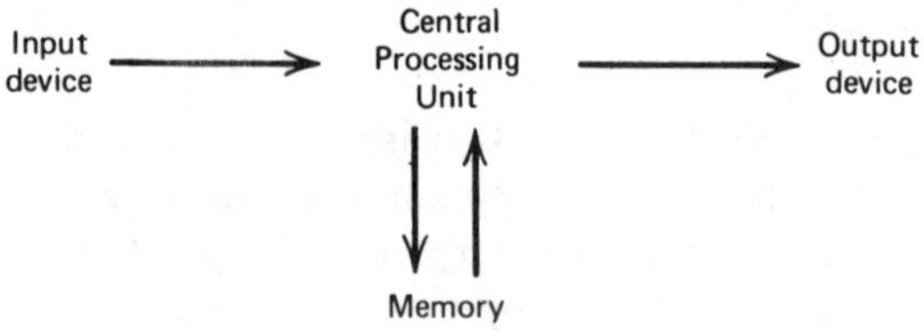

FIGURE 14.1 THE BASIC COMPONENTS OF COMPUTER HARDWARE. THE ARROWS INDICATE FLOW OF DATA.

little consequence to most users. Word lengths are usually expressed in "bits" (which we will define later). Suffice it to say at this point that memory is composed of words, each having an address or location. These memory words can be used to store data or lists of instructions (programs). Depending upon the design of the computer, a word could hold one or more letters or numbers of data, and one instruction might take one or more words to store. Memory size is usually expressed in thousands of words or "kilowords" or "K." A "K" of computer words is actually 1,024 words (2 to the tenth power; computer arithmetic is in powers of 2, as we shall see later). Thus, a small computer with 8K of memory has 8,192 words of memory.

We mentioned earlier that the computer is an electronic device, but we so far have avoided saying how the computer represents data internally. Data inside the computer is represented by electric signals. There are two primary ways to represent numeric data by an electric signal: *analog* and *digital*. Analog representation of data means that the magnitude of the signal is proportional to the size of the number. An analog signal is a continuous variable. Analog computers exist, but normally when we speak of a computer we mean a digital computer. Digital computers represent numbers by discrete signals, usually using only two numbers, 0 and 1, representing "off" and "on" electrically. Representing data only as 0 or 1 in a digital computer makes for the maximum simplicity in circuitry and cost, and maximum speed. Since there are only two numbers, we have what is known as the *binary number system*. Our normal number system has 10 integers, 0 through 9, hence the name decimal number system. One integer of data is a "*bit*." (We said earlier that computer word lengths are expressed in bits.) Table 14.1 shows some binary numbers and their decimal equivalents. But how then are alphabetic characters represented in a computer? Alphabetic characters are represented by a code of binary bits. One of the most common codes is ASCII, which is an acronym for American Standard Code for Information Interchange (pronounced "as′ key"). Table 14.2

TABLE 14.1 DECIMAL AND BINARY NUMBERS

Binary	*Decimal*	*Binary*	*Decimal*
0	0	101	5
1	1	110	6
10	2	111	7
11	3	1000	8
100	4	1001	9

shows some of the ASCII codes. Note that there are ASCII codes for letters, numbers, punctuation marks, and several other types of marks. Fortunately, the computer user usually does not have to be concerned with binary numbers or ASCII codes since most programs or "software" are written in "higher level languages," as we shall see shortly.

In summary, the computer hardware consists of an input device that converts data to electrical signals usable by the CPU. The CPU may manipulate this data, store it in memory, retrieve it from memory, and perhaps send it to the output device, which converts the electrical signals to human readable form. Remember, however, that the CPU does only what its list of instructions or program tells it to do. The program is written and placed in the memory of the computer by the programmer. We shall now consider the nature of this list of instructions or program.

SOFTWARE

Computer programs are known as *software*. As we said earlier, the computer basically can do only very simple tasks and does only what it

TABLE 14.2 ASCII CODES

Character	*ASCII Code*	*Character*	*ASCII Code*
A	1000001	7	0110111
B	1000010	8	0111000
C	1000011	9	0111001
D	1000100	:	0111010
E	1000101	;	0111011

is instructed or "programmed" to do. The list of instructions must be written in minute detail and stored in the memory of the computer. Ultimately, the codes for the various instructions are stored in the computer memory as binary bits. Programs can actually be written in these binary codes. Such programs are said to be written in "machine language." Obviously, machine language programming is a very laborious and tedious undertaking. At some point in the life of a computer, it must be programmed in machine language since this is the form that is actually stored in the computer memory. However, most computers have "compilers" or translators that permit the use of "high level languages," that is, languages that are more like English or mathematical statements. These compilers are actually programs with which the computer translates the high level language to machine language instructions it can execute. There are many of these high level languages, each with features making it particularly suited for a particular application. Some of these languages are BASIC, Fortran, Cobol, Algol, and APL. There are even "dialects" of these languages that reflect particular characteristics of particular brands of computers. BASIC is probably the easiest language for the beginning programmer to use. We shall look at a simple BASIC program to illustrate how a computer is instructed to perform a task.

The list below is a simple program in BASIC.

```
1  INPUT A
2  INPUT B
3  C = A + B
4  PRINT C
5  END
```

This program instructs the computer to read in a value for A from its input device in Statement 1. Statement 2 tells it to read in a value for B. Statement 3 tells it to add A to B to get C. Statement 4 tells it to print the sum C. The END statement is there to tell the compiler that it is the end of the list of statements to be translated to machine language.

A slightly more complicated BASIC program that instructs the computer to find the mean ($\overline{Y}$) of a set of five numbers is shown in Table 14.3. Before writing more complex programs such as the one given in this table, one should first diagram the "flow" of the program by means of a *flowchart*. A flowchart is a diagram of the sequence of steps taken to perform a task. Figure 14.2 presents a flowchart of the program shown in Table 14.3.

Recall that the computer cannot execute BASIC statements di-

TABLE 14.3 BASIC PROGRAM SHOWING CALCULATION OF $\bar{Y}$

Computer Instruction	*English Translation*
10 YSUM = 0	Initialize variable called YSUM.
20 N = 0	Initialize N (N is a counter used to "count" number of Y values read into computer).
30 INPUT Y	Read one value of variable Y from input device.
40 YSUM = YSUM + Y	Add value of Y to YSUM (this is called a running total).
50 N = N + 1	Increment counter (N is increased by 1 each time the computer reads a value for Y).
60 IF N < 5 GO TO 30	Decision: If statement true (fewer than 5 Y values read), repeat steps beginning at statement 30; if false, go to next statement (statement 70).
70 YBAR = YSUM/5	Divide total of Y's (YSUM) by 5 and label as YBAR.
80 PRINT YBAR	Print value of YBAR on output device.
90 END	

rectly; they must be translated to machine language instructions. This is done in the machine by means of the translator. Although BASIC statements are easier to write than machine language instructions, the task must still be defined in minute detail. The computer only does what it is told, and the instructions must be complete and absolutely correct. A human can interpret an instruction with a typographical error, the computer cannot. When the computer encounters an error in an instruction, it does the wrong thing or it "bombs" (computereze for halting with an error or being sent into a series of erroneous operations). Such errors in a computer program are referred to as "bugs" and the process of finding and correcting them is called "debugging" the program. Thus, a computer program must be both typographically perfect and operationally correct.

Computer hardware instructed by a proper program (software) can perform a multitude of tasks on data with high speed and extreme accuracy. Most computer "errors" are actually the result of erroneous input or faulty programming. Hardware malfunctions are usually not the cause. From the above discussion it becomes obvious that the computer cannot itself "think." A computer can make comparisons and arrive at

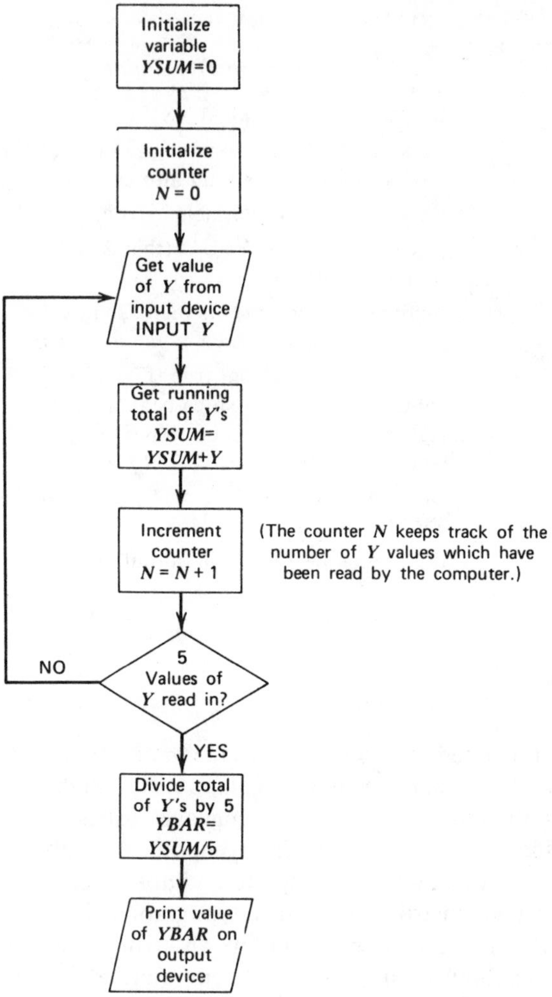

FIGURE 14.2 FLOWCHART OF PROGRAM IN TABLE 14.3

decisions as instructed by the program but does not itself make any decisions. It only applies criteria dictated by the programmer and determines if the data do or do not meet those criteria. The machine-subservient bureaucrat who insists that "the computer did it!" or "the computer will only do it this way" would do well to learn that the computer does only what the programmer tells it to do.

COMPUTER TECHNOLOGY IN THE HEALTH SCIENCES

Figure 14.3 presents a summary of the wide-ranging applications of computer technology to health care. As can be seen from the figure, these applications may be broadly classified according to four general categories: (1) administrative, (2) teaching, (3) service, and (4) research.

Administrative

Due to their high costs, computers have in the past been used only in very large hospitals or other health agencies to maintain personnel files, accounting and billing systems, and to issue management reports. However, as more and more less costly, smaller computer systems are being made available, physicians and others in private health care delivery are now able to take advantage of computer technology not only for administrative purposes but for direct patient care as well.

Service

Computer technology is being used in almost every phase of patient care. These functions include systems that monitor patient medication regimens, monitor patients in a critical care environment, and maintain patient data files for storage and retrieval of the facts about the care of individual patients. In addition, specialized microcomputers are being increasingly used to analyze data obtained from laboratory instruments such as are found in clinical chemistry, pathology, and radiology laboratories.

Teaching

The use of computer-assisted instruction as a tool for teaching in the health sciences is a relatively new but rapidly expanding application of computer technology. Computer programs have been designed to present simulated examples of clinical problems that students are asked to solve in an "interactive" mode. These "on-line" computer-simulated situations provide students with the opportunity to participate in clinical problem-solving, thus gaining valuable experience before being faced with the real-life situations.

Research

The computer is an invaluable instrument in the conduct of health-related research. Large bodies of research data may be processed swiftly

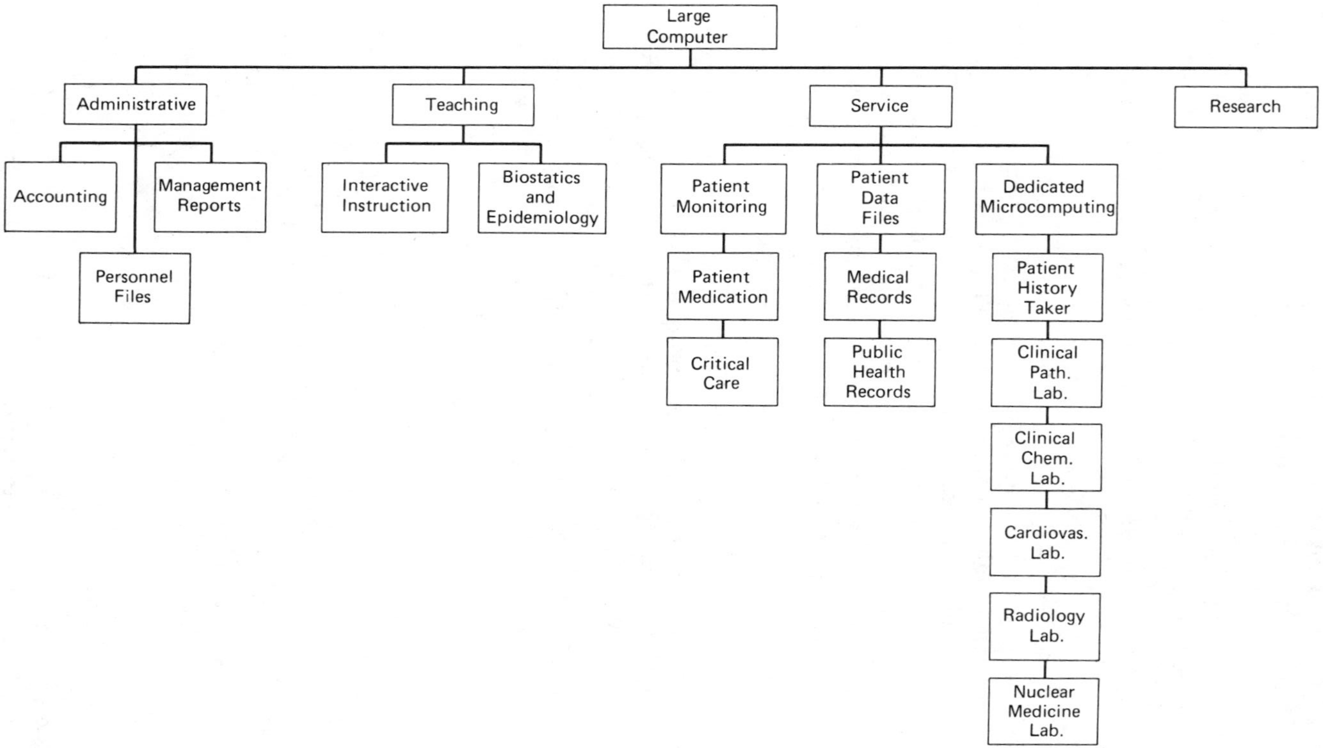

FIGURE 14.3. APPLICATIONS OF COMPUTER TECHNOLOGY IN THE HEALTH SCIENCES (ADAPTED IN PART FROM *COMPUTER TECHNOLOGY IN THE HEALTH SCIENCES* BY DAVID B. SHIRES, SPRINGFIELD, ILL., CHARLES C THOMAS, PUBLISHER, 1974, P. 6. USED BY PERMISSION.)

and accurately, thus facilitating more rapid incorporation of research information into the service area.

The increasing proliferation of computers in health care delivery and research makes it mandatory for health professionals to have some understanding of what a computer is, and most important, what it can do.

APPENDIX 1

TABLES OF STATISTICAL TESTS

Note: In the following tables, the tests specified for a given level of measurement are cumulative down each column.

DESCRIPTIVE STATISTICS

			Numerical	
	Tabular	*Graphical*	*Location*	*Spread*
Nominal	Frequency tables (discrete categories)	Bar charts Circle graphs	Mode	
Ordinal	Frequency tables (ordered categories)	Bar charts Circle graphs	Median	
Equal Interval	Frequency tables (continuous categories)	Histograms Frequency polygons	Mean Median Mode	Range Variance Standard deviation

TESTS OF SIGNIFICANCE

		Two-sample		
	One-sample	*Related*	*Independent*	*k-sample*
Nominal	Binomial test X^2 one-sample test	McNemar test	X^2	X^2
Ordinal	*	Sign test Signed-ranks test	Rank-sum test Median test	Median test *Kruskall-Wallis ANOVA *Friedman two-way ANOVA
Equal Interval	Simple t-test	Paired t-test	Pooled t-test	Analysis of Variance

* Tests not discussed or only mentioned briefly.

CORRELATION

Nominal	Contingency Coefficient*
Ordinal	Spearman's Rank Order Correlation Coefficient Kendall's Tau* Coefficient of Concordance*
Equal Interval	Pearson's Product Moment Correlation Coefficient

* Tests not discussed or only mentioned briefly.

APPENDIX 2

SOLUTIONS TO EXERCISES

CHAPTER 1

EXERCISE 1.1

a. $\Sigma Y_i = Y_1 + Y_2 + Y_3 + Y_4 + Y_5$

$= 2 + 1 + 1 + 2 + 4$

$= 10$

b. $\Sigma Y_i^2 = Y_1^2 + Y_2^2 + Y_3^2 + Y_4^2 + Y_5^2$

$= 2^2 + 1^2 + 1^2 + 2^2 + 4^2$

$= 26$

c. $\frac{(\Sigma Y_i)^2}{5} = \frac{(Y_1 + Y_2 + Y_3 + Y_4 + Y_5)^2}{5}$

$= \frac{10^2}{5} = 20$

d. $\Sigma(Y_i - 2) = (Y_1 - 2) + (Y_2 - 2) + (Y_3 - 2) + (Y_4 - 2) + (Y_5 - 2)$

$= (2 - 2) + (1 - 2) + (1 - 2) + (2 - 2) + (4 - 2)$

$= 0 + (-1) + (-1) + 0 + 2$

$= 0$

EXERCISE 1.2

a. $\frac{\Sigma Y_i}{5} = \frac{Y_1 + Y_2 + Y_3 + Y_4 + Y_5}{5} = \frac{10}{5} = 2$

b. $\Sigma(Y_i - 2)^2 = (Y_1 - 2)^2 + (Y_2 - 2)^2 + (Y_3 - 2)^2 + (Y_4 - 2)^2 + (Y_5 - 2)^2$

$= (2 - 2)^2 + (1 - 2)^2 + (1 - 2)^2 + (2 - 2)^2 + (4 - 2)^2$

$= 0 + (-1)^2 + (-1)^2 + 0 + 2^2$

$= 1 + 1 + 4 = 6$

c. $$\frac{(\Sigma Y_i)^2}{5} = \frac{(Y_1 + Y_2 + Y_3 + Y_4 + Y_5)^2}{5} = \frac{(10)^2}{5} = 20$$

d. $$\frac{\Sigma Y_i^2 - \dfrac{(\Sigma Y_i)^2}{5}}{n - 1} = \frac{26 - 20}{4}$$

$$= \frac{6}{4} = 1.5$$

EXERCISE 1.3

a. Variable: weight of newborns on special diet. Measurement scale: ratio.
b. Variable: type of defect (i.e., in hearing, sight, etc.). Measurement scale: nominal.
c. Variable: clinical expertise of senior nursing students. Measurement scale: ordinal.
d. Variable: body temperature of infants. Measurement scale: interval.

EXERCISE 1.4

a. Continuous
b. Continuous
c. Discrete
d. Continuous
e. Continuous
f. Discrete
g. Discrete

CHAPTER 2

EXERCISE 2.1

a.

FREQUENCY TABLE OF WAITING TIMES RECORDED BY 10 PATIENTS

Time	*Frequency*
4	1
6	2
7	3
8	4
Total	10

b.

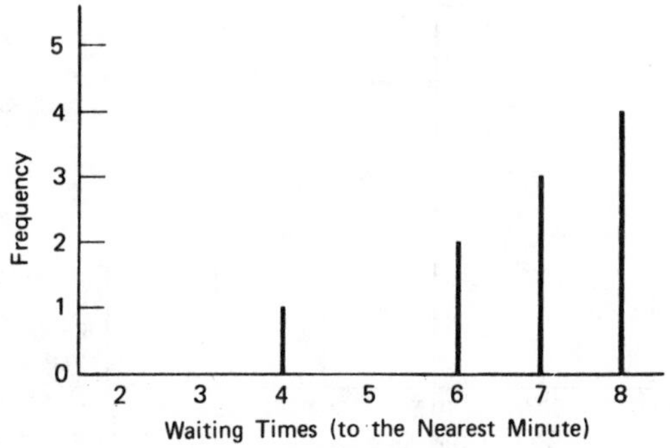

FREQUENCY DIAGRAM OF WAITING TIMES RECORDED BY 10 PATIENTS

c. The distribution is skewed in the direction of long waiting times. Based on this sample of 10 patients, the efficiency of the nursing staff may be questioned.

EXERCISES 2.2, 2.3

FREQUENCY DISTRIBUTION OF AGE (IN YEARS) OF PATIENTS IN A CORONARY CARE FACILITY

Age Intervals	*Frequency*	*Relative Frequency*
20.5–30.5	2	.13
30.5–40.5	5	.33
40.5–50.5	4	.27
50.5–60.5	2	.13
60.5–70.5	2	.13
Total	15	

EXERCISE 2.4

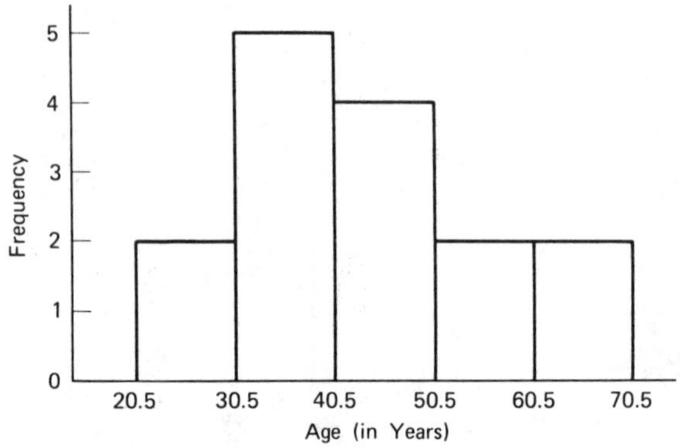

HISTOGRAM OF AGE (IN YEARS) OF 15 PATIENTS IN A CORONARY CARE FACILITY

EXERCISE 2.5

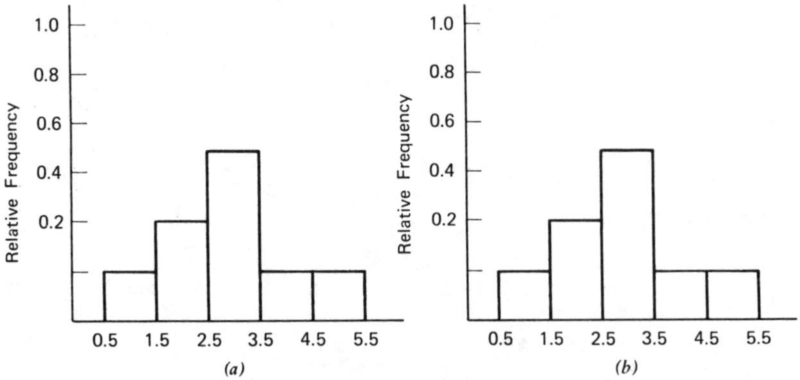

(*A*) ATTITUDES OF 50 JUNIOR NURSING STUDENTS TOWARD NURSING SCHOOL (*B*) ATTITUDES OF 20 SENIOR NURSING STUDENTS TOWARD NURSING SCHOOL

The relative frequency is more meaningful when comparing two groups of scores.

EXERCISE 2.6

FREQUENCY DISTRIBUTION OF SYSTOLIC BLOOD PRESSURES (mmHg) IN INFANTS AFTER THREE AND SEVEN MINUTES OF CRYING

	After Three Minutes		*After Seven Minutes*	
Intervals	*Frequency*	*Relative Frequency*	*Frequency*	*Relative Frequency*
85.5– 95.5	1	.1	1	.1
95.5–105.5	2	.2	0	0
105.5–115.5	2	.2	2	.2
115.5–125.5	3	.3	3	.3
125.5–135.5	1	.1	3	.3
135.5–145.5	1	.1	1	.1
Total	10		10	

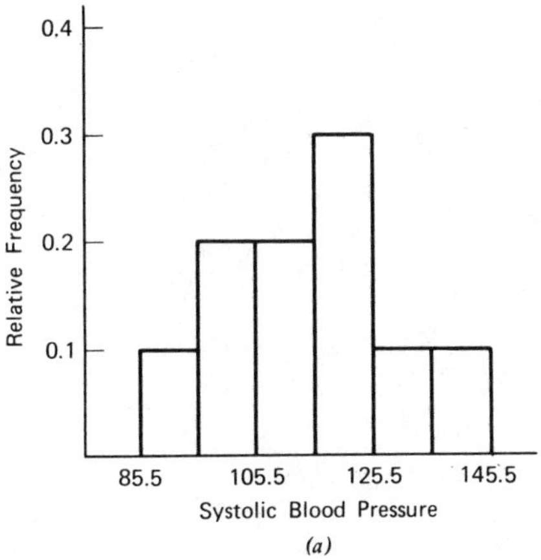

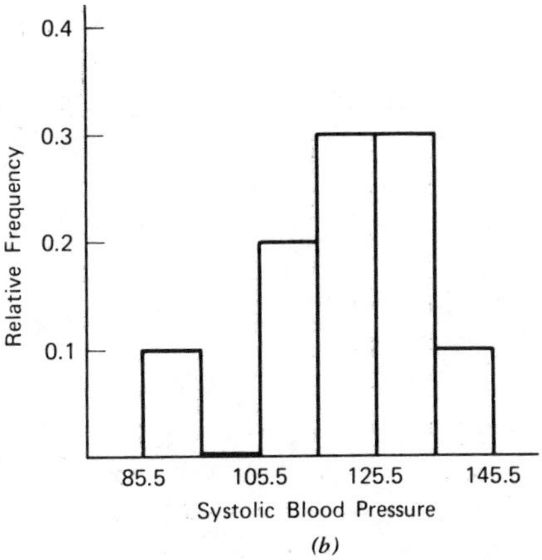

(*A*) HISTOGRAM OF SYSTOLIC BLOOD PRESSURE AFTER THREE MINUTES OF CRYING (*B*) HISTOGRAM OF SYSTOLIC BLOOD PRESSURE AFTER SEVEN MINUTES OF CRYING

EXERCISE 2.7

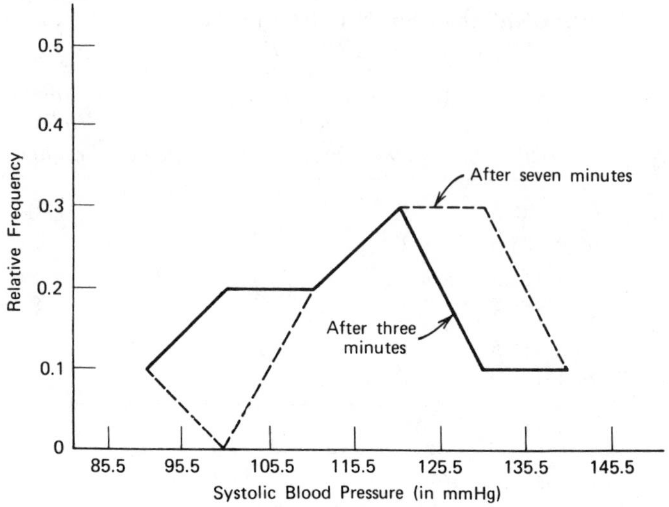

FREQUENCY POLYGON OF SYSTOLIC BLOOD PRESSURE AFTER THREE MINUTES OF CRYING AND AFTER SEVEN MINUTES OF CRYING

EXERCISE 2.8

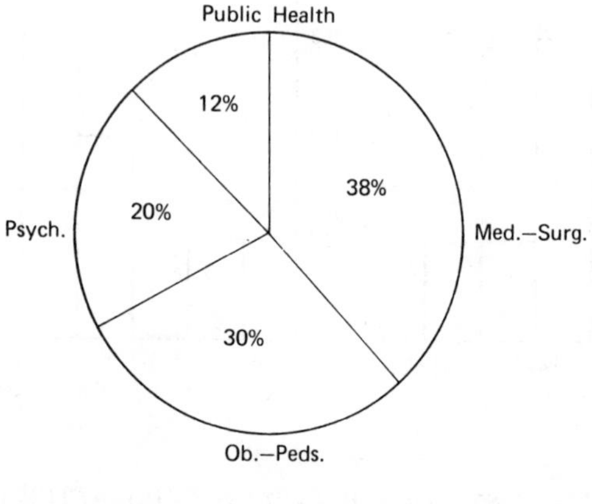

PIE CHART SHOWING PERCENTAGE OF NURSING STUDENTS ACCORDING TO SPECIALTY

EXERCISE 2.9

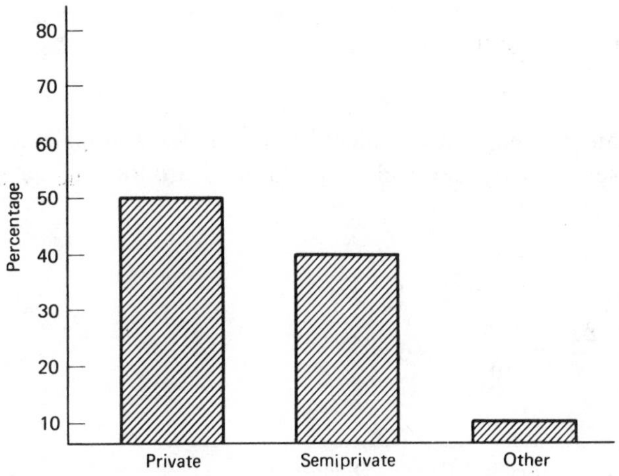

BAR GRAPH OF ROOM PREFERENCE OF 50 PATIENTS

CHAPTER 3

EXERCISE 3.1

The data in this example do not contain extreme values; therefore, the most useful measure of central tendency would probably be the mean. The median may also be reported.

EXERCISE 3.2

Since there is one very small survival time in the data, the mean could possibly give a misleadingly small idea of survival time. In this instance, the median would give a better picture of the "center" of the data.

EXERCISE 3.3

The investigator is interested in the most frequently worn size rather than an average size. The mode in the case would be appropriate.

EXERCISE 3.4

In this study, the measurements are made on an ordinal scale; therefore, the most appropriate measure of central tendency would be the median.

EXERCISE 3.5

The largest score is 5 and the smallest is 1.

$$\text{Range} = 5 - 1 = 4$$

EXERCISE 3.6

Histogram (*b*) has the smallest standard deviation since the data values seem to be more closely clumped around the center point than in (*a*).

EXERCISE 3.7

$$\text{Mean} = \frac{\Sigma Y_i}{n} = \frac{180}{10} = 18.0 = \overline{Y}$$

Patient	*Survival Time* Y_i	*Deviation* $(Y_i - \overline{Y})$	$(Y_i - \overline{Y})^2$
1	24	+6	36
2	8	−10	100
3	12	−6	36
4	3	−15	225
5	20	+2	4
6	18	0	0
7	24	+6	36
8	19	+1	1
9	27	+9	81
10	25	+7	49
Total	180	0	568

$$s = \sqrt{\frac{\Sigma(Y_i - \overline{Y})^2}{n - 1}} = \sqrt{\frac{568}{9}} = \sqrt{63.1} = 7.9$$

EXERCISE 3.8

a. $\text{Mean} = \overline{Y} = \frac{\Sigma Y_i}{6} = \frac{24}{6} = 4$

b. Data arranged in ascending order

2, 2, 3, 4, 5, 8

$$\text{Median} = \frac{3 + 4}{2} = 3.5$$

c. Mode = 2
d. Range = 8 − 2 = 6
e. Deviations

Y_i	$(Y_i - \bar{Y})$	$(Y_i - \bar{Y})^2$	Y_i^2
4	0	0	16
3	−1	1	9
8	+4	16	64
5	+1	1	25
2	−2	4	4
2	−2	4	4
24	0	26	122

$$\text{Variance} = s^2 = \frac{\Sigma(Y_i - \bar{Y})^2}{n - 1} = \frac{26}{5} = 5.2$$

Using computational formula

$$s^2 = \frac{\Sigma Y^2 - \dfrac{(\Sigma Y)^2}{n}}{n - 1} = \frac{122 - \dfrac{(24)^2}{6}}{5} = 5.$$

f. Standard deviation $= s = \sqrt{s^2} = \sqrt{5.2} = 2.28$

EXERCISE 3.9

Method A has much less variability in measurements ana, assuming the mean blood pressure readings for both methods are "true," Method A would be the best.

EXERCISE 3.10

Patient	(Y_i)	(Y_i^2)
1	4	16
2	3	9
3	8	64
4	5	25
5	2	4
6	2	4
	$\Sigma Y_i = 24$	$\Sigma Y_i^2 = 122$

$$s = \sqrt{\frac{\Sigma Y^2 - \frac{(\Sigma Y)^2}{n}}{n-1}}$$

$$= \sqrt{\frac{122 - \frac{(24)^2}{6}}{5}} = \sqrt{\frac{26}{5}} = 2.28$$

For convenience, the subscript i is usually omitted from the summation notation.

EXERCISE 3.11

Data from Exercise 3.1: 86, 70, 75, 82, 68

$$\sum_{i=1}^{5} Y_i = 381 \qquad \sum_{i=1}^{5} Y_i^2 = 29269$$

a. Mean:

$$\bar{Y} = \frac{\sum_{i=1}^{5} Y_i}{5} = \frac{381}{5} = 76.2$$

b. Median:

Ordered array: 68, 70, (75), 82, 86

Median = 75

c. Mode: No mode

d. Range:

$$86 - 68 = 18$$

e. Standard deviation

	Y_i	$(Y_i - \overline{Y})$	$(Y_i - \overline{Y})^2$	Y_i^2
	86	9.8	96.04	7396
	70	−6.2	38.44	4900
	75	−1.2	1.44	5625
	82	5.8	33.64	6724
	68	−8.2	67.24	4624
Totals:	381	0	236.8	29269
	ΣY_i	$\Sigma(Y_i - \overline{Y})$	$\Sigma(Y_i - \overline{Y})^2$	ΣY_i^2

Theoretical formula:

$$s = \sqrt{\frac{\Sigma(Y_i - \overline{Y})^2}{n - 1}} = \sqrt{\frac{236.8}{4}} = 7.69$$

Computational formula:

$$s = \sqrt{\frac{\Sigma Y_i^2 - \dfrac{(\Sigma Y_i)^2}{n}}{n - 1}} = \sqrt{\frac{29269 - \dfrac{(381)^2}{5}}{4}}$$

$$= \sqrt{\frac{236.8}{4}} = 7.69$$

EXERCISE 3.12

Data from Exercise 3.2: 24, 8, 12, 3, 20, 18, 24, 19, 27, 25

$$\Sigma Y_i = 180 \qquad \Sigma Y_i^2 = 3808$$

a. Mean:

$$\overline{Y} = \frac{\Sigma Y_i}{10} = \frac{180}{10} = 18$$

b. Median: 19.5

Ordered array 3, 8, 12, 18, 19|20, 24, 24, 25, 27

c. Mode: 24

d. Range:

$$27 - 3 = 24$$

e. Standard deviation:

Y_i	$(Y_i - \overline{Y})$	$(Y_i - \overline{Y})^2$	Y_i^2
24	6	36	576
8	−10	100	64
12	−6	36	144
3	−15	225	9
20	2	4	400
18	0	0	324
24	6	36	576
19	1	1	361
27	9	81	729
25	7	49	625
180	0	568	3808
ΣY_i	$\Sigma(Y_i - \overline{Y})$	$\Sigma(Y_i - \overline{Y})^2$	ΣY_i^2

Theoretical formula:

$$s = \sqrt{\frac{\Sigma(Y_i - \overline{Y})^2}{n - 1}} = \sqrt{\frac{568}{9}} = 7.94$$

Computational formula:

$$s = \sqrt{\frac{\Sigma Y_i^2 - \frac{(\Sigma Y_i)^2}{n}}{n - 1}} = \sqrt{\frac{3808 - \frac{(180)^2}{10}}{9}}$$

$$= \sqrt{\frac{568}{9}} = 7.94$$

CHAPTER 4

EXERCISE 4.1

The *parameter* of interest is the mean score μ on the anxiety exam for all mothers visiting sick babies. The statistic is $\overline{Y}$, the mean score for the 10 mothers in the sample.

EXERCISE 4.2

The *parameter* is μ, the mean change in blood pressure following a 10-minute back rub for all geriatric patients treated under similar conditions. The *statistic* is $\overline{Y}$, the mean change in blood pressure following back rub for the 30 patients in the sample.

EXERCISE 4.3

Sixty-eight percent of the scores will lie within one standard deviation in either direction of the mean. One standard deviation in this example is 10 units; therefore, 68% of the scores would be between $50 - 10 = 40$ and $50 + 10 = 60$. Ninety-five percent of the scores will fall within 1.96 standard deviations of the mean, i.e.,

$$50 - 1.96(10) = 30.4$$

$$50 + 1.96(10) = 69.6$$

Ninety-nine percent of the scores will lie within 2.58 standard deviations of the mean, i.e.,

$$50 - 2.58(10) = 24.2$$

$$50 + 2.58(10) = 75.8$$

EXERCISE 4.4

The target population is all infants in South Carolina. The population sampled is infants in a particular county in South Carolina.

EXERCISE 4.5

The female weighed 143 pounds. In units of standard deviation, this is:

$$\frac{143 - 130}{5} = \frac{13}{5} = 2.6$$

She is 2.6 standard deviation units above the mean optimum weight and would therefore be classified as obese.

EXERCISE 4.6

The mean optimum weight is 130 pounds, with a standard deviation (σ) of 5 pounds.

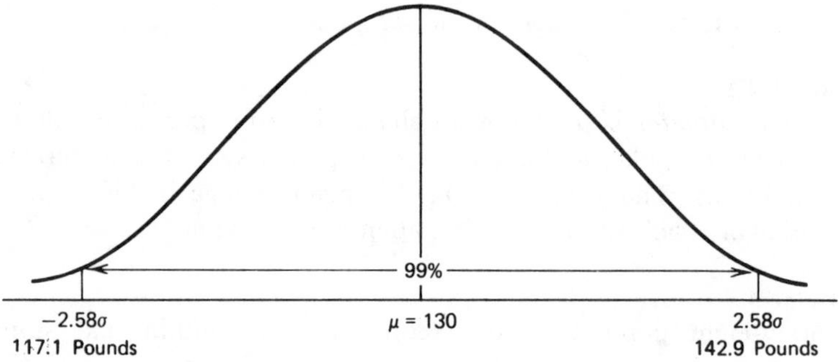

Ninty-nine percent of all weights will lie within 2.58 standard deviations of the mean μ. Since 1 standard deviation is equal to 5 pounds ($\sigma = 5$), 2.58 standard deviations would be equal to:

$$(2.58)5 = 12.9 \text{ pounds:}$$

Therefore, 99% of the weights would be between

$$130 - (2.58)(5) = 117.1 \text{ pounds}$$

and

$$130 + (2.58)(5) = 142.9 \text{ pounds}$$

EXERCISE 4.7

To classify student performance on the exam, we must first convert the students' scores into units of standard deviation. This is done by finding the difference between the raw score and the mean score and dividing by standard deviation. Thus,

$$z = \frac{Y - \mu}{\sigma}$$

where Y = individual score on exam, μ = mean score on exam = 70, and σ = standard deviation of exam scores = 10.

a. $$z = \frac{Y - \mu}{\sigma} = \frac{95 - 70}{10} = 2.5$$

A score of 95 is 2.5 standard deviations *above* the class mean. Classification: Excellent.

b. $z = \frac{Y - \mu}{\sigma} = \frac{65 - 70}{10} = -0.5$

A score of 65 is 0.5 standard deviations *below* the class mean. Classification: Average.

c. $z = \frac{Y - \mu}{\sigma} = \frac{72 - 70}{10} = 0.2$

A score of 72 is 0.2 standard deviations *above* the class mean. Classification: Average.

d. $z = \frac{Y - \mu}{\sigma} = \frac{52 - 70}{10} = -1.8$

A score of 52 is 1.8 standard deviations *below* the class mean. Classification: Poor.

e. $z = \frac{Y - \mu}{\sigma} = \frac{82 - 70}{10} = 1.2$

A score of 82 is 1.2 standard deviations *above* the class mean. Classification: Good.

EXERCISE 4.8

The mean score on the exam is 70 ($\mu = 70$) and the standard deviation of the scores is 10 ($\sigma = 10$).

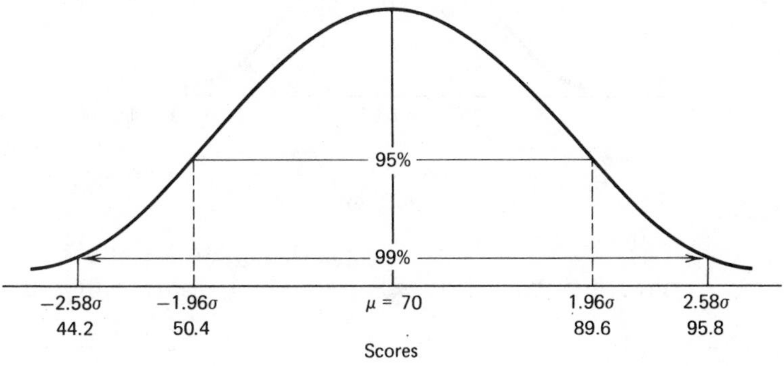

Ninety-five percent of all scores on this exam will fall within 1.96 standard deviations on either side of the mean. One standard deviation is equal to 10 ($\sigma = 10$); 1.96 standard deviations would be equal to:

$$1.96(10) = 19.6$$

Ninety-five percent of all scores on the exam will lie between:

$$70 - (1.96)(10) = 50.4$$

and

$$70 + (1.96)(10) = 89.6$$

Similarly, 99% of all scores will lie between:

$$70 - (2.58)(10) = 44.2$$

and

$$70 + (2.58)(10) = 95.8$$

EXERCISE 4.9

a.

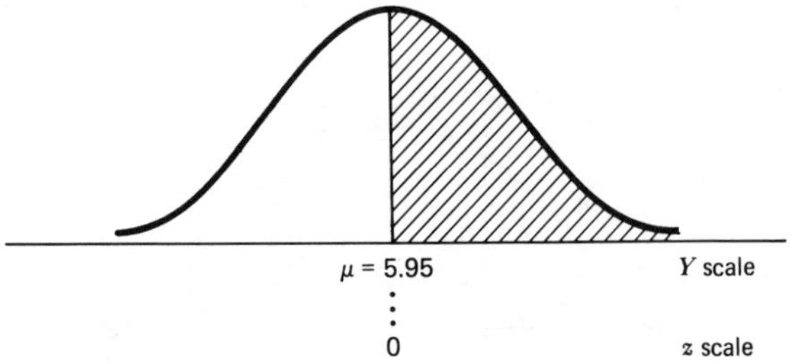

$$\Pr(Y \geq 5.95) = \Pr\left(z \geq \frac{5.95 - 5.95}{2}\right) = \Pr(z \geq 0) = 0.5$$

b.

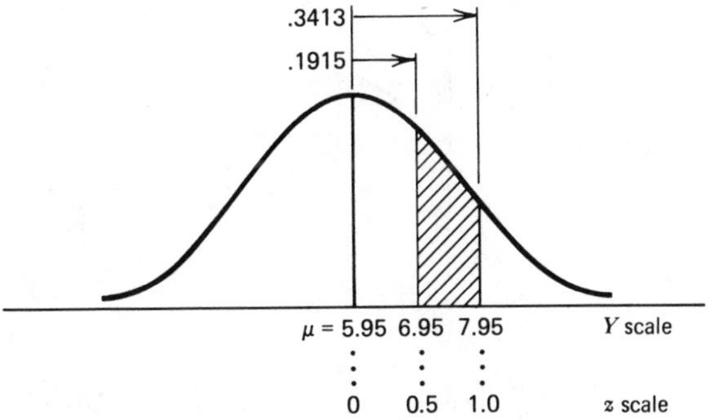

$$z_1 = \frac{Y_1 - \mu}{\sigma} = \frac{6.95 - 5.95}{2} = 0.5$$

$$z_2 = \frac{Y_2 - \mu}{\sigma} = \frac{7.95 - 5.95}{2} = 1.0$$

$$\begin{aligned} \Pr(6.95 \le Y \le 7.95) &= \Pr(0.5 \le z \le 1.0) \\ &= .3413 - .1915 \\ &= .1498 \end{aligned}$$

EXERCISE 4.9

c.

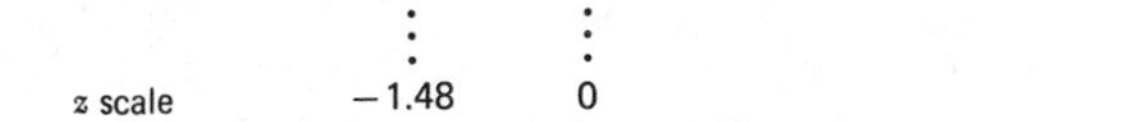

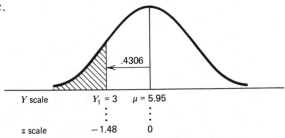

$$z_1 = \frac{Y_1 - \mu}{\sigma} = \frac{3 - 5.95}{2} = -1.475 \quad (-1.48)$$

$$\begin{aligned} \Pr(Y_1 \le 3) = \Pr(z \le -1.48) &= 0.5 - .4306 \\ &= .0694 \end{aligned}$$

d.

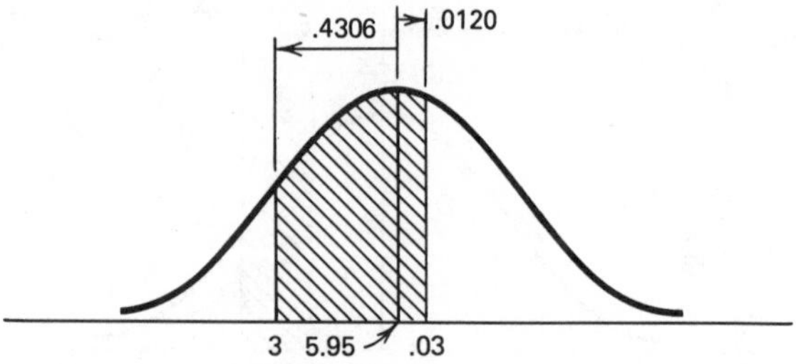

$$z_1 = \frac{Y_1 - \mu}{\sigma} = \frac{3 - 5.95}{2} = -1.48$$

$$z_2 = \frac{Y_2 - \mu}{\sigma} = \frac{6 - 5.95}{2} = .025 \quad (.03)$$

$$\begin{aligned} \Pr(3 \le Y \le 6) &= \Pr(-1.48 \le z \le .03) \\ &= .4306 + .0120 \\ &= .4426 \end{aligned}$$

EXERCISE 4.10

a. $P(Y \le 150) = P(Z \le 1.5) = 0.5 + 0.4332 = 0.9332$

$P(Y \le 110) = P(Z \le -0.5) = 0.5 - 0.1915 = 0.3085$

b. $Z = \dfrac{Y - \mu}{\sigma} = \dfrac{160 - 120}{20} = 2$

c. $Z = \dfrac{Y - \mu}{\sigma}$

$$1.96 = \frac{Y - 120}{20} \qquad -1.96 = \frac{Y - 120}{20}$$

$$Y = 159.2 \qquad Y = 80.8$$

$$P(80.8 \le Y \le 159.2) = .95$$

90% of the observations lie within ±1.65 standard deviations.

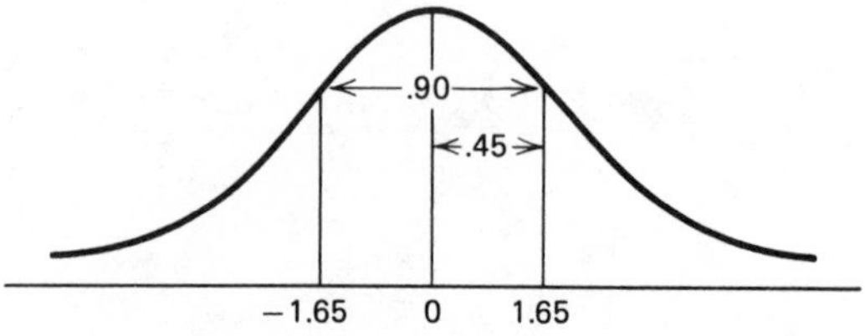

$$+1.65 = \frac{Y - 120}{20} \qquad -1.65 = \frac{Y - 120}{20}$$

$$Y = 153 \qquad\qquad Y = 87$$

$$P(87 \leq Y \leq 153) = .90$$

d. $P(100 \leq Y \leq 140) = P(-1 \leq Z \leq +1) = .3413 + .3413 = .6826 = 68.3\%$

e. $P(60 \leq Y \leq 180) = P(-3 \leq Z \leq +3) = .4987 + .4987 = .9974 = 99.7\%$

$<1\%$ of the observations lie outside the interval.

EXERCISE 4.11

The mean of the population of mean values is equal to 120 (mmHg) and the standard deviation is

$$\frac{\sigma}{\sqrt{n}} = \frac{20}{\sqrt{100}} = 2$$

a. $P(\overline{Y} \geq 126) = P\left[Z \geq \frac{126 - 120}{2}\right]$

$$= P(Z \geq 3.0) = .5 - .49865 = .001$$

b. $Z = \frac{\overline{Y} - \mu}{\sigma/\sqrt{n}} = \frac{126 - 120}{2} = 3.0$

c. 95% of all sample means fall within ± 1.96 standard deviations of μ.

$$Z = \frac{\overline{Y} - \mu}{\sigma/\sqrt{n}} \qquad -1.96 = \frac{\overline{Y} - 120}{2}$$

$$\overline{Y} = 116.08$$

$$1.96 = \frac{\overline{Y} - 120}{2}$$

$$\overline{Y} = 123.92$$

$$P(116.08 \leq \overline{Y} \leq 123.92) = .95$$

d. The Z value above which 10% of the means will fall is $+1.28$.

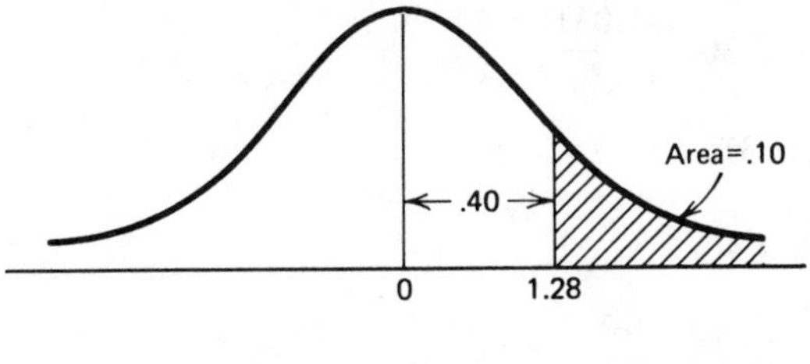

$$Z = \frac{\overline{Y} - \mu}{\sigma/\sqrt{n}}$$

10% of the mean values will lie above $\overline{Y} = 122.56$.

$$1.28 = \frac{\overline{Y} - 120}{2}$$

$$\overline{Y} = 122.56$$

e. $$P(\overline{Y} \leq 115) = P\left(Z \leq \frac{115 - 120}{2}\right)$$

$$= P(Z \leq -2.5) = .5 - .4938 = .0062$$

CHAPTER 5

EXERCISE 5.1

Given: $n = 16$

$\overline{Y} = 150$

$\sigma = 20$

$\overline{Y} \pm 1.96\sigma/\sqrt{n}$ Use z table because σ known

$$150 \pm 1.96 \frac{20}{\sqrt{16}}$$

150 ± 9.8

(140.2, 159.8)

Interpretation: We are 95% confident that the interval 140.2 to 159.8 mmHg contains the true mean systolic blood pressure for all senior nursing students.

EXERCISE 5.2

Given: $n = 9$

$\overline{Y} = 6.5$

$s = 1.031$

$\overline{Y} \pm t(s/\sqrt{n})$

$$6.5 \pm 2.306 \left(\frac{1.031}{\sqrt{9}}\right)$$

(5.7, 7.3)

Interpretation: We are 95% confident that the interval 5.7 to 7.3 pounds contains the true mean birth weight of infants for the population of mothers on the special diet.

EXERCISE 5.3

Sample values:

Premature	*Full-Term*
$n_1 = 5$	$n_2 = 5$
$\overline{Y}_1 = 2.0$	$\overline{Y}_2 = 4.0$

$$\sigma_1 = \sigma_2 = 1.0$$

$$\sigma\sqrt{\frac{1}{n_1} + \frac{1}{n_2}} = 1.0\sqrt{\frac{1}{5} + \frac{1}{5}} = .632$$

$$(\overline{Y}_1 - \overline{Y}_2) \pm Z\left(\sigma\sqrt{\frac{1}{n_1} + \frac{1}{n_2}}\right)$$

$$(2.0 - 4.0) \pm 2.58(.632)$$

$$(-.37, -3.63)$$

Interpretation: We are 99% confident that the interval −.37 to −3.63 mg/100 cc contains the true difference in mean serum indirect bilirubin levels for premature and full-term infants for the population studied.

EXERCISE 5.4

Sample values:

Exposed	*Not Exposed*
$n_1 = 100$	$n_2 = 100$
$\overline{Y}_1 = 145$	$\overline{Y}_2 = 120$
$s_1 = 20$	$s_2 = 15$

The pooled standard deviation is:

$$s_p = \sqrt{\frac{s_1^2(n_1 - 1) + s_2^2(n_2 - 1)}{n_1 + n_2 - 2}} = \sqrt{\frac{(20)^2(99) + (15)^2(99)}{198}}$$

$$= 17.7$$

$$(\overline{Y}_1 - \overline{Y}_2) \pm (1.6525)(17.7)\sqrt{\frac{1}{100} + \frac{1}{100}}$$

25 ± 4.136

$(20.9, 29.1)$

Note: We use $df = 200$, since this is the closest we have to $df = 198$ in the table.

Interpretation: We are 90% confident that the interval 20.9 to 29.1 mmHg contains the true difference in mean systolic blood pressure for the exposed and not exposed groups.

EXERCISE 5.6

Sample values:

$n = 9$

$\bar{d} = 2.111$

$s_d = 4.197 \qquad s_d/\sqrt{n} = 1.399$

$\bar{d} \pm t(s_d/\sqrt{n})$

$2.111 \pm 2.306(1.399)$

$(-1.11, 5.34)$

Interpretation: We are 95% confident that the interval −1.11 to 5.34 pounds contains the true mean difference in weight gain before and at the end of the three-month period for the population studied.

EXERCISE 5.7

Sample values:

$n = 200$

$\hat{p} = \frac{90}{200} = 0.45 \qquad$ proportion responding favorably

$$\hat{p} \pm 1.96\sqrt{\frac{\hat{p}(1-\hat{p})}{n}}$$

$$.45 \pm 1.96\sqrt{\frac{(.45)(.55)}{200}}$$

$(.38, .52)$

Interpretation: We are 95% confident that the interval 38% to 52%

contains the true proportion of favorable responders in the community.

EXERCISE 5.8

Sample values:

Experimental	*Control*
$n_1 = 100$	$n_2 = 90$
$\hat{p}_1 = \dfrac{75}{100} = .75$	$\hat{p}_2 = \dfrac{50}{90} = .56$

$$(\hat{p}_1 - \hat{p}_2) \pm 1.96\sqrt{\frac{\hat{p}_1(1 - \hat{p}_1)}{n_1} + \frac{\hat{p}_2(1 - \hat{p}_2)}{n_2}}$$

$$(.75 - .56) \pm 1.96\sqrt{\frac{(.75)(.25)}{100} + \frac{(.56)(.44)}{90}}$$

$$(.06, .32)$$

Interpretation: We are 95% confident that the interval 6% to 32% contains the true difference in proportion of satisfactory responses for the experimental and control groups.

EXERCISE 5.9

Sample values:

Expelled	*Retained*
$n_E = 10$	$n_R = 15$
$\overline{Y}_E = 28$	$\overline{Y}_R = 33$
$s_E = 10$	$s_R = 12$

$$s_p = \sqrt{\frac{s_E^2(n_E - 1) + s_R^2(n_R - 1)}{n_E + n_R - 2}} = \sqrt{\frac{(10)^2(9) + (12)^2(14)}{10 + 15 - 2}}$$

$$= 11.3$$

$$(\overline{Y}_E - \overline{Y}_R) \pm t\left(s_p\sqrt{\frac{1}{n_R} + \frac{1}{n_E}}\right)$$

$$(28 - 33) \pm (2.0687)(11.3)\sqrt{\frac{1}{15} + \frac{1}{10}}$$

$$(-14.5, 4.5)$$

Interpretation: We are 95% confident that the interval −14.5 to 4.5 years contains the true difference in mean ages for the group who expelled IUD's and the group who retained IUD's.

EXERCISE 5.10

Sample values:

$$\bar{d} = \frac{48}{9} = 5.33$$

$$n = 9$$

$$s_d = \sqrt{\frac{734 - \frac{(48)^2}{9}}{8}} = 7.73$$

$$\bar{d} \pm t(s_d/\sqrt{n})$$

$$5.33 \pm 3.3554\left(\frac{7.73}{\sqrt{9}}\right)$$

$$(-3.3,\ 14.0)$$

Interpretation: We are 99% confident that the interval −3.3 to 14.0 contains the true mean difference in pre- and posttest scores for the population of nursing students studied.

EXERCISE 5.11

Sample values:

$$n = 100$$

$$\bar{Y} = 98.02$$

$$s = 0.9$$

$$\bar{Y} \pm t(s/\sqrt{n})$$

$$98.02 \pm 1.984\,\frac{0.9}{\sqrt{100}}$$

$$(97.8^\circ,\quad 98.2^\circ)$$

Use $df = 100$, since this is the closest table value to $df = n - 1 = 99$. *Interpretation:* We are 95% confident that the interval 97.8°F to 98.2°F covers the true mean rectal temperature of the population of infants studied.

EXERCISE 5.12

Sample values:

Hospital 1	*Hospital 2*
$n_1 = 15$	$n_2 = 15$
$\overline{Y}_1 = 85$	$\overline{Y}_2 = 70$
$s_1 = 10$	$s_2 = 10$

$$s_p = \sqrt{\frac{(10)^2(14) + (10)^2(14)}{15 + 15 - 2}} = 10$$

$$(\overline{Y}_1 - \overline{Y}_2) \pm t\left(s_p\sqrt{\frac{1}{n_1} + \frac{1}{n_2}}\right)$$

$$(85 - 70) \pm (2.7633)(10)\sqrt{\frac{1}{15} + \frac{1}{15}}$$

$$(4.9, 25.1)$$

Interpretation: We are 99% confident that the interval 4.9 to 25.1 covers the true difference in mean audit performance scores for the two hospitals studied.

EXERCISE 5.13

Sample values:

Freshmen	*Seniors*
$n_1 = 20$	$n_2 = 20$
$\hat{p}_1 = \frac{12}{20} = 0.6$	$\hat{p}_2 = \frac{8}{20} = 0.4$

a. 95% confidence interval on true proportion of smokers in freshman class:

$$\hat{p}_1 \pm 1.96\sqrt{\frac{\hat{p}_1(1 - \hat{p}_1)}{n_1}}$$

$$0.6 \pm 1.96\sqrt{\frac{(.6)(.4)}{20}}$$

$$(.39, .81)$$

Interpretation: We are 95% confident that the interval 39% to 81% contains the true proportion of smokers in the freshman class at the college studied.

b. 95% confidence interval on true proportion of smokers in senior class:

$$\hat{p}_2 \pm 1.96\sqrt{\frac{\hat{p}_2(1-\hat{p}_2)}{n_2}}$$

$$0.4 \pm 1.96\sqrt{\frac{(.4)(.6)}{20}}$$

$$(.19, .61)$$

Interpretation: We are 95% confident that the interval 19% to 61% contains the true proportion of smokers in the senior class at the college studied.

c. 95% confidence interval on true difference in proportion of smokers in freshman and senior classes:

$$(\hat{p}_1 - \hat{p}_2) \pm 1.96\sqrt{\frac{\hat{p}_1(1-\hat{p}_1)}{n_1} + \frac{\hat{p}_2(1-\hat{p}_2)}{n_2}}$$

$$(.6 - .4) \pm 1.96\sqrt{\frac{(.6)(.4)}{20} + \frac{(.4)(.6)}{20}}$$

$$(-.10, .50)$$

Interpretation: We are 95% confident that the interval −.10 to .50 contains the true difference in proportion of smokers in the freshman and senior classes at the college studied.

CHAPTER 6

EXERCISE 6.1

The statistic $\overline{Y}$, the mean blood pressure of the 30 patients, is an *estimate* of the true mean blood pressure of all geriatric patients. How good an estimate $\overline{Y}$ is depends on the representativeness of the sample. The statistic $\overline{Y}$ is rarely, if ever, equal to the unknown mean μ. A new sample of 30 patients may yield an entirely different $\overline{Y}$. All we know is that the values of the $\overline{Y}$'s cluster around the true μ. The failure of a $\overline{Y}$ to be exactly equal to μ is called *sampling error*.

EXERCISE 6.2

a. The null hypothesis in this example would be that pulse rate is unaffected by the presence of a nurse with a hypodermic syringe, i.e., pulse rate is equal to the normal value. The alternate hypothesis is that mean pulse rate of male patients who are approached by a nurse carrying a hypodermic syringe is higher than the norm, 72 beats/min.

b. The test statistic is a measure of the discrepancy between the hypothesized pulse rate ($\mu = 72$) and the mean pulse rate determined from the sample ($\overline{Y} = 90$). The null hypothesis is rejected when the discrepancy between $\overline{Y}$ and μ is "large."

c. The specified significance level determines what values will be considered "large." These large values constitute the rejection region for the test.

d. To reject the null hypothesis at the $\alpha = .01$ level of significance states that the investigator is willing to risk a 1% chance of rejecting H_0 when it is true.

EXERCISE 6.3

a. The discrepancy between the normal mean urine chloride level and that observed for the sample of premature infants could be explained by:

   (1) The true mean urine chloride level for premature infants is the same as the true mean urine chloride level for full-term infants, but due to random chance (sampling error), the mean of the sample of premature infants is less than the true mean for full-term infants.

   (2) The mean of the sample of premature infants is less than the true mean for full-term infants because the true mean urine chloride level for premature infants is *not* equal to the true mean urine chloride level for full-term infants.

b. The null hypothesis is that there is *no difference* in mean urine chloride levels for premature and for full-term infants. That is, we are hypothesizing that the true mean urine chloride level for premature infants is equal to the true mean for full-term infants ($\mu = 210$ mEq/24 hr). The alternate hypothesis is that mean urine chloride levels for premature infants is *lower* than for full-term infants.

c. The test statistic provides a measure of the discrepancy between what we hypothesize as the true mean urine chloride level for premature infants ($\mu = 210$) and what we actually observe to be the mean urine chloride level for the *sample* of 25 premature infants ($\overline{Y} = 170$).

The test of significance decides which of the alternatives in *a* above is the "most likely" explanation of the difference between $\overline{Y}$, the sample mean, and μ, the hypothesized population mean. "Large" values of the test statistic suggest that *a2* is the explanation, while "small" values of the test statistic suggest that *a1* is the reason for the difference between $\overline{Y}$ and μ. The rejection region for the test of significance specifies what values we will consider to be large enough to warrant rejection of the null hypothesis. The level of significance chosen by the investigator specifies the cutoff points for the rejection region. A test reported as significant at the $\alpha = .01$ level of significance means that the test statistic fell in the rejection region, the null hypothesis was rejected, and in rejecting the null hypothesis the investigator is willing to accept a 1% chance of rejecting H_0 when it is true.

CHAPTER 7

EXERCISE 7.1

Branch IIC: Pearson's Product Moment Correlation Coefficient accompanied by a test of its statistical significance.

EXERCISE 7.2

Branch IC1: Single sample t test.

EXERCISE 7.3

Branch IA1: Single sample X^2 test.

EXERCISE 7.4

Branch IC2a: Pooled t test.

EXERCISE 7.5

Branch IC2b: Paired t test.

EXERCISE 7.6

Branch IC3a: ANOVA–CRD.

EXERCISE 7.7

Branch IB2a: Rank sum test or median test.

EXERCISE 7.8

Branch IA1: Single sample X^2 test.

EXERCISE 7.9

Branch IIA: X^2 analysis.

EXERCISE 7.10

Branch IIC: Pearson's Product Moment Correlation Coefficient along with test of its statistical significance. (*Note:* Some may argue that the scores are *ordinal.* If so, Branch IIB should be followed.)

EXERCISE 7.11

Branch IIB: Spearman's Rank Order Correlation Coefficient along with test of statistical significance.

CHAPTER 8

EXERCISE 8.1

1. *Data*. The data consist of glucose-6-dehydrogenase measurements made on nine adult male patients admitted to an alcohol treatment center. The measurements are continuous in nature and possess at least an interval scale.
2. *Assumptions*. It is assumed that the underlying population of measurements from which the nine patients were randomly selected has a normal distribution.
3. *Hypotheses*. The question of interest is whether the production of glucose-6-dehydrogenase in the blood is different for alcoholics. Stated in null form, the hypothesis is that the mean level of glucose-6-dehydrogenase for alcoholics is the same as for normal adults, i.e.,

   $$H_0: \quad \mu = 375$$

   The researcher has hypothesized that production of glucose-6-dehydrogenase in the blood is different for alcoholics and normal (nondiseased) adults. Hence, the alternate hypothesis of interest is

   $$H_A: \quad \mu \neq 375$$

4. *Test statistic*. The mean glucose-6-dehydrogenase level for the nine alcoholic patients was found to be 450 units/10^9 cells with a standard deviation s of 60 units/10^9 cells, i.e.,

   $$\overline{Y} = 450 \qquad s = 60$$

The test statistic is given by:

$$t = \frac{\bar{Y} - \mu_0}{s/\sqrt{n}}$$

$$= \frac{450 - 375}{60/\sqrt{9}} = \frac{75}{20}$$

$$= 3.75$$

5. *Decision rule.* Since the investigator is interested in testing whether or not a difference (in either direction) exists between normals and alcoholics, the test is a *two-tailed* test of significance. The cutoff point is located in Appendix Table A at the intersection of the column $\alpha/2 = .025$ and $df = n - 1 = 8$. This value is 2.306. The rejection region consists of values of the test statistic under the shaded areas below.

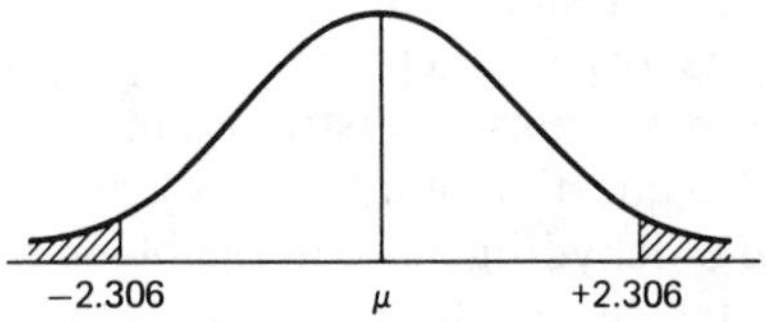

6. *Conclusion.* The test statistic (Step 4) falls in the rejection region. It may be concluded that the mean glucose-6-dehydrogenase level for alcoholics is significantly different from the normal value (at the $\alpha = 0.05$ level of significance).

EXERCISE 8.2

1. *Data.* The data, which consist of the nine temperature readings, are continuous in nature and possess an underlying interval scale.
2. *Assumptions.* It is assumed that the nine measurements are a random sample from a population of measurements that is normally distributed.
3. *Hypotheses.* The null hypothesis is that there is no difference in temperature for infants who are fondled and the normal body temperature of infants.

H_0: $\mu = 99.6$

H_A: $\mu \neq 99.6$

4. *Test statistic*. The mean temperature for the nine infants is:

$$\overline{Y} = \frac{\Sigma Y}{n} = \frac{903.6}{9} = 100.4$$

The standard deviation s is:

$$s = \sqrt{\frac{\Sigma Y^2 - \frac{(\Sigma Y)^2}{n}}{n - 1}} = \sqrt{\frac{90725.06 - \frac{(903.6)^2}{9}}{8}}$$

$$= .67$$

The test statistic is:

$$t = \frac{\overline{Y} - \mu_0}{s/\sqrt{n}} = \frac{100.4 - 99.6}{.67/\sqrt{9}} = \frac{0.8}{.22} = 3.6$$

5. *Decision rule*. The cutoff point is found in Appendix Table A at the intersection of row $df = n - 1 = 8$ and column $\alpha/2 = .025$. The value is 2.306 and the rejection region is under the shaded areas below.

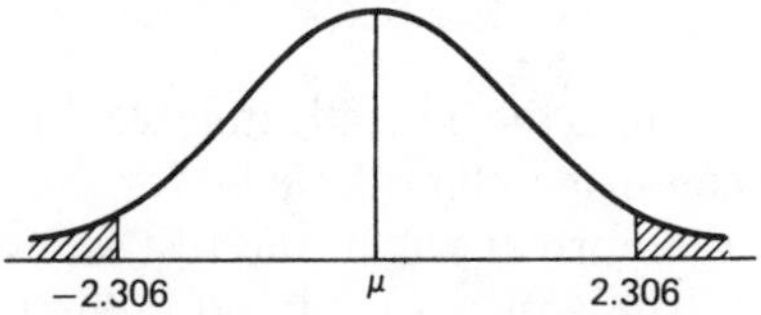

6. *Conclusion*. Since the value of the test statistic falls in the rejection region, the null hypothesis may be rejected at the $\alpha = .05$ level of significance. It may be concluded that the body temperature of infants who are fondled is different from the normal infant body temperature. There is a 5% chance of rejecting H_0 when it is true.

EXERCISE 8.3

1. *Data*. The data consist of urine chloride measurements made on 9 premature infants. The measurements possess at

least an interval scale and are continuous in nature.

2. *Assumptions*. It is assumed that the nine measurements are randomly selected from a population of measurements whose distribution is normal.
3. *Hypotheses*. The null hypothesis is that the mean urine chloride level for premature infants is the same as the normal value.

$$H_0: \quad \mu = 210$$

$$H_A: \quad \mu < 210$$

The alternative hypothesis is that the mean is lower for premature infants.

4. *Test statistic*. The mean urine chloride level calculated from the sample of nine premature infants is 205 mEq/24hr with a standard deviation s of 12 mEq/24hr.

$$\bar{Y} = 205 \qquad s = 12$$

The test statistic is:

$$t = \frac{\bar{Y} - \mu_0}{s/\sqrt{n}}$$

$$= \frac{205 - 210}{12/\sqrt{9}} = -1.25$$

5. *Decision rule*. Since the investigator would like to determine if the mean urine chloride level for premature infants is lower than the normal value, the test is a one-tailed test of significance. The entire α is placed in the lower tail. The cutoff value, found in Appendix Table A at the intersection of the column $\alpha = .01$ and $df = n - 1 = 8$, is 2.896.

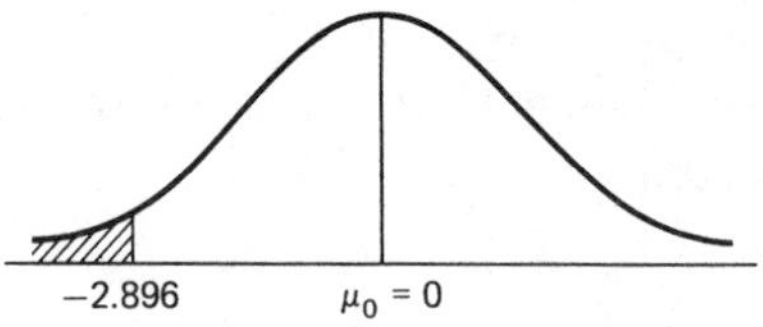

6. *Conclusion*. The test statistic does not fall in the rejection

region. It must, therefore, be concluded that there is not enough evidence to show that mean urine chloride level of premature infants is lower than the normal value.

EXERCISE 8.4

The investigator would like to show that counseling patients for 20 minutes would result in a lowering of their blood pressure. The null hypothesis is that there is no difference in mean systolic pressure for the counseled group and the known mean systolic blood pressure for noncounseled presurgery patients. In other words, the null hypothesis states that mean systolic blood pressure for the counseled group is equal to the known mean systolic pressure for the presurgery patients who receive no counseling. Symbolically,

$$H_0: \quad \mu = 150$$

The alternate hypothesis is

$$H_A: \quad \mu < 150$$

A t-test was performed and the null hypothesis was not rejected. However, from this it *cannot* be concluded that counseling does not lower blood pressure. All that may be concluded when H_0 is not rejected is that there is not enough evidence to show that counseling results in a lowering of blood pressure. The investigator's statement implies "acceptance" of the null hypothesis as absolutely true. This is not correct since failure to reject H_0 may not be taken as "proof" that H_0 is true.

EXERCISE 8.5

a. The curriculum committee wish to determine whether the students who participate in the self-instructional experience will score significantly higher than the national average on the standardized proficiency exam. In null form, the hypothesis to be tested is that the mean score on the exam for the self-instruction group is equal to the national average. The alternative to the null hypothesis is that these students score higher than the national average. Symbolically,

$$H_0: \quad \mu_{SI} = 70$$

$$H_A: \quad \mu_{SI} > 70$$

The sample consists of students at a particular college of nursing who participate in the self-instructional class. These students may be selected at random from among their classmates or the entire class may be asked to participate. At the end of the 4-week period, the students are given the standardized exam and their mean exam score $\overline{Y}$ is calculated. The mean score of the sample $\overline{Y}$ is compared to the hypothesized mean $\mu = 70$. If the discrepancy between $\overline{Y}$ and μ is large, i.e., the test statistic is large, then the null hypothesis is rejected. (The rejection region, which specifies how large the test statistic must be in order to reject H_0, depends upon the desired level of significance α.)

b. To say that the students scored significantly higher than the national average at $\alpha = .20$ means that the null hypothesis has been rejected in favor of the alternative hypothesis. With the given level of significance, there is a 20% chance that the investigator is incorrect in saying that the self-instructional group performs better than the national average. Most investigators are unwilling to accept such a large probability for error when instituting a new procedure.

EXERCISE 8.6

a. If the committee is interested in assessing the performance of the self-instructional group as compared with the performance of students taught using traditional methods, it may select a group of students to participate in the treatment (self-instruction) group and another group of students for the control (traditional method) group. For a pooled t-test, it is assumed that the students would be randomly assigned to either the treatment or the control group. At the end of the four-week period both groups would be given the standardized proficiency exam. The null hypothesis would be that there is no difference in performance on the exam for the two groups, i.e., the mean exam score for the treatment group is equal to the mean exam score for the control group. Symbolically,

$$H_0: \quad \mu_T = \mu_C$$

or

$$\mu_T - \mu_C = 0$$

$$H_A: \quad \mu_T > \mu_C \quad \text{or} \quad \mu_T - \mu_C > 0$$

b. The use of a paired t-test would require that students in the two groups be matched on the basis of characteristics such as age, IQ, previous performance, sex, etc. That is, Student No. 1 in the self-instruction group would have a matched counterpart in the traditional group. Since the participants have been matched, the differences in individual scores between the pairs may be obtained. The null hypothesis is that there is no difference in scores for the two groups, i.e., the true mean of the differences is zero.

$$H_0: \quad \mu_T - \mu_C = 0$$

or

$$H_0: \quad D = 0$$

$$H_A: \quad D > 0$$

The alternative hypothesis is that the scores for the treatment group are greater than the scores for the control group.

EXERCISE 8.7

a. The investigator is interested in determining if prolonged crying affects blood pressure. Specifically, it is of interest to show that prolonged periods of crying result in elevated blood pressure. The null hypothesis for this study would be that there is no difference in blood pressure following three minutes of crying and after seven minutes of crying. The alternate hypothesis is that blood pressure after 7 minutes of crying is higher than before the crying. Since this study is a "before and after" design, that is, the blood pressure for each infant was measured at the end of three minutes and again after seven minutes of crying, the paired t-test would be the appropriate method of analysis.

b. The statement that there is a significant increase in blood pressure after seven minutes of crying at $\alpha = .05$ says that the null hypothesis of no difference in blood pressure readings has been rejected at the .05 level of significance. The test was a one-tailed test since the investigator was interested in the alternative hypothesis that the mean blood pressure after seven minutes is *greater than* the mean blood pressure after three minutes of crying. To reject the null hypothesis in favor of the above alternative at the .05

level of significance means that the investigator risks a 5% chance of error in making this statement.

EXERCISE 8.8

a. The investigator is interested in determining if counseling for presurgery patients significantly lowers their blood pressure. To carry out this study using a pooled t-test, the investigator would select a group of presurgery patients and randomly assign them to either the treatment (counseling) group or the control (no-counseling) group. The treatment group would receive a 20-minute counseling session and the control group would be treated similarly in all respects but would receive no counseling. The null hypothesis is that the true mean blood pressure for the treatment group is the same as the true mean blood pressure for the control group. Symbolically,

$$H_0: \quad \mu_T = \mu_C$$

or

$$\mu_T - \mu_C = 0$$

The alternative to H_0 is that the mean for the treatment group is less than the mean for the control group, i.e.,

$$H_A: \quad \mu_T - \mu_C < 0$$

The means for both groups would be calculated and the difference $\overline{Y}_T - \overline{Y}_C$ obtained. Large differences in sample means support rejection of the null hypothesis, while small differences do not support rejection of H_0. The test statistic for the pooled t-test provides a measure of the discrepancy between the hypothesized difference ($\mu_T - \mu_C = 0$) and the observed difference ($\overline{Y}_T - \overline{Y}_C$). The level of significance chosen by the investigator specifies the rejection region for the test.

b. To carry out this investigation using a paired t-test, the subjects in both the treatment and control group must be paired or matched on the basis of relevant factors such as age, sex, etc., or each patient may serve as his own control. This "before and after" situation would involve making a blood pressure reading on a patient prior to counseling and then again after counseling. The null hypothesis would be that the mean difference in before-and-after readings is equal to zero, i.e., there is no difference in before-and-after blood pressure readings. The paired t-test would be carried out on the observed differences in readings.

EXERCISE 8.9

1. *Data*. The data, which consist of serum indirect bilirubin levels for two groups of infants (premature and full-term), are continuous and have at least an interval scale.
2. *Assumptions*. It is assumed that the two samples were independently and randomly selected and that the population of serum bilirubin measurements is normally distributed. It is further assumed that the standard deviations for the two populations of measurements are equal ($\sigma_1 = \sigma_2 = 1.0$ mg/100cc).
3. *Hypothesis*. The null hypothesis is that there is no difference in the mean serum indirect bilirubin level for the two groups, i.e., the true mean bilirubin level for premature infants is equal to that for full-term infants. Symbolically,

H_0: $\mu_1 - \mu_2 = 0$ 1 = Premature

H_A: $\mu_1 - \mu_2 \neq 0$ 2 = Full-term

4. *Test statistic*.

Sample Values

Premature	*Full-term*
$n_1 = 5$	$n_2 = 5$
$\bar{Y}_1 = 2.0$	$\bar{Y}_2 = 4.0$

$\sigma_1 = \sigma_2 = 1.0 = s_p$

$$s_p \sqrt{\frac{1}{n_1} + \frac{1}{n_2}} = 1.0 \sqrt{\frac{1}{5} + \frac{1}{5}} = 0.632$$

Note: We usually calculate s_1 and s_2, the standard deviation for the two samples. However, since we were given the population values $\sigma_1 = \sigma_2 = 1$, we will use them instead. The test statistic for the pooled t-test is:

$$t = \frac{\bar{Y}_1 - \bar{Y}_2}{s_p \sqrt{\frac{1}{n_1} + \frac{1}{n_2}}} = \frac{2.0 - 4.0}{1\sqrt{\frac{1}{5} + \frac{1}{5}}} = \frac{-2}{.632} = -3.16$$

5. *Decision rule*. The rejection region for the two-tailed test is found by locating in Appendix Table A the column $\alpha/2 = .025$ and the row $df = n_1 + n_2 - 2 = 5 + 5 - 2 = 8$. The value is 2.306. The rejection region is shown under the shaded portions of the curve below.

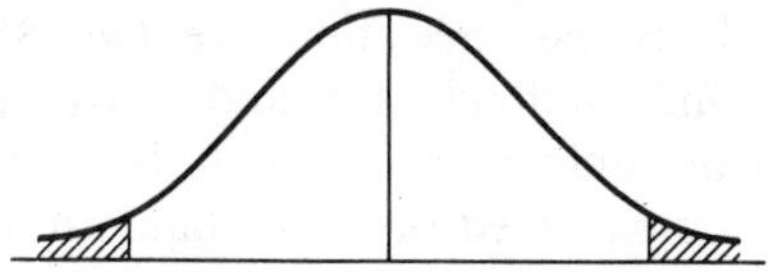

6. *Conclusion*. The value of the test statistic ($t = -3.16$) falls in the rejection region. The null hypothesis is rejected at $\alpha = .05$ level of significance. It may therefore be concluded that there is a significant difference in serum indirect bilirubin levels for premature and full-term infants (at $\alpha = .05$).

EXERCISE 8.10

WEIGHT GAIN IN NINE FEMALES FOLLOWING CONTRACEPTIVE USAGE

Subject	*Initial*	*Three Months*	*Difference*
1	120	123	+3
2	141	143	+2
3	130	140	+10
4	150	145	−5
5	135	140	+5
6	140	143	+3
7	120	118	−2
8	140	141	+1
9	130	132	+2

1. *Data*. The data consist of weight measurements on nine females before using oral contraceptives and again three months after an oral contraceptive has been used. Each subject serves as her own control.
2. *Assumptions*. The observed differences constitute a random sample from a population of differences that are normally distributed.
3. *Hypotheses*. The null hypothesis is that the mean weights for the before-and-after measurements are equal. This is equivalent to saying that the mean of the differences is equal to zero:

$$H_0:\quad D = 0$$

where D is the true mean of the entire population of differences. The alternative of interest is that the mean weight following three months of oral contraceptive use is greater than the mean weight prior to taking the medication.
4. *Test statistic*. Sample values:

$$n = 9,\ \Sigma d_i = 19,\ \Sigma d_i^2 = 181$$

$$\bar{d} = 2.111 \text{ (the mean of the differences)}$$

$$s = 4.197 \text{ (the standard deviation of the differences)}$$

The test statistic is:

$$t = \frac{\bar{d} - 0}{s/\sqrt{n}} = \frac{2.111}{4.197/\sqrt{9}} = \frac{2.111}{1.399} = 1.51$$

5. *Decision rule*. The rejection region for the one-tailed test is found by locating in Appendix Table A the column $\alpha = 0.05$ and the row $df = n - 1 = 8$. The value is 1.8595. The rejection region is under the shaded area shown below.

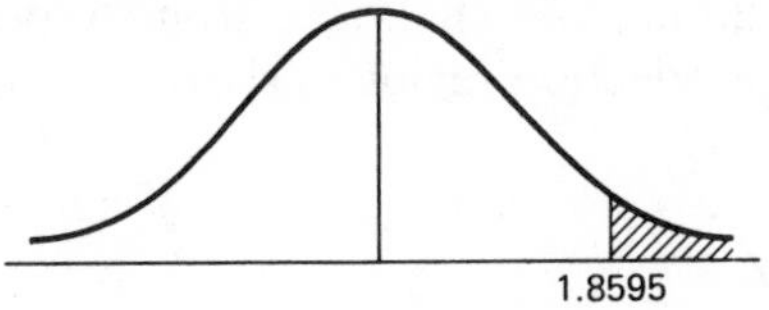

6. *Conclusion*. The test statistic does not fall in the rejection region; thus the null hypothesis cannot be rejected. It

cannot be shown that there is an increase in weight following oral contraceptive usage.

EXERCISE 8.11

1. *Data*. The data consist of ages for two samples of females. The variable age is continuous and possesses at least an interval scale of measurement.
2. *Assumptions*. It is assumed that the two samples were randomly and independently selected and that the population of ages is normally distributed. It is further assumed that the standard deviations for the two populations are equal.
3. *Hypotheses*. The null hypothesis is that there is no difference in mean ages for the expelled and retained groups.

$$H_0: \quad \mu_E - \mu_R = 0$$

$$H_A: \quad \mu_E - \mu_R \neq 0$$

4. *Test statistic*.

$$t = \frac{\bar{Y}_E - \bar{Y}_R}{s_p \sqrt{\frac{1}{n_E} + \frac{1}{n_R}}} = \frac{28 - 33}{11.3 \sqrt{\frac{1}{10} + \frac{1}{15}}} = -1.08$$

where

$$s_p = \sqrt{\frac{s_E{}^2(n_E - 1) + s_R{}^2(n_R - 1)}{n_E + n_R - 2}}$$

$$= \sqrt{\frac{(10)^2(9) + (12)^2(14)}{10 + 15 - 2}}$$

$$= 11.3$$

5. *Decision rule*. For $df = n_1 + n_2 - 2 = 23$ and $\alpha/2 = .025$ (two-tailed alternative), the value from Appendix Table A is 2.0687. The rejection region is shown below.

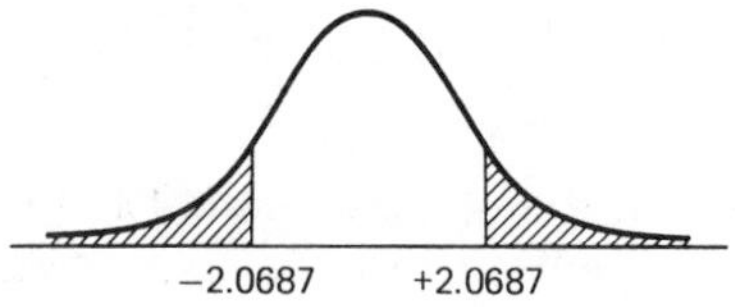

6. *Conclusion*. Since the value of the test statistic ($t = -1.08$) does not fall in the rejection region, it cannot be shown that the mean age for the expelled group is different from the mean age of the retained group (H_0 not rejected).

EXERCISE 8.12

1. *Data*. The data consist of 15 nursing audit scores from Hospital 1 and 15 scores from Hospital 2. The scores are assumed to be continuous and will be treated as if they possess an underlying interval scale of measurement.
2. *Assumptions*. It is assumed that the two samples represent independent random samples from the two populations of scores. The populations are assumed to be normally distributed with equal standard deviations ($\sigma_1 = \sigma_2$).
3. *Hypotheses*.

$$H_0: \mu_1 - \mu_2 = 0$$

$$H_A: \mu_1 - \mu_2 \neq 0$$

4. *Test statistic*.

$$t = \frac{\overline{Y}_1 - \overline{Y}_2}{s_p \sqrt{\dfrac{1}{n_1} + \dfrac{1}{n_2}}} = \frac{85 - 70}{10 \sqrt{\dfrac{1}{15} + \dfrac{1}{15}}} = 4.1$$

$$s_p = \sqrt{\frac{(10)^2(14) + (10)^2(14)}{15 + 15 - 2}} = 10$$

5. *Decision rule*. For $df = n_1 + n_2 - 2 = 28$ and $\alpha/2 = .005$ (two-tailed alternative), the value from Appendix Table A is 2.7633. The rejection region is:

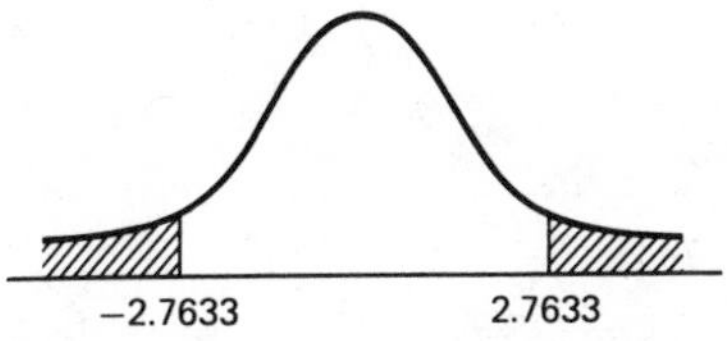

6. *Conclusion*. Since the value of the test statistic ($t = 4.1$) falls in the rejection region, it can be concluded that the

mean nursing performance scores for the two hospitals are different (H_0 rejected at the .01 level of significance).

EXERCISE 8.13

1. *Data*. The data consist of pre- and posttest scores for 9 randomly selected senior nursing students.
2. *Assumptions*. It is assumed that the observed differences are a random sample of differences from a population that is normally distributed.
3. *Hypotheses*. The null hypothesis is that there is no difference in pre- and posttest scores.

$$H_0: \quad D = 0$$

The alternate hypothesis is that the posttest scores are higher than the pretest scores, i.e.:

$$H_A: \quad D > 0$$

4. *Test statistic*.

Student	*Pretest*	*Posttest*	*Differences*
1	80	85	+5
2	75	90	+15
3	85	85	0
4	60	75	+15
5	95	98	+3
6	70	75	+5
7	65	70	+5
8	75	85	+10
9	90	80	−10

$$\bar{d} = \frac{48}{9} = 5.33$$

$$s_d = \sqrt{\frac{734 - \dfrac{(48)^2}{9}}{8}} = 7.73$$

$$t = \frac{\bar{d}}{s_d/\sqrt{n}} = \frac{5.33}{7.73/\sqrt{9}} = 2.07$$

5. *Decision rule*. For $df = n - 1 = 8$ and $\alpha = .05$ (one-tailed alternative), the value from Appendix Table A is 1.8595.

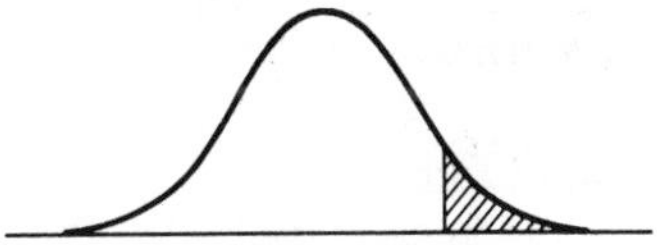

6. *Conclusion*. Since the value of the test statistic falls in the rejection region, it may be concluded that the posttest scores are significantly higher than the pretest scores at the .05 level of significance (reject H_0).

CHAPTER 9

EXERCISE 9.1

1. *Data*. The data consist of three sets of serum cholesterol measurements as shown below. The measurements are continuous and possess at least an interval scale.

CHOLESTEROL LEVELS

Diet A	*Diet B*	*Diet C*
a_1	b_1	c_1
a_2	b_2	c_2
a_3	b_3	c_3
a_4	b_4	c_4
a_5	b_5	c_5
	b_6	c_6
$\overline{Y}_A$	$\overline{Y}_B$	$\overline{Y}_C$

2. *Assumptions*. It is assumed that the three samples were randomly selected from normal populations having equal standard deviations.
3. *Hypotheses*. The null hypothesis is that the true mean serum cholesterol levels are the same for the three diets.

$$H_0: \quad \mu_A = \mu_B = \mu_C$$

The alternate hypothesis is that at least one of the pairs of means is not equal.

4. *Test statistic.*

$$F = \frac{\text{Among-diet Variation}}{\text{Within-diet Variation}} = 2.75$$

The ANOVA table would be:

ANOVA

Source of Variation	*df*	*SS*	*MS*	*F*
Among diets	3 − 1 = 2	*SST*	*SST*/2	2.75
Within diets	17 − 3 = 14	*SSE*	*SSE*/14	
Total	17 − 1 = 16			

5. *Decision rule.* The cutoff point for the rejection region is located at the intersection of column 2 (numerator *df*) and row 14 (denominator *df*) in the table headed $\alpha = .05$. The value is 3.74.
6. *Conclusion.* The calculated F value (Step 4) does not fall in the rejection region; therefore the null hypothesis cannot be rejected at the $\alpha = 0.05$ level of significance. There is not enough evidence to show that serum cholesterol level is different for any of the three diets studied.

† EXERCISE 9.2

1. *Data.* The data, which consist of 20 total audit scores, are continuous in nature and will be treated as if the underlying scale of measurement is equal interval.
2. *Assumptions.* It is assumed that the four samples were randomly selected from populations whose distributions are normal with equal standard deviations.
3. *Hypotheses.* The null hypothesis is that the mean audit scores are the same for all four services. Symbolically,

$$H_0: \quad \mu_1 = \mu_2 = \mu_3 = \mu_4$$

The alternative to H_0 is that at least one of the pairs is not equal.

4. *Test statistic*. The test statistic for the Analysis of Variance is the F-ratio:

$$F = \frac{MST}{MSE} = 3.37$$

5. *Decision rule*. The cutoff point for the rejection region is located at the intersection of column 3 (numerator df) and row 16 (denominator df) in the table headed $\alpha = .01$. The value is 5.29. The rejection region is shown below.

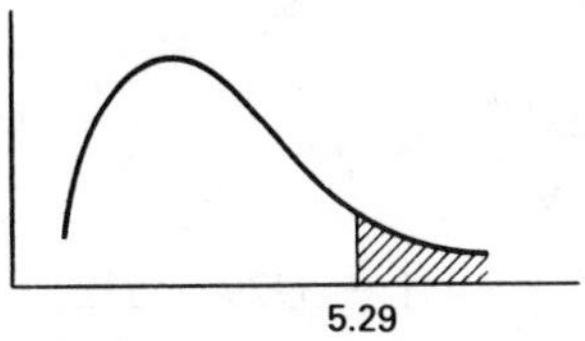

6. *Conclusion*. Since the calculated F value ($F = 3.37$) does not fall in the rejection region, the null hypothesis may not be rejected at the $\alpha = .01$ level of significance. There is insufficient evidence to conclude that at least one of the pairs of means is not equal (with a 1% risk of error).

† EXERCISE 9.3

1. *Data*. The data consist of response times in minutes for nine patients. The data are continuous in nature and possess at least an interval scale.
2. *Assumptions*. It is assumed that the nine patients were selected at random and assigned to the three methods. The populations from which the samples were taken are assumed to be normally distributed with equal standard deviations.
3. *Hypotheses*. The null hypothesis is that the mean response times for the three populations are equal.

$$H_0: \quad \mu_1 = \mu_2 = \mu_3$$

The alternative is that at least one pair of means is not equal.

4. *Test statistic.*

	Method 1	*Method 2*	*Method 3*	
	30	25	21	
	20	15	20	
	40	35	25	
Totals (T_i)	90	75	66	$\sum_{\text{all}} Y = 231$
(n_i)	3	3	3	$\sum_{\text{all}} Y^2 = 6441$
($\bar{Y}_i$)	30	25	22	

$$CF = \frac{(\sum_{\text{all}} Y)^2}{N} = \frac{(231)^2}{9} = 5929$$

$$SS_{\text{TOTAL}} = \sum_{\text{all}} Y^2 - CF = 6441 - 5929 = 512$$

$$SST = \sum \frac{T_i^2}{n_i} - CF = \frac{(90)^2}{3} + \frac{(75)^2}{3} + \frac{(66)^2}{3} - 5929 = 98$$

$$SSE = SS_{\text{TOTAL}} - SST = 512 - 98 = 414$$

The ANOVA table is:

ANOVA

Source	*df*	*SS*	*MS*	*F*
Among	2	98	49	.71
Within	6	414	69	
Total	8	512		

5. *Decision rule*. For $\alpha = .05$ and degrees of freedom equal to 2 (numerator) and 6 (denominator), the cutoff point for the rejection region is 5.14.

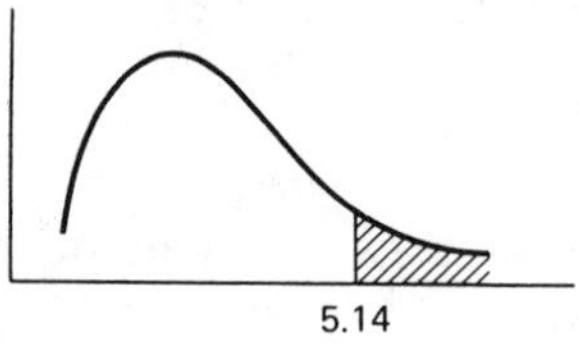

6. *Conclusion*. Since the value of F ($F = .71$) does not fall into the rejection region, it cannot be concluded that there is a difference in mean response times for the three methods at the $\alpha = .05$ level of significance.

EXERCISE 9.4

1. *Data*. The data layout would be as follows:

Cholesterol Levels

Initial Levels	*Diet A*	*Diet B*	*Diet C*
1	Y_{1A}	Y_{1B}	Y_{1C}
2	Y_{2A}	Y_{2B}	Y_{2C}
3	Y_{3A}	Y_{3B}	Y_{3C}

The measurements of serum cholesterol values are continuous and have an equal-interval scale.

2. *Assumptions*. Each cholesterol measurement Y_{ij} represents an independent random sample of size 1 from the population of patients in Block (initial level) i who receive Treatment (diet) j. It is assumed that each of the nine populations is normally distributed with the same standard deviation σ. It is also assumed that there is no interaction between initial levels (blocks) and diets (treatments).
3. *Hypotheses*. The null hypothesis is that the cholesterol levels are the same for the three diets. The alternate hypothesis is that cholesterol levels are not equal for all diets.
4. *Test statistic*.

$$F = 5.88$$

The ANOVA Table would be

ANOVA

Source of Variation	*df*	*SS*	*MS*	*F*
Among treatments (diets)	2	*SST*	*MST*	*MST/MSE*
Among blocks (initial levels)	2	*SSB*		
Error	4	*SSE*	*MSE*	
Total	8			

5. *Decision rule.*

$$df_{\text{numerator}} = 2$$
$$df_{\text{denominator}} = 4$$

From Table C for $\alpha = .05$, the appropriate table value is 6.94.

6. *Conclusion.* Since the calculated test statistic ($F = 5.88$) is less than the tabulated value (6.94). We cannot reject the null hypothesis at the $\alpha = .05$ level of significance. There is, therefore, insufficient evidence to show that cholesterol levels are different for the three diets.

EXERCISE 9.5

Randomized complete block design.

	Treatments				
	Blocks Patient	*Diet Alone*	*Chlorpropamide (100 mg/day)*	*Chlorpropamide (250 mg/day)*	*Block Totals* (B_j)
	1	8	5	5	18
	2	7	6	5	18
	3	9	8	7	24
	4	7	5	5	17
	5	8	6	7	21
Treatment totals (T_i)		39	30	29	98
$\overline{Y}_i$		7.8	6.0	5.8	—

Calculation of sums of squares:

$$CF = \frac{(\Sigma Y)^2}{N} = \frac{(98)^2}{15} = 640.267$$

$$SS_{\text{Total}} = \sum_{\text{all}} Y^2 - CF = 666 - 640.267 = 25.733$$

$$SS_{\text{Treatments}} = SST = \frac{(39)^2 + (30)^2 + (29)^2}{5} - 640.267$$

$$= 652.4 - 640.267 = 12.133$$

$$SS_{\text{Blocks}} = SSB = \frac{(18)^2 + (18)^2 + (24)^2 + (17)^2 + (21)^2}{3} - 640.267$$

$$= 651.333 - 640.267 = 11.067$$

$$SSE = 25.7333 - 12.133 - 11.067 = 2.533$$

ANOVA

Source	*df*	*SS*	*MS*	*F*
Treatments	2	12.133	6.067	19.16
Blocks	4	11.067	2.767	
Error	8	2.533	.317	
Total	14	25.733		

The tabulated value is $F_{2.8,.01} = 8.65$. Since the calculated $F = 19.16$ is greater than the tabulated value, we reject the null hypothesis of equality of treatment effects at the $\alpha = .01$ level of significance. We may conclude that at least one pair of population means is different. The data do provide sufficient evidence to indicate a difference in HB A_{1c} (percentage) among the three different treatments.

EXERCISE 9.6

Hypothesis	*t Statistic*
$\mu_1 = \mu_2$	$\dfrac{7.8 - 6.0}{\sqrt{.317\left(\frac{1}{5} + \frac{1}{5}\right)}} = 5.05$
$\mu_1 = \mu_3$	$\dfrac{7.8 - 5.8}{.356} = 5.62$
$\mu_2 = \mu_3$	$\dfrac{6.0 - 5.8}{.356} = .562$

$$t_{(.01/3)/2} = t_{.002}$$

$$t_{.002,8} = z_{.498} + \frac{z_{.498}^3 + z_{.498}}{4(6)} = 2.88 + \frac{(2.88)^3 + 2.88}{24} = 4.0$$

We may conclude that diet alone is different from chlorpropamide (100 mg/day) and from chlorpropamide (250 mg/day) at the overall per experiment error rate of $\alpha = .01$.

EXERCISE 9.7

Hypotheses	*t-Statistic*
Medical vs. surgical $\mu_M = \mu_S$	$\dfrac{64 - 80}{\sqrt{75.625\left(\frac{1}{5} + \frac{1}{5}\right)}} = \dfrac{-16}{5.5} = -2.91$
Medical vs. pediatrics $\mu_M = \mu_P$	$\dfrac{64 - 71}{\sqrt{75.625\left(\frac{1}{5} + \frac{1}{5}\right)}} = \dfrac{-7}{5.5} = -1.27$
Medical vs. ob-gyn $\mu_M = \mu_O$	$\dfrac{64 - 66}{\sqrt{75.625\left(\frac{1}{5} + \frac{1}{5}\right)}} = \dfrac{-2}{5.5} = -0.36$
Surgical vs. pediatrics $\mu_S = \mu_P$	$\dfrac{80 - 71}{\sqrt{75.625\left(\frac{1}{5} + \frac{1}{5}\right)}} = \dfrac{9}{5.5} = 1.64$
Surgical vs. ob-gyn $\mu_S = \mu_O$	$\dfrac{80 - 66}{\sqrt{75.625\left(\frac{1}{5} + \frac{1}{5}\right)}} = 2.55$
Pediatrics vs. ob-gyn $\mu_P = \mu_O$	$\dfrac{71 - 66}{\sqrt{75.625\left(\frac{1}{5} + \frac{1}{5}\right)}} = 0.91$

$$t_{(.01/6)/2} = t_{.001}$$

$$t_{.001,16} = z_{.499} + \frac{z_{.499}^3 + z_{.499}}{4(14)} = 3.09 + \frac{(3.09)^3 + 3.09}{56} = 3.67$$

None of the six t-statistics exceeds 3.67; thus, we cannot say that any of the pairs of means are different. *Note:* Since the ANOVA was not statistically significant (i.e., we did not reject H_0 in Example 9.2), we would not, in practice, carry out these multiple comparisons. However, we did so here in order to illustrate the procedure.

CHAPTER 10

EXERCISE 10.1

$$Syy = \Sigma Y^2 - \frac{(\Sigma Y)^2}{n} = 95 - \frac{(21)^2}{5} = 6.8$$

$$Sxx = \Sigma X^2 - \frac{(\Sigma X)^2}{n} = 90 - \frac{(20)^2}{5} = 10$$

$$Sxy = \Sigma XY - \frac{(\Sigma X)(\Sigma Y)}{n} = 91 - \frac{(20)(21)}{5} = 7$$

$$r = \frac{Sxy}{\sqrt{SxxSyy}} = \frac{7.}{\sqrt{(6.8)(10)}} = \frac{7.}{\sqrt{68}} = \frac{7.}{8.25} = .85$$

EXERCISE 10.2

a.

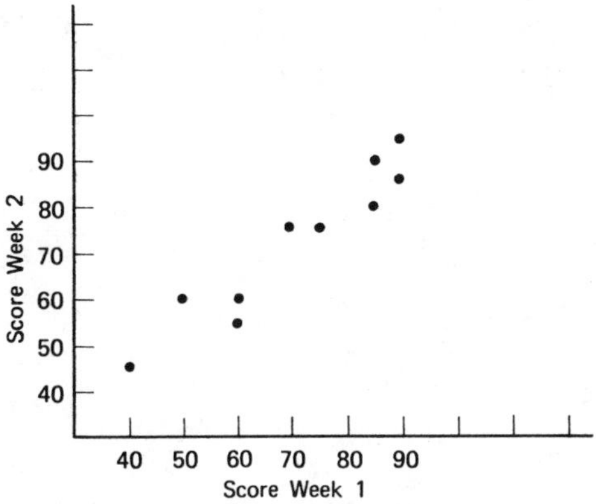

b. There appears to be a positive linear relationship between scores for Week 1 and scores for Week 2.

c. For large values of X (scores for Week 1), we have large values for Y (scores for Week 2), and for small X values we have small Y values.

EXERCISE 10.3

A correlation of +.95 indicates a strong positive association between the scores for Week 1 and the scores for Week 2. Since reliability is the degree of "reproducibility" of results, a positive correlation of +.95 would indicate good reliability.

EXERCISE 10.4

In correlation studies, one can assume causation only when all conceivable extraneous variables have been controlled for. This is rarely, if ever, achievable in practice. Therefore, a correlation of .90 between vocabulary proficiency and grades in a senior level research course does not imply that a good vocabulary will cause a student to do well in the course. It could as easily be said that a good grade in the course will cause a student to have a good vocabulary.

EXERCISE 10.5

Grades in Ob-Peds are probably used in computing overall clinical grade point average. Therefore, the Ob-Peds grades have been, in a sense, correlated with themselves, resulting in the high positive correlation.

EXERCISE 10.6

a.

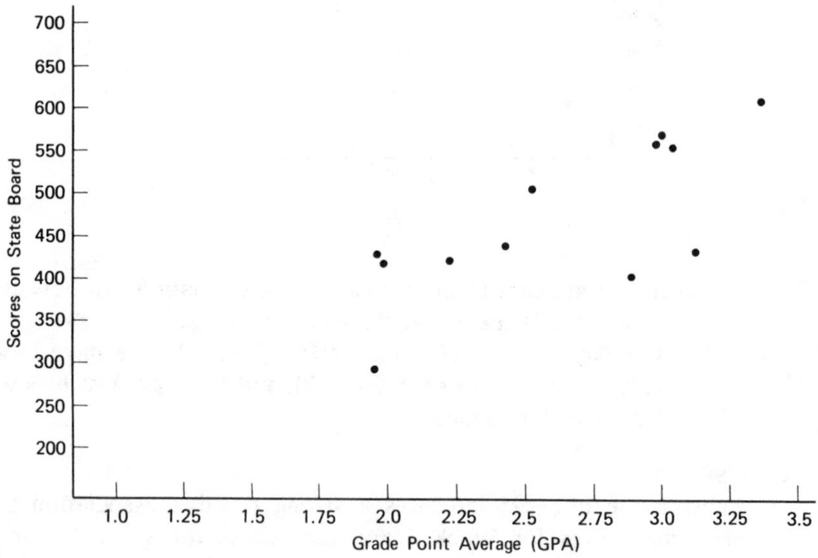

The scatter diagram suggests a positive linear relationship between grade point average and state board scores.

b. For $\alpha = .05$ and $df = n - 2 = 10$, the value in Table F is .5760. The calculated r value is 0.75. Since the calculated value of r ($r = .75$) is greater than the value found in the table (.5760), it may be concluded that the correlation is significantly different from zero. This means that there is a significant positive relationship between grade point average and scores on state boards. High grade average values have high values for state board scores, and low grade averages have low state board scores.

EXERCISE 10.7

a.

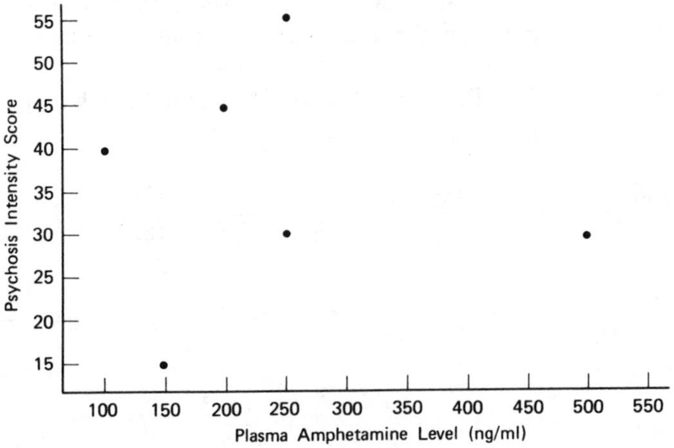

The points on the diagram seem widely scattered with no discernible underlying pattern.

b. For $\alpha = .05$ and $df = n - 2 = 4$, the critical value from Table F is 0.8114. To be significant, the calculated correlation ($r = -.047$) must be greater than .8114 (sign ignored). Thus the correlation coefficient is not significantly different from zero (at $\alpha = .05$). It cannot be shown that a significant linear relationship exists between the plasma amphetamine level and the psychosis intensity score.

EXERCISE 10.8

a.

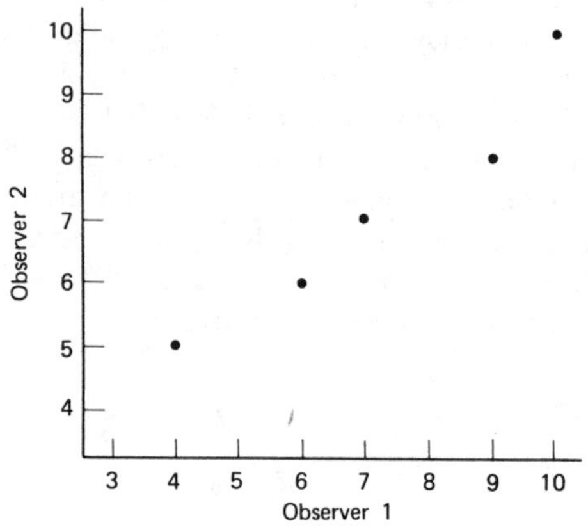

The scatter diagram suggests a positive linear relationship between the readings for Observer 1 and those for Observer 2.

b. The Pearson's Product Moment Correlation Coefficient is given by:

$$r = \frac{Sxy}{\sqrt{SxxSyy}} = \frac{17.8}{\sqrt{(22.8)(14.8)}} = \frac{17.8}{18.4} = .97$$

$$Sxx = \Sigma X^2 - \frac{(\Sigma X)^2}{n} = 282 - \frac{(36)^2}{5} = 282 - 259.2 = 22.8$$

$$Syy = \Sigma Y^2 - \frac{(\Sigma Y)^2}{n} = 274 - \frac{(36)^2}{5} = 274 - 259.2 = 14.8$$

$$Sxy = \Sigma XY - \frac{(\Sigma X)(\Sigma Y)}{n} = 277 - \frac{(36)(36)}{5} = 277 - 259.2 = 17.8$$

c. A correlation of .97 indicates a significant linear relationship between the readings of the two observers (at the .01 level of significance). In terms of reliability, this means that the two observers are obtaining close to the same results. This would indicate good interobserver reliability of the instrument.

EXERCISE 10.9

To establish the validity of the new tool, the investigator would compare proficiency scores obtained using the new tool with those obtained using the established tool. The correlation coefficient would be computed for the scores obtained with the new tool and those obtained with the traditional one. A high correlation between these two sets of scores would imply strong agreement between the two instruments. The validity of the new instrument may thus be established.

EXERCISE 10.10

a. Since the measurements are in the form of ranks, Spearman's Rank Order Correlation Coefficient would be used.

Ranks		*Difference (d)*
Child	*Parent*	*Between Ranks*
4	5	+1
2	2	0
1	3	2
3	1	−2
5	6	+1
6	4	−2

The Rank Order Correlation is given by:

$$r_s = 1 - \frac{6(\Sigma d^2)}{n(n^2 - 1)} = 1 - \frac{6(14)}{6(36 - 1)}$$

$$= 1 - \frac{84}{210} = 1 - .4 = 0.6$$

b. For $n = 6$ and $\alpha = .05$, the value from Table G at the intersection of column $\alpha/2 = .025$ and row $n = 6$ is .8286. The calculated $r_s = .6$ does not exceed this value; therefore, it cannot be concluded that the correlation between the creativity scores is significantly different from zero ($\alpha = .05$). There does not appear to be a significant linear relationships between the two variables.

EXERCISE 10.11

Nurse	*Observer 1*	*Observer 2*	*Difference (d)*
1	3	2	+1
2	2	3	−1
3	5	4	+1
4	1	1	0
5	4	5	−1
			$\Sigma d^2 = 4$

$$r_s = 1 - \frac{6(\Sigma d^2)}{n(n^2 - 1)}$$

$$= 1 - \frac{6(4)}{5(24)} = 1 - \frac{24}{120}$$

$$= 0.8$$

For $n = 5$ pairs of observations and $\alpha = .05$, the value from Table G for $\alpha/2 = .025$ is .90. Therefore, we cannot say that a significant linear relationship exists between the two variables (at the .05 level of significance).

† **EXERCISE 10.12**

a.

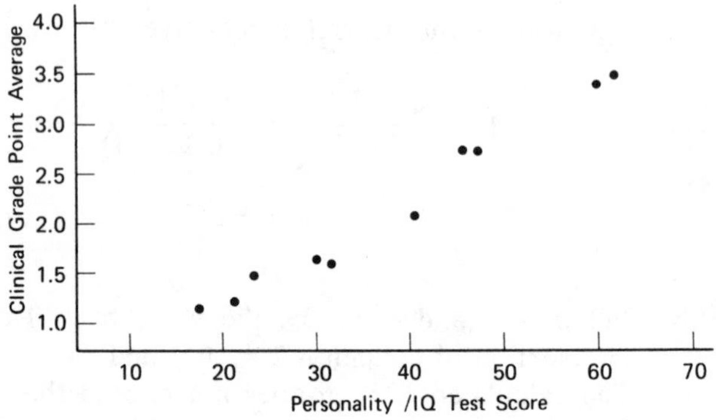

b.

$$\overline{X} = \frac{383}{10} = 38.3 \qquad \overline{Y} = \frac{21.8}{10} = 2.18$$

$$S_{xx} = \Sigma X^2 - \frac{(\Sigma X)^2}{n} = 16767 - \frac{(383)^2}{10} = 2098.1$$

$$S_{yy} = \Sigma Y^2 - \frac{(\Sigma Y)^2}{n} = 54.06 - \frac{(21.8)^2}{10} = 6.536$$

$$S_{xy} = \Sigma XY - \frac{(\Sigma X)(\Sigma Y)}{n} = 951.1 - \frac{(383)(21.8)}{10} = 116.16$$

$$\hat{\beta}_1 = \frac{S_{xy}}{S_{xx}} = \frac{116.16}{2098.1} = 0.0554$$

$$\hat{\beta}_0 = \overline{Y} - \hat{\beta}_1\overline{X} = (2.18) - (.0554)(38.3) = .0582$$

Therefore, the best fitting straight line relating personality/IQ scores to grade point average is:

$$\hat{Y} = 0.0582 + 0.0554X$$

c. Next, we wish to test whether $\hat{\beta}_1$ is significantly different from zero. That is, we wish to determine if a linear relationship exists between Y and X or if the value $\hat{\beta}_1 = 0.0554$ would have occurred by chance when, in reality, the true value of β_1 is zero. To perform this test of hypothesis we must first determine the variation around the regression line.

$$S_{yy} - \hat{\beta}_1 S_{xy} = 6.536 - (0.0554)(116.16)$$

$$= 0.1007$$

$$s_{y.x} = \sqrt{\frac{0.1007}{8}} = 0.1122$$

We will use the t statistic to test the hypothesis of no linear relationship between variables.

(1) *Hypotheses.*

H_0: $\beta_1 = 0$

H_A: $\beta_1 \neq 0$

(2) *Sample values.*

$\hat{\beta}_1 = 0.0554$

(3) *Test statistic.*

$$t = \frac{\hat{\beta}_1 - 0}{s_{y.x}/\sqrt{S_{xx}}} = \frac{0.0554}{.1122/\sqrt{2098.1}} = \frac{0.0554}{.0024} = 23.1$$

(4) *Rejection region.* For $\alpha = .01$ and $df = n - 2 = 8$, the value from Table A is 3.3554.

(5) *Conclusion.* Since the calculated t lies in the rejection region, we reject the null hypothesis of no linear relationship between variables. We may conclude that there is a significant linear relationship between personality/IQ test score and clinical grade point average (at the $\alpha = .01$ level of significance).

† EXERCISE 10.13

b.

$$S_{xx} = 108 - \frac{(26)^2}{8} = 23.5 \qquad \bar{Y} = \frac{30.5}{8} = 3.81$$

$$S_{yy} = 133.75 - \frac{(30.5)^2}{8} = 17.469 \qquad \bar{X} = \frac{26}{8} = 3.25$$

$$S_{xy} = 118.5 - \frac{(26)(30.5)}{8} = 19.375$$

$$\hat{\beta}_1 = \frac{S_{xy}}{S_{xx}} = \frac{19.375}{23.5} = .82$$

$$\hat{\beta}_0 = \bar{Y} - \hat{\beta}_1\bar{X}$$

$$= (3.81) - (.82)(3.25)$$

$$= 1.15$$

The equation for the regression line is:

$$\hat{Y} = 1.15 + .82X$$

c. (1) *Hypotheses.*

$$H_0: \ \beta_1 = 0$$

(i.e., there is no linear relationship between X and Y)

$$H_A: \ \beta_1 \neq 0$$

(2) *Sample values.*

$$\hat{\beta}_1 = .82$$

$$s_{y.x} = \sqrt{\frac{S_{yy} - \hat{\beta}_1 S_{xy}}{n-2}}$$

$$= \sqrt{\frac{17.469 - (.82)(19.375)}{6}} = 0.51$$

(3) *Test statistic.*

$$t = \frac{\hat{\beta}_1 - \beta_1}{s_{y.x}/\sqrt{S_{xx}}}$$

$$= \frac{.82 - 0}{0.51/\sqrt{23.5}} = 7.8$$

(4) *Rejection region.* For $df = 6$ and $\alpha/2 = .025$, the value from Appendix Table A is 2.4469.

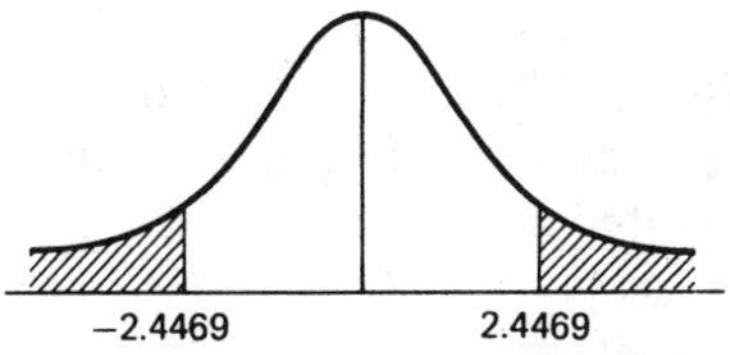

(5) *Conclusion.* Since the value of the test statistic falls in the rejection region, we may reject H_0 at the $\alpha = .05$ level of significance and conclude that there is a significant linear relationship between age of infant and weight.

d. For age = 3 weeks, weight is:

$$\hat{Y} = 1.15 + .82(3)$$

$$= 3.6 \text{ pounds}$$

CHAPTER 11

EXERCISE 11.1

		Eduation Level Completed by Parents			
		Elementary	Secondary	College	Total
Children	Yes	15 (21)	30 (28)	25 (21)	70
vaccinated	No	15 (9)	10 (12)	5 (9)	30
Total		30	40	30	100

1,2. *Data and assumptions.* The data consist of observed frequencies falling into the different categories in a 2×3 contingency table. It is assumed that the response of one individual in no way affects the response of any other individual.

3. *Hypotheses.* There is no association between education level completed and whether or not children are vaccinated against polio. Stated another way, the proportion of children who are vaccinated against polio is the same for all three levels of parents' education. The alternate hypothesis is that there is an association between educational level completed and whether or not children are vaccinated.

4. *Test statistic.* The observed and expected (in parentheses) frequencies are displayed in the contingency table. The calculated X^2 value is:

$$X^2 = \Sigma \frac{(O - E)^2}{E}$$

$$= \frac{(15-21)^2}{21} + \frac{(30-28)^2}{28} + \frac{(25-21)^2}{21}$$

$$+ \frac{(15-9)^2}{9} + \frac{(10-12)^2}{12} + \frac{(5-9)^2}{9}$$

$$= 8.7$$

5. *Decision rule.* The cutoff point for the rejection region is found by locating in Table D the intersection of the row

$df = (r - 1)(c - 1)$ and the column $1 - \alpha = .95$. The value is 5.991. The rejection region consists of values of the test statistic under the shaded area shown below.

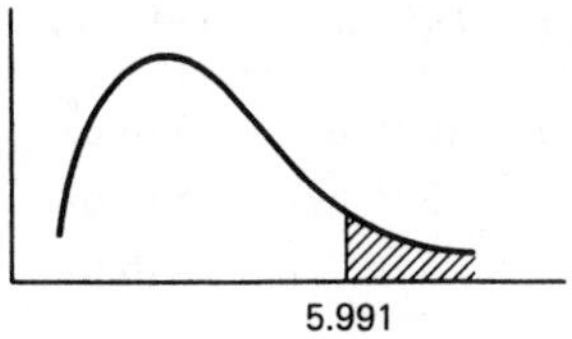

6. *Conclusion.* Since the calculated test statistic (step 4) does fall in the rejection region, we can reject the null hypothesis. We may conclude that there is an association between parents' educational level and whether or not children are vaccinated against polio (at the .05 level of significance).

EXERCISE 11.2

For this investigation, the data would consist of the number of individuals who experience relief for the various pain relievers being studied. For example, the data may be laid out as follows:

	Brand of Pain Reliever		
	Brand X	*Brand Y*	*Brand Z*
Relief	*A*	*B*	*C*
No relief	*D*	*E*	*F*

where *A* would be the number of people who experience relief of headache pain with Brand *X*, *B* the number experiencing relief with Brand *Y*, etc. The null hypothesis is that the proportion of individuals experiencing relief from pain is the same for all brands of pain reliever. A significant X^2 implies that the null hypothesis has been rejected at the given level of significance. It may be concluded that the proportion of people experiencing relief from pain is different for the three brands.

EXERCISE 11.3

	Perceived Difficulty of Course			
Class Rank	*Not Difficult*	*Intermediate Difficulty*	*Very Difficult*	*Total*
Upper third	5 (6.67)	10 (6.67)	5 (6.67)	20
Middle third	10 (6.67)	5 (6.67)	5 (6.67)	20
Lower third	5 (6.67)	5 (6.67)	10 (6.67)	20
Total	20	20	20	60

a. The null hypothesis is that there is no association between course difficulty rating and class rank. Stated another way, the proportions of students falling into the three perceived difficulty classifications are constant across the three class rankings.

b. $$X^2 = \Sigma \frac{(O - E)^2}{E}$$

$$= \frac{(5 - 6.67)^2}{6.67} + \frac{(10 - 6.67)^2}{6.67} + \frac{(5 - 6.67)^2}{6.67}$$

$$+ \frac{(10 - 6.67)^2}{6.67} + \frac{(5 - 6.67)^2}{6.67} + \frac{(5 - 6.67)^2}{6.67}$$

$$+ \frac{(5 - 6.67)^2}{6.67} + \frac{(5 - 6.67)^2}{6.67} + \frac{(10 - 6.67)^2}{6.67}$$

$$= 0.418 + 1.663 + 0.418 + 1.663 + 0.418$$

$$+ 0.418 + 0.418 + 0.418 + 1.663$$

$$= 7.497$$

c. The rejection region for the test is found by locating in Appendix Table D the intersection of column $(1 - \alpha) = .95$ and the row $df = (r - 1)(c - 1) = 4$. The value is 9.488.

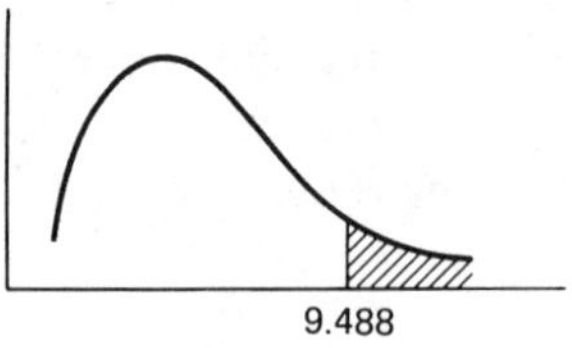

The calculated test statistic ($X^2 = 7.497$) does not fall in the rejection region (shaded area). Therefore the null hypothesis of no association cannot be rejected. It must be concluded that there is not enough evidence to show that students' perceived difficulty of the course is related to class rank.

EXERCISE 11.4

a. The null hypothesis is that the proportion of mistakes is constant for the three treatments.

b. The calculated test statistic ($X^2 = 27.25$) falls in the rejection region. (The cutoff point found in Appendix Table D for the rejection region for $df = 4$ and $X^2_{.99}$ is 13.277.) The null hypothesis may therefore be rejected and it may be concluded that there is an association between number of mistakes and type of treatment ($\alpha = .01$).

EXERCISE 11.5

a.

	Performance		
	Satisfactory	*Not Satisfactory*	*Total*
Antidepressant	12	18	30
Antidepressant and alcohol	8	22	30
Total	20	40	60

b. *Hypotheses.* H_0: The proportion of patients with satisfactory response is the same for the two treatment groups. H_A: The proportion of patients with satisfactory response is different for the two treatment groups.

c. The rejection region is found by locating the intersection of the column $(1 - \alpha) = .95$ and $df = (r - 1)(c - 1) = 1$. The value is 3.841.

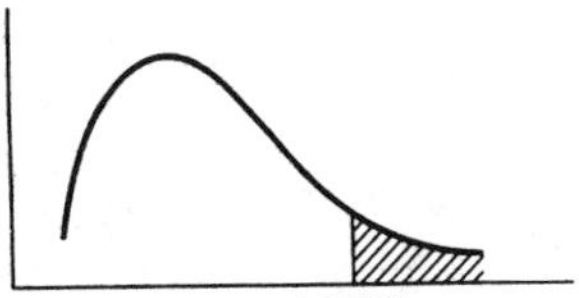

The calculated test statistic ($X^2 = .675$) does not fall in the rejection region. It must be concluded that there is not enough evidence to show that the proportion of patients with satisfactory response is different for the two treatment groups.

EXERCISE 11.6

a. *Null hypothesis*. There is no association between reasons given for not sleeping well and type of treatment received. The alternate hypothesis is that there is an association between reasons given for not sleeping well and type of medication received.

b. The rejection region is found by locating the intersection of $(1 - \alpha) = .99$ and the row $df = 6$. The value from Table D is 16.812. The calculated test statistic ($X^2 = 6.5$) does not fall in the rejection region. The null hypothesis of no association between reason for sleeplessness and treatment cannot be rejected. There is insufficient evidence to show that perceived reason for not sleeping well is related to type of medication.

EXERCISE 11.7

1. *Hypotheses*. There is no difference in proportions of births according to seasons for institutionalized schizophrenics:

$$H_0: \quad P_1 = P_2 = P_3 = P_4 = .25$$

where P = true proportion for each season.

H_A: The proportions are not all equal.

2. *Contingency table*.

Season of Birth	*Observed*	*Expected*
Fall	20	25
Winter	35	25
Spring	20	25
Summer	25	25
	100	100

3. *Test statistic.*

$$X^2 = \Sigma \frac{(O - E)^2}{E}$$

$$= \frac{(20 - 25)^2}{25} + \frac{(35 - 25)^2}{25} + \frac{(20 - 25)^2}{25} + \frac{(25 - 25)^2}{25}$$

$$= 6.0$$

4. *Rejection region.* From Table D, the cutoff point for the rejection region is located at the intersection of the column $(1 - \alpha) = .95$ and the row $df = 3$. The value is 7.815.

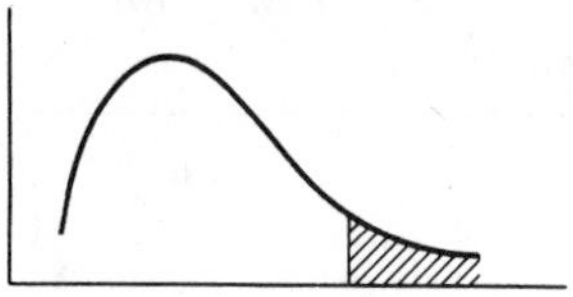

5. *Conclusion.* Since the calculated test statistic ($X^2 = 6.0$) does not fall in the rejection region, the null hypothesis of equal proportion of births according to seasons cannot be rejected. It cannot be shown that there is a difference in number of births according to different seasons of the year for institutionalized schizophrenics.

CHAPTER 12

EXERCISE 12.1

Sign Test:

Patient	*Ulcer Severity Score* *Before*	*After*	*Sign**
1	15	10	−
2	20	14	−
3	10	13	+
4	7	5	−
5	12	9	−
6	8	6	−
7	6	10	+

* + indicates higher score after treatment.

The null hypothesis is that the median difference for the matched pairs is 0. The minority sign is "+" and it occurred twice. From Table J for $N = 7$ and $\alpha = .05$, we see that 0 plus signs would have had to occur before the null hypothesis could be rejected. It must therefore be concluded that there is not enough evidence to show that the samples are from populations with different medians (i.e., it cannot be shown that there is a statistically significant difference in the "before" and "after" scores at the .05 level of significance).

Signed-Rank Test:

Patient	*Ulcer Severity Score* *Before*	*After*	*Differences*	*Rank Disregarding Sign*	*Signed Rank*
1	15	10	−5	6	−6
2	20	14	−6	7	−7
3	10	13	+3	3.5	+3.5
4	7	5	−2	1.5	−1.5
5	12	9	−3	3.5	−3.5
6	8	6	−2	1.5	−1.5
7	6	10	+4	5	+5

Sum of positive signed ranks = 8.5
Sum of negative signed ranks = 19.5
Sample size $n = 7$

The data consist of ulcer severity scores for seven patients taken before and after a treatment has been applied. The null hypothesis is that there is no difference in scores for the before and after time periods (i.e., the two populations have equal medians). To test this hypothesis, the smaller of the sum of signed ranks (8.5) is compared to a critical value found in Table H. For $n = 7$ and $\alpha = .05$, the critical value is 2. In order to reject the null hypothesis, the smaller sum of signed ranks (8.5) must be less than 2. Clearly, this is not the case. Therefore, it cannot be shown that there is a statistically significant difference in the before and after ulcer severity scores.

EXERCISE 12.2

Median Test: The null hypothesis is that the two independent samples have been drawn from populations with equal medians. The median score for the two groups combined is 20.5.

	Family Visits	*No Family Visits*	*Total*
No. scores above median	4	6	10
No. scores below median	6	4	10
Total	10	10	20

$$X^2 = \frac{N\left(|AD - BC| - \frac{N}{2}\right)^2}{(A + B)(C + D)(A + C)(B + D)}$$

$$= \frac{20\left(|(4)(4) - (6)(6)| - \frac{20}{2}\right)^2}{(10)(10)(10)(10)} = 0.2$$

The value from Table D for $\alpha = .01$ and $df = 1$ is 6.635. Since the calculated value of X^2 is not greater than the value obtained from Table D, the null hypothesis cannot be rejected at the .01 level of significance. There is not enough evidence to show that the samples are from populations whose medians are different.

Rank Sum Test: Ordering and Rank for the 20 combined scores are as follows (the scores for patients whose families visited are marked by an asterisk):

Order	8	9	10	10*	10*	14*	14*	17*	19	20*
Rank	1	2	4	4	4	6.5	6.5	8	9	10
Order	21*	27	52*	54*	54	60	64	74*	79	84
Rank	11	12	13	14.5	14.5	16	17	18	19	20

The null hypothesis is that there is no difference in length of hospitalization for patients whose families visit frequently and for patients whose families do not visit. Since there is an equal number in both groups, either group of ranks may be summed. The sum of the ranks for Group I (denoted by *) is 95.5. From Table I for $n_1 = 10$, $n_2 = 10$, and $\alpha = .01$, the critical values are 71 and 139. To be significant, the rank sum (95.5) must be less than 71 or greater than 139. Therefore, the null hypothesis cannot be rejected at the $\alpha = .01$ level of significance. It cannot be shown that there is a difference in length of hospitalization for the two groups.

EXERCISE 12.3

The data consist of greeting behavior scores for two independent samples. The null hypothesis is that there is no difference in greeting behavior scores for mothers who received counseling and for mothers who did not. The appropriate nonparametric test is the Wilcoxon Rank Sum Test. (The Median Test may also be used.) Ordering and Rank for the 10 combined scores are as follows: (the scores for mothers who received counseling are marked by an asterisk):

Order	2	3	3*	4	5*	6	7*	8*	8	10*
Rank	1	2.5	2.5	4	5	6	7	8.5	8.5	10

Since there is an equal number in both groups, either group of ranks may be summed. The sum of the ranks for the group receiving counseling (denoted by *) is 33. From Table I for $n_1 = 5$, $n_2 = 5$ and $\alpha = .05$, the critical values are 17 and 38. Since the rank sum (33) is not less than 17 or greater than 38, the null hypothesis cannot be rejected ($\alpha = .05$). It cannot be shown that there is a difference in greeting behavior scores for mothers who received counseling and mothers who did not receive counseling.

EXERCISE 12.4

	Dental Hygiene Score				
Child	*Before*	*After*	*Difference*	*Rank Disregarding Sign*	*Signed Rank*
1	4	2	−2	4	−4
2	2	3	+1	1.5	+1.5
3	3	5	+2	4	+4
4	6	4	−2	4	−4
5	5	6	+1	1.5	+1.5
6	1	1	0	—	—
7	1	4	+3	6.5	+6.5
8	4	7	+3	6.5	+6.5

Sum of positive signed ranks = 20
Sum of negative signed ranks = 8
Sample size $n = 7$ (1 zero discarded)

The null hypothesis is that there is no difference in dental hygiene scores before and after instruction (i.e., the median of the differences is zero). The smaller of the sum of signed ranks (8) is compared to a critical value found in Table H. For $n = 7$ and $\alpha = .05$, the table value is 2. Since the smaller sum of signed ranks (8) is not less than 2, the null hypothesis cannot be rejected ($\alpha = .05$). It cannot be shown that there is a difference in dental hygiene scores before and after instruction. (The above test could also have been carried out using the Sign Test. However, since the Signed Rank Test takes into account the magnitude of the differences, it is a more powerful test and is probably, therefore, preferable.)

EXERCISE 12.5

Student	*Before*	*After*	*Sign**
1	8	10	+
2	5	10	+
3	14	15	+
4	22	20	−
5	17	20	+
6	8	16	+
7	16	24	+
8	10	23	+
9	20	25	+
10	14	24	+
11	10	23	+
12	20	18	−

* + indicates a higher score following the workshop.

The null hypothesis is that there is no difference in before and after scores (i.e., the medians for the two populations are equal). The minority sign ("−") occurs twice. From Table J for $N = 12$ and $\alpha = .05$, we see that the null hypothesis may be rejected if the minority sign occurs *two* or *fewer* times. The null hypothesis of equal medians may, therefore, be rejected and it may be concluded that there is a difference in student attitudes toward alcoholics following the workshop (at an .05 level of significance).

EXERCISE 12.6

The null hypothesis is that the two random samples are from populations with equal medians (i.e., there is no difference in anxiety scores for the two groups). The combined median for the two groups is 37.5.

	Treatment	*Control*	*Total*
No. scores above median	2	8	10
No. scores below median	8	2	10
Total	10	10	20

$$X^2 = \frac{N\left(|AD - BC| - \frac{N}{2}\right)^2}{(A + B)(C + D)(A + C)(B + D)}$$

$$= \frac{20\left(|(2)(2) - (8)(8)| - \frac{20}{2}\right)^2}{(10)(10)(10)(10)} = 5$$

The value from Table D for $df = 1$ and $\alpha = .05$ is 3.841. Since the calculated X^2 is greater than the value from Table D, we may reject the null hypothesis at the $\alpha = .05$ level of significance and conclude that the medians for the two populations are not equal; that is, we may conclude that the anxiety scores for the treatment and control groups are different (at the .05 level of significance).

CHAPTER 13

EXERCISE 13.1

Type of sample: Convenience. Generalizability: A convenience sample is a nonprobability or nonrandom sample. Conclusions may not be generalized beyond those students in the sample.

EXERCISE 13.2

Type of sample: Cluster sample. The hospitals are the primary sampling units (the cluster) and the diabetic patients within a given hospital are the secondary sampling units. Generalizability: A cluster sample is a probability sample since a random mechanism was used to select the hospitals. Since hospitals were randomly selected within a given city, conclusions may be generalized statistically to all diabetics in the given city. To generalize further from the population sampled (diabetics in the city) to the target population (all diabetics) is subjective.

EXERCISE 13.3

Type of sample: Volunteer. The 60% of the people who responded to the questionnaire did so on a voluntary basis. The respondants as a group may have characteristics which are different from those of the nonrespondants. Generalizability: A volunteer sample is a nonrandom sample and the conclusions based on this sample are statistically nongeneralizable.

EXERCISE 13.4

Type of sample: Systematic random sample. Generalizability: Conclusions may be statistically generalized to the population sampled (all patients in the particular hospital). To generalize to a larger target population must be justified subjectively.

EXERCISE 13.5

Type of sample: Random sample (random allocation). The 15 patients were randomly assigned to the two groups. Generalizability: Conclusions based on random assignment may not be generalized beyond those in the sample.

EXERCISE 13.6

Type of sample: Quota sample. Generalizability: A quota sample is a nonprobability sample since it does not involve the use of a mechanism to insure randomness. Conclusions may not be statistically generalized beyond those individuals in the sample.

EXERCISE 13.7

Type of sample: Purposive sample. Generalizability: A purposive sample is a nonprobability sample. Conclusions are only valid for those individuals in the study.

EXERCISE 13.8

Type of sample: Purposive sample. Generalizability: Same as for Exercise 13.7.

EXERCISE 13.9

Type of sample: Cluster sample. The colleges are the primary sampling unit (the clusters) and the students within the colleges are the secondary sampling units. Generalizability: A cluster sample is a probability sample. Conclusions may be generalized to the population sampled (all colleges of nursing in the United States). Note: If the return on the questionnaire had been less than 100%, then the sample would be considered volunteer.

EXERCISE 13.10

Type of sample: Simple random sample. Generalizability: Conclusions may be generalized to the population sampled, i.e., all colleges of nursing in the United States.

EXERCISE 13.11

Retrospective study

EXERCISE 13.12

Prospective study

EXERCISE 13.13

Cross-sectional study

EXERCISE 13.14

Cross-sectional study

EXERCISE 13.15

Prospective study

APPENDIX 3

TABLES

TABLE A. THE t DISTRIBUTION

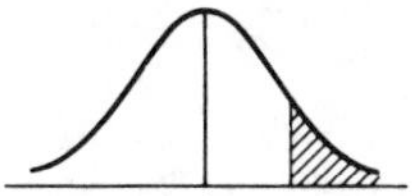

df	$t_{.10}$	$t_{.05}$	$t_{.025}$	$t_{.01}$	$t_{.005}$
1	3.078	6.3138	12.706	31.821	63.657
2	1.886	2.9200	4.3027	6.965	9.9248
3	1.638	2.3534	3.1825	4.541	5.8409
4	1.533	2.1318	2.7764	3.747	4.6041
5	1.476	2.0150	2.5706	3.365	4.0321
6	1.440	1.9432	2.4469	3.143	3.7074
7	1.415	1.8946	2.3646	2.998	3.4995
8	1.397	1.8595	2.3060	2.896	3.3554
9	1.383	1.8331	2.2622	2.821	3.2498
10	1.372	1.8125	2.2281	2.764	3.1693
11	1.363	1.7959	2.2010	2.718	3.1058
12	1.356	1.7823	2.1788	2.681	3.0545
13	1.350	1.7709	2.1604	2.650	3.0123
14	1.345	1.7613	2.1448	2.624	2.9768
15	1.341	1.7530	2.1315	2.602	2.9467
16	1.337	1.7459	2.1199	2.583	2.9208
17	1.333	1.7396	2.1098	2.567	2.8982
18	1.330	1.7341	2.1009	2.552	2.8784
19	1.328	1.7291	2.0930	2.539	2.8609
20	1.325	1.7247	2.0860	2.528	2.8453

TABLE A. CONTINUED

df	$t_{.10}$	$t_{.05}$	$t_{.025}$	$t_{.01}$	$t_{.005}$
21	1.323	1.7207	2.0796	2.518	2.8314
22	1.321	1.7171	2.0739	2.508	2.8188
23	1.319	1.7139	2.0687	2.500	2.8073
24	1.318	1.7109	2.0639	2.492	2.7969
25	1.316	1.7081	2.0595	2.485	2.7874
26	1.315	1.7056	2.0555	2.479	2.7787
27	1.314	1.7033	2.0518	2.473	2.7707
28	1.313	1.7011	2.0484	2.467	2.7633
29	1.311	1.6991	2.0452	2.462	2.7564
30	1.310	1.6973	2.0423	2.457	2.7500
35	1.3062	1.6896	2.0301	2.438	2.7239
40	1.3031	1.6839	2.0211	2.423	2.7045
45	1.3007	1.6794	2.0141	2.412	2.6896
50	1.2987	1.6759	2.0086	2.403	2.6778
60	1.2959	1.6707	2.0003	2.390	2.6603
70	1.2938	1.6669	1.9945	2.381	2.6480
80	1.2922	1.6641	1.9901	2.374	2.6388
90	1.2910	1.6620	1.9867	2.368	2.6316
100	1.2901	1.6602	1.9840	2.364	2.6260
120	1.2887	1.6577	1.9799	2.358	2.6175
140	1.2876	1.6558	1.9771	2.353	2.6114
160	1.2869	1.6545	1.9749	2.350	2.6070
180	1.2863	1.6534	1.9733	2.347	2.6035
200	1.2858	1.6525	1.9719	2.345	2.6006
∞	1.282	1.645	1.96	2.326	2.576

SOURCE: From *Documenta Geigy, Scientific Tables,* 7th Edition, 1970, pp. 32–35. Courtesy of Ciba-Geigy Limited, Basle, Switzerland.

TABLE B. THE STANDARD NORMAL DISTRIBUTION

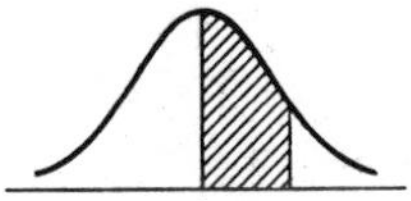

z	*.00*	*.01*	*.02*	*.03*	*.04*	*.05*	*.06*	*.07*	*.08*	*.09*
.0	.0000	.0040	.0080	.0120	.0160	.0199	.0239	.0279	.0319	.0359
.1	.0398	.0438	.0478	.0517	.0557	.0596	.0636	.0675	.0714	.0753
.2	.0793	.0832	.0871	.0910	.0948	.0987	.1026	.1064	.1103	.1141
.3	.1179	.1217	.1255	.1293	.1331	.1368	.1406	.1443	.1480	.1517
.4	.1554	.1591	.1628	.1664	.1700	.1736	.1772	.1808	.1844	.1879
.5	.1915	.1950	.1985	.2019	.2054	.2088	.2123	.2157	.2190	.2224
.6	.2257	.2291	.2324	.2357	.2389	.2422	.2454	.2486	.2517	.2549
.7	.2580	.2611	.2642	.2673	.2704	.2734	.2764	.2794	.2823	.2852
.8	.2881	.2910	.2939	.2967	.2995	.3023	.3051	.3078	.3106	.3133
.9	.3159	.3186	.3212	.3238	.3264	.3289	.3315	.3340	.3365	.3389
1.0	.3413	.3438	.3461	.3485	.3508	.3531	.3554	.3577	.3599	.3621
1.1	.3643	.3665	.3686	.3708	.3729	.3749	.3770	.3790	.3810	.3830
1.2	.3849	.3869	.3888	.3907	.3925	.3944	.3962	.3980	.3997	.4015
1.3	.4032	.4049	.4066	.4082	.4099	.4115	.4131	.4147	.4162	.4177
1.4	.4192	.4207	.4222	.4236	.4251	.4265	.4279	.4292	.4306	.4319
1.5	.4332	.4345	.4357	.4370	.4382	.4394	.4406	.4418	.4429	.4441
1.6	.4452	.4463	.4474	.4484	.4495	.4505	.4515	.4525	.4535	.4545
1.7	.4554	.4564	.4573	.4582	.4591	.4599	.4608	.4616	.4625	.4633
1.8	.4641	.4649	.4656	.4664	.4671	.4678	.4686	.4693	.4699	.4706
1.9	.4713	.4719	.4726	.4732	.4738	.4744	.4750	.4756	.4761	.4767
2.0	.4772	.4778	.4783	.4788	.4793	.4798	.4803	.4808	.4812	.4817
2.1	.4821	.4826	.4830	.4834	.4838	.4842	.4846	.4850	.4854	.4857
2.2	.4861	.4864	.4868	.4871	.4875	.4878	.4881	.4884	.4887	.4890
2.3	.4893	.4896	.4898	.4901	.4904	.4906	.4909	.4911	.4913	.4916
2.4	.4918	.4920	.4922	.4925	.4927	.4929	.4931	.4932	.4934	.4936
2.5	.4938	.4940	.4941	.4943	.4945	.4946	.4948	.4949	.4951	.4952
2.6	.4953	.4955	.4956	.4957	.4959	.4960	.4961	.4962	.4963	.4964
2.7	.4965	.4966	.4967	.4968	.4969	.4970	.4971	.4972	.4973	.4974
2.8	.4974	.4975	.4976	.4977	.4977	.4978	.4979	.4979	.4980	.4981
2.9	.4981	.4982	.4982	.4983	.4984	.4984	.4985	.4985	.4986	.4986
3.0	.4987	.4987	.4987	.4988	.4988	.4989	.4989	.4989	.4990	.4990

SOURCE: From John E. Freund and Frank J. Williams, *Elementary Business Statistics: The Modern Approach,* 2nd Edition, © 1972, p. 473. Reprinted by permission of Prentice-Hall, Inc., Englewood Cliffs, New Jersey.

TABLE C. THE *F* DISTRIBUTION ($\alpha = 0.05$)

Denominator Degrees of Freedom	*Numerator Degrees of Freedom* 1	2	3	4	5	6	7	8	9	10	12	15	*Numerator Degrees of Freedom* 20	24	30	40	60	120	∞
1	161.4	199.5	215.7	224.6	230.2	234.0	236.8	238.9	240.5	241.9	243.9	245.9	248.0	249.1	250.1	251.1	252.2	253.3	254.3
2	18.51	19.00	19.16	19.25	19.30	19.33	19.35	19.37	19.38	19.40	19.41	19.43	19.45	19.45	19.46	19.47	19.48	19.49	19.50
3	10.13	9.55	9.28	9.12	9.01	8.94	8.89	8.85	8.81	8.79	8.74	8.70	8.66	8.64	8.62	8.59	8.57	8.55	8.53
4	7.71	6.94	6.59	6.39	6.26	6.16	6.09	6.04	6.00	5.96	5.91	5.86	5.80	5.77	5.75	5.72	5.69	5.66	5.63
5	6.61	5.79	5.41	5.19	5.05	4.95	4.88	4.82	4.77	4.74	4.68	4.62	4.56	4.53	4.50	4.46	4.43	4.40	4.36
6	5.99	5.14	4.76	4.53	4.39	4.28	4.21	4.15	4.10	4.06	4.00	3.94	3.87	3.84	3.81	3.77	3.74	3.70	3.67
7	5.59	4.74	4.35	4.12	3.97	3.87	3.79	3.73	3.68	3.64	3.57	3.51	3.44	3.41	3.38	3.34	3.30	3.27	3.23
8	5.32	4.46	4.07	3.84	3.69	3.58	3.50	3.44	3.39	3.35	3.28	3.22	3.15	3.12	3.08	3.04	3.01	2.97	2.93
9	5.12	4.26	3.86	3.63	3.48	3.37	3.29	3.23	3.18	3.14	3.07	3.01	2.94	2.90	2.86	2.83	2.79	2.75	2.71
10	4.96	4.10	3.71	3.48	3.33	3.22	3.14	3.07	3.02	2.98	2.91	2.85	2.77	2.74	2.70	2.66	2.62	2.58	2.54
11	4.84	3.98	3.59	3.36	3.20	3.09	3.01	2.95	2.90	2.85	2.79	2.72	2.65	2.61	2.57	2.53	2.49	2.45	2.40
12	4.75	3.89	3.49	3.26	3.11	3.00	2.91	2.85	2.80	2.75	2.69	2.62	2.54	2.51	2.47	2.43	2.38	2.34	2.30
13	4.67	3.81	3.41	3.18	3.03	2.92	2.83	2.77	2.71	2.67	2.60	2.53	2.46	2.42	2.38	2.34	2.30	2.25	2.21
14	4.60	3.74	3.34	3.11	2.96	2.85	2.76	2.70	2.65	2.60	2.53	2.46	2.39	2.35	2.31	2.27	2.22	2.18	2.13
15	4.54	3.68	3.29	3.06	2.90	2.79	2.71	2.64	2.59	2.54	2.48	2.40	2.33	2.29	2.25	2.20	2.16	2.11	2.07
16	4.49	3.63	3.24	3.01	2.85	2.74	2.66	2.59	2.54	2.49	2.42	2.35	2.28	2.24	2.19	2.15	2.11	2.06	2.01
17	4.45	3.59	3.20	2.96	2.81	2.70	2.61	2.55	2.49	2.45	2.38	2.31	2.23	2.19	2.15	2.10	2.06	2.01	1.96
18	4.41	3.55	3.16	2.93	2.77	2.66	2.58	2.51	2.46	2.41	2.34	2.27	2.19	2.15	2.11	2.06	2.02	1.97	1.92
19	4.38	3.52	3.13	2.90	2.74	2.63	2.54	2.48	2.42	2.38	2.31	2.23	2.16	2.11	2.07	2.03	1.98	1.93	1.88
20	4.35	3.49	3.10	2.87	2.71	2.60	2.51	2.45	2.39	2.35	2.28	2.20	2.12	2.08	2.04	1.99	1.95	1.90	1.84
21	4.32	3.47	3.07	2.84	2.68	2.57	2.49	2.42	2.37	2.32	2.25	2.18	2.10	2.05	2.01	1.96	1.92	1.87	1.81
22	4.30	3.44	3.05	2.82	2.66	2.55	2.46	2.40	2.34	2.30	2.23	2.15	2.07	2.03	1.98	1.94	1.89	1.84	1.78
23	4.28	3.42	3.03	2.80	2.64	2.53	2.44	2.37	2.32	2.27	2.20	2.13	2.05	2.01	1.96	1.91	1.86	1.81	1.76
24	4.26	3.40	3.01	2.78	2.62	2.51	2.42	2.36	2.30	2.25	2.18	2.11	2.03	1.98	1.94	1.89	1.84	1.79	1.73
25	4.24	3.39	2.99	2.76	2.60	2.49	2.40	2.34	2.28	2.24	2.16	2.09	2.01	1.96	1.92	1.87	1.82	1.77	1.71
26	4.23	3.37	2.98	2.74	2.59	2.47	2.39	2.32	2.27	2.22	2.15	2.07	1.99	1.95	1.90	1.85	1.80	1.75	1.69
27	4.21	3.35	2.96	2.73	2.57	2.46	2.37	2.31	2.25	2.20	2.13	2.06	1.97	1.93	1.88	1.84	1.79	1.73	1.67
28	4.20	3.34	2.95	2.71	2.56	2.45	2.36	2.29	2.24	2.19	2.12	2.04	1.96	1.91	1.87	1.82	1.77	1.71	1.65
29	4.18	3.33	2.93	2.70	2.55	2.43	2.35	2.28	2.22	2.18	2.10	2.03	1.94	1.90	1.85	1.81	1.75	1.70	1.64
30	4.17	3.32	2.92	2.69	2.53	2.42	2.33	2.27	2.21	2.16	2.09	2.01	1.93	1.89	1.84	1.79	1.74	1.68	1.62
40	4.08	3.23	2.84	2.61	2.45	2.34	2.25	2.18	2.12	2.08	2.00	1.92	1.84	1.79	1.74	1.69	1.64	1.58	1.51
60	4.00	3.15	2.76	2.53	2.37	2.25	2.17	2.10	2.04	1.99	1.92	1.84	1.75	1.70	1.65	1.59	1.53	1.47	1.39
120	3.92	3.07	2.68	2.45	2.29	2.17	2.09	2.02	1.96	1.91	1.83	1.75	1.66	1.61	1.55	1.50	1.43	1.35	1.25
∞	3.84	3.00	2.60	2.37	2.21	2.10	2.01	1.94	1.88	1.83	1.75	1.67	1.57	1.52	1.46	1.39	1.32	1.22	1.00

SOURCE: From *Biometrika Tables for Statisticians*, 3rd Edition, Vol. 1, London, 1966. Reprinted by permission of Mrs. E. J. Snell on behalf of the Biometrika Trustees.

TABLE C. THE F DISTRIBUTION ($\alpha = 0.01$)

Denominator Degrees of Freedom	Numerator Degrees of Freedom 1	2	3	4	5	6	7	8	9	10	12	15	20	24	30	40	60	120	∞
1	4052	4999.5	5403	5625	5764	5859	5928	5981	6022	6056	6106	6157	6209	6235	6261	6287	6313	6339	6366
2	98.50	99.00	99.17	99.25	99.30	99.33	99.36	99.37	99.39	99.40	99.42	99.43	99.45	99.46	99.47	99.47	99.48	99.49	99.50
3	34.12	30.82	29.46	28.71	28.24	27.91	27.67	27.49	27.35	27.23	27.05	26.87	26.69	26.60	26.50	26.41	26.32	26.22	26.13
4	21.20	18.00	16.69	15.98	15.52	15.21	14.98	14.80	14.66	14.55	14.37	14.20	14.02	13.93	13.84	13.75	13.65	13.56	13.46
5	16.26	13.27	12.06	11.39	10.97	10.67	10.46	10.29	10.16	10.05	9.89	9.72	9.55	9.47	9.38	9.29	9.20	9.11	9.02
6	13.75	10.92	9.78	9.15	8.75	8.47	8.26	8.10	7.98	7.87	7.72	7.56	7.40	7.31	7.23	7.14	7.06	6.97	6.88
7	12.25	9.55	8.45	7.85	7.46	7.19	6.99	6.84	6.72	6.62	6.47	6.31	6.16	6.07	5.99	5.91	5.82	5.74	5.65
8	11.26	8.65	7.59	7.01	6.63	6.37	6.18	6.03	5.91	5.81	5.67	5.52	5.36	5.28	5.20	5.12	5.03	4.95	4.86
9	10.56	8.02	6.99	6.42	6.06	5.80	5.61	5.47	5.35	5.26	5.11	4.96	4.81	4.73	4.65	4.57	4.48	4.40	4.31
10	10.04	7.56	6.55	5.99	5.64	5.39	5.20	5.06	4.94	4.85	4.71	4.56	4.41	4.33	4.25	4.17	4.08	4.00	3.91
11	9.65	7.21	6.22	5.67	5.32	5.07	4.89	4.74	4.63	4.54	4.40	4.25	4.10	4.02	3.94	3.86	3.78	3.69	3.60
12	9.33	6.93	5.95	5.41	5.06	4.82	4.64	4.50	4.39	4.30	4.16	4.01	3.86	3.78	3.70	3.62	3.54	3.45	3.36
13	9.07	6.70	5.74	5.21	4.86	4.62	4.44	4.30	4.19	4.10	3.96	3.82	3.66	3.59	3.51	3.43	3.34	3.25	3.17
14	8.86	6.51	5.56	5.04	4.69	4.46	4.28	4.14	4.03	3.94	3.80	3.66	3.51	3.43	3.35	3.27	3.18	3.09	3.00
15	8.68	6.36	5.42	4.89	4.56	4.32	4.14	4.00	3.89	3.80	3.67	3.52	3.37	3.29	3.21	3.13	3.05	2.96	2.87
16	8.53	6.23	5.29	4.77	4.44	4.20	4.03	3.89	3.78	3.69	3.55	3.41	3.26	3.18	3.10	3.02	2.93	2.84	2.75
17	8.40	6.11	5.18	4.67	4.34	4.10	3.93	3.79	3.68	3.59	3.46	3.31	3.16	3.08	3.00	2.92	2.83	2.75	2.65
18	8.29	6.01	5.09	4.58	4.25	4.01	3.84	3.71	3.60	3.51	3.37	3.23	3.08	3.00	2.92	2.84	2.75	2.66	2.57
19	8.18	5.93	5.01	4.50	4.17	3.94	3.77	3.63	3.52	3.43	3.30	3.15	3.00	2.92	2.84	2.76	2.67	2.58	2.49
20	8.10	5.85	4.94	4.43	4.10	3.87	3.70	3.56	3.46	3.37	3.23	3.09	2.94	2.86	2.78	2.69	2.61	2.52	2.42
21	8.02	5.78	4.87	4.37	4.04	3.81	3.64	3.51	3.40	3.31	3.17	3.03	2.88	2.80	2.72	2.64	2.55	2.46	2.36
22	7.95	5.72	4.82	4.31	3.99	3.76	3.59	3.45	3.35	3.26	3.12	2.98	2.83	2.75	2.67	2.58	2.50	2.40	2.31
23	7.88	5.66	4.76	4.26	3.94	3.71	3.54	3.41	3.30	3.21	3.07	2.93	2.78	2.70	2.62	2.54	2.45	2.35	2.26
24	7.82	5.61	4.72	4.22	3.90	3.67	3.50	3.36	3.26	3.17	3.03	2.89	2.74	2.66	2.58	2.49	2.40	2.31	2.21
25	7.77	5.57	4.68	4.18	3.85	3.63	3.46	3.32	3.22	3.13	2.99	2.85	2.70	2.62	2.54	2.45	2.36	2.27	2.17
26	7.72	5.53	4.64	4.14	3.82	3.59	3.42	3.29	3.18	3.09	2.96	2.81	2.66	2.58	2.50	2.42	2.33	2.23	2.13
27	7.68	5.49	4.60	4.11	3.78	3.56	3.39	3.26	3.15	3.06	2.93	2.78	2.63	2.55	2.47	2.38	2.29	2.20	2.10
28	7.64	5.45	4.57	4.07	3.75	3.53	3.36	3.23	3.12	3.03	2.90	2.75	2.60	2.52	2.44	2.35	2.26	2.17	2.06
29	7.60	5.42	4.54	4.04	3.73	3.50	3.33	3.20	3.09	3.00	2.87	2.73	2.57	2.49	2.41	2.33	2.23	2.14	2.03
30	7.56	5.39	4.51	4.02	3.70	3.47	3.30	3.17	3.07	2.98	2.84	2.70	2.55	2.47	2.39	2.30	2.21	2.11	2.01
40	7.31	5.18	4.31	3.83	3.51	3.29	3.12	2.99	2.89	2.80	2.66	2.52	2.37	2.29	2.20	2.11	2.02	1.92	1.80
60	7.08	4.98	4.13	3.65	3.34	3.12	2.95	2.82	2.72	2.63	2.50	2.35	2.20	2.12	2.03	1.94	1.84	1.73	1.60
120	6.85	4.79	3.95	3.48	3.17	2.96	2.79	2.66	2.56	2.47	2.34	2.19	2.03	1.95	1.86	1.76	1.66	1.53	1.38
∞	6.63	4.61	3.78	3.32	3.02	2.80	2.64	2.51	2.41	2.32	2.18	2.04	1.88	1.79	1.70	1.59	1.47	1.32	1.00

SOURCE: From *Biometrika Tables for Statisticians*, 3rd Edition, Vol. 1, London, 1966. Reprinted by permission of Mrs. E. J. Snell on behalf of the Biometrika Trustees.

TABLE D. THE CHI-SQUARED DISTRIBUTION

df	$\chi^2_{.005}$	$\chi^2_{.025}$	$\chi^2_{.05}$	$\chi^2_{.90}$	$\chi^2_{.95}$	$\chi^2_{.975}$	$\chi^2_{.99}$	$\chi^2_{.995}$
1	.0000393	.000982	.00393	2.706	3.841	5.024	6.635	7.879
2	.0100	.0506	.103	4.605	5.991	7.378	9.210	10.597
3	0.717	.216	.352	6.251	7.815	9.348	11.345	12.838
4	.207	.484	.711	7.779	9.488	11.143	13.277	14.860
5	.412	.831	1.145	9.236	11.070	12.832	15.086	16.750
6	.676	1.237	1.635	10.645	12.592	14.449	16.812	18.548
7	.989	1.690	2.167	12.017	14.067	16.013	18.475	20.278
8	1.344	2.180	2.733	13.362	15.507	17.535	20.090	21.955
9	1.735	2.700	3.325	14.684	16.919	19.023	21.666	23.589
10	2.156	3.247	3.940	15.987	18.307	20.483	23.209	25.188
11	2.603	3.816	4.575	17.275	19.675	21.920	24.725	26.757
12	3.074	4.404	5.226	18.549	21.026	23.336	26.217	28.300
13	3.565	5.009	5.892	19.812	22.362	24.736	27.688	29.819
14	4.075	5.629	6.571	21.064	23.685	26.119	29.141	31.319
15	4.601	6.262	7.261	22.307	24.996	27.488	30.578	32.801
16	5.142	6.908	7.962	23.542	26.296	28.845	32.000	34.267
17	5.697	7.564	8.672	24.769	27.587	30.191	33.409	35.718
18	6.265	8.231	9.390	25.989	28.869	31.526	34.805	37.156
19	6.844	8.907	10.117	27.204	30.144	32.852	36.191	38.582
20	7.434	9.591	10.851	28.412	31.410	34.170	37.566	39.997
21	8.034	10.283	11.591	29.615	32.671	35.479	38.932	41.401
22	8.643	10.982	12.338	30.813	33.924	36.781	40.289	42.796
23	9.260	11.688	13.091	32.007	35.172	38.076	41.638	44.181
24	9.886	12.401	13.848	33.196	36.415	39.364	42.980	45.558
25	10.520	13.120	14.611	34.382	37.652	40.646	44.314	46.928
26	11.160	13.844	15.379	35.563	38.885	41.923	45.642	48.290
27	11.808	14.573	16.151	36.741	40.113	43.194	46.963	49.645
28	12.461	15.308	16.928	37.916	41.337	44.461	48.278	50.993
29	13.121	16.047	17.708	39.087	42.557	45.722	49.588	52.336
30	13.787	16.791	18.493	40.256	43.773	46.979	50.892	53.672
35	17.192	20.569	22.465	46.059	49.802	53.203	57.342	60.275
40	20.707	24.433	26.509	51.805	55.758	59.342	63.691	66.766
45	24.311	28.366	30.612	57.505	61.656	65.410	69.957	73.166
50	27.991	32.357	34.764	63.167	67.505	71.420	76.154	79.490
60	35.535	40.482	43.188	74.397	79.082	83.298	88.379	91.952
70	43.275	48.758	51.739	85.527	90.531	95.023	100.425	104.215
80	51.172	57.153	60.391	96.578	101.879	106.629	112.329	116.321
90	59.196	65.647	69.126	107.565	113.145	118.136	124.116	128.299
100	67.328	74.222	77.929	118.498	124.342	129.561	135.807	140.169

SOURCE: From A. Hald and S. A. Sinkbaek, "A Table of Percentage Points of the X^2 Distribution," *Skandinavisk Aktuarietidskrift*, *33*(1950):168–175. Used by permission.

TABLE E. SQUARES AND SQUARE ROOTS

n	n^2	$\sqrt{n}$	$\sqrt{10n}$	$(10n)^2$
1.0	1.00	1.00000	3.16228	100
1.1	1.21	1.04881	3.31662	121
1.2	1.44	1.09545	3.46410	144
1.3	1.69	1.14018	3.60555	169
1.4	1.96	1.18322	3.74166	196
1.5	2.25	1.22474	3.87298	225
1.6	2.56	1.26491	4.00000	256
1.7	2.89	1.30384	4.12311	289
1.8	3.24	1.34164	4.24264	324
1.9	3.61	1.37840	4.35890	361
2.0	4.00	1.41421	4.47214	400
2.1	4.41	1.44914	4.58258	441
2.2	4.84	1.48324	4.69042	484
2.3	5.29	1.51658	4.79583	529
2.4	5.76	1.54919	4.89898	576
2.5	6.25	1.58114	5.00000	625
2.6	6.76	1.61245	5.09902	676
2.7	7.29	1.64317	5.19615	729
2.8	7.84	1.67332	5.29150	784
2.9	8.41	1.70294	5.38516	841
3.0	9.00	1.73205	5.47723	900
3.1	9.61	1.76068	5.56776	961
3.2	10.24	1.78885	5.65685	1024
3.3	10.89	1.81659	5.74456	1089
3.4	11.56	1.84391	5.83095	1156
3.5	12.25	1.87083	5.91608	1225
3.6	12.96	1.89737	6.00000	1296
3.7	13.69	1.92354	6.08276	1369
3.8	14.44	1.94936	6.16441	1444
3.9	15.21	1.97484	6.24500	1521
4.0	16.00	2.00000	6.32456	1600
4.1	16.81	2.02485	6.40312	1681
4.2	17.64	2.04939	6.48074	1764
4.3	18.49	2.07364	6.55744	1849
4.4	19.36	2.09762	6.63325	1936
4.5	20.25	2.12132	6.70820	2025
4.6	21.16	2.14476	6.78233	2116
4.7	22.09	2.16795	6.85565	2209
4.8	23.04	2.19089	6.92820	2304
4.9	24.01	2.21359	7.00000	2401
5.0	25.00	2.23607	7.07107	2500
5.1	26.01	2.25832	7.14143	2601
5.2	27.04	2.28035	7.21110	2704
5.3	28.09	2.30217	7.28011	2809

SOURCE: From Wayne W. Daniel, *Biostatistics: A Foundation for Analysis in the Health Sciences*, © 1974, by John Wiley & Sons, Inc., p. 382. Used by permission.

TABLE E. CONTINUED

n	n^2	$\sqrt{n}$	$\sqrt{10n}$	$(10n)^2$
5.4	29.16	2.32379	7.34847	2916
5.5	30.25	2.34521	7.41620	3025
5.6	31.36	2.36643	7.48331	3136
5.7	32.49	2.38747	7.54983	3249
5.8	33.64	2.40832	7.61577	3364
5.9	34.81	2.42899	7.68115	3481
6.0	36.00	2.44949	7.74597	3600
6.1	37.21	2.46982	7.81025	3721
6.2	38.44	2.48998	7.87401	3844
6.3	39.69	2.50998	7.93725	3969
6.4	40.96	2.52982	8.00000	4096
6.5	42.25	2.54951	8.06226	4225
6.6	43.56	2.56905	8.12404	4356
6.7	44.89	2.58844	8.18535	4489
6.8	46.24	2.60768	8.24621	4624
6.9	47.61	2.62679	8.30662	4761
7.0	49.00	2.64575	8.36660	4900
7.1	50.41	2.66458	8.42615	5041
7.2	51.84	2.68328	8.48528	5184
7.3	53.29	2.70185	8.54400	5329
7.4	54.76	2.72029	8.60233	5476
7.5	56.25	2.73861	8.66025	5625
7.6	57.76	2.75681	8.71780	5776
7.7	59.29	2.77489	8.77496	5929
7.8	60.84	2.79285	8.83176	6084
7.9	62.41	2.81069	8.88819	6241
8.0	64.00	2.82843	8.94427	6400
8.1	65.61	2.84605	9.00000	6561
8.2	67.24	2.86356	9.05539	6724
8.3	68.89	2.88097	9.11043	6889
8.4	70.56	2.89828	9.16515	7056
8.5	72.25	2.91548	9.21954	7225
8.6	73.96	2.93258	9.27362	7396
8.7	75.69	2.94958	9.32738	7569
8.8	77.44	2.96648	9.38083	7744
8.9	79.21	2.98329	9.43398	7921
9.0	81.00	3.00000	9.48683	8100
9.1	82.81	3.01662	9.53939	8281
9.2	84.64	3.03315	9.59166	8464
9.3	86.49	3.04959	9.64365	8649
9.4	88.36	3.06594	9.69536	8836
9.5	90.25	3.08221	9.74679	9025
9.6	92.16	3.09839	9.79796	9216
9.7	94.09	3.11448	9.84886	9409
9.8	96.04	3.13050	9.89949	9604
9.9	98.01	3.14643	9.94987	9801

TABLE F. PEARSON'S PRODUCT MOMENT CORRELATION

*df**	*.05*	*.01*
1	.996917	.9998766
2	.95000	.990000
3	.8783	.95873
4	.8114	.91720
5	.7545	.8745
6	.7067	.8343
7	.6664	.7977
8	.6319	.7646
9	.6021	.7348
10	.5760	.7079
11	.5529	.6835
12	.5324	.6614
13	.5139	.6411
14	.4973	.6226
15	.4821	.6055
16	.4683	.5897
17	.4555	.5751
18	.4438	.5614
19	.4329	.5487
20	.4227	.5368
25	.3809	.4869
30	.3494	.4487
35	.3246	.4182
40	.3044	.3932
45	.2875	.3721
50	.2732	.3541
60	.2500	.3248
70	.2319	.3017
80	.2172	.2830
90	.2050	.2673
100	.1946	.2540

SOURCE: Reprinted with permission of Macmillan Publishing Co., Inc., from *Statistical Methods for Research Workers*, 14th Edition, p. 209, by R. A. Fisher. Copyright © 1970 University of Adelaide.

* The degrees of freedom (*df*) are 2 less than the number of pairs in the sample.

TABLE G. SPEARMAN'S RANK ORDER CORRELATION (SIGNIFICANCE LEVEL, α)

n	.001	.005	.010	.025	.050	.100
4	—	—	—	—	.8000	.8000
5	—	—	.9000	.9000	.8000	.7000
6	—	.9429	.8857	.8286	.7714	.6000
7	.9643	.8929	.8571	.7450	.6786	.5357
8	.9286	.8571	.8095	.7143	.6190	.5000
9	.9000	.8167	.7667	.6833	.5833	.4667
10	.8667	.7818	.7333	.6364	.5515	.4424
11	.8364	.7545	.7000	.6091	.5273	.4182
12	.8182	.7273	.6713	.5804	.4965	.3986
13	.7912	.6978	.6429	.5549	.4780	.3791
14	.7670	.6747	.6220	.5341	.4593	.3626
15	.7464	.6536	.6000	.5179	.4429	.3500
16	.7265	.6324	.5824	.5000	.4265	.3382
17	.7083	.6152	.5637	.4853	.4118	.3260
18	.6904	.5975	.5480	.4716	.3994	.3148
19	.6737	.5825	.5333	.4579	.3895	.3070
20	.6586	.5684	.5203	.4451	.3789	.2977
21	.6455	.5545	.5078	.4351	.3688	.2909
22	.6318	.5426	.4963	.4241	.3597	.2829
23	.6186	.5306	.4852	.4150	.3518	.2767
24	.6070	.5200	.4748	.4061	.3435	.2704
25	.5962	.5100	.4654	.3977	.3362	.2646
26	.5856	.5002	.4564	.3894	.3299	.2588
27	.5757	.4915	.4481	.3822	.3236	.2540
28	.5660	.4828	.4401	.3749	.3175	.2490
29	.5567	.4744	.4320	.3685	.3113	.2443
30	.5479	.4665	.4251	.3620	.3059	.2400

Note: The corresponding lower-tail critical value for r_s is $-r_s^*$.

SOURCE: From Gerald J. Glasser and Robert F. Winter, "Critical Values of the Coefficient of Rank Correlation for Testing Hypothesis of Independence," *Biometrika* *48*(1962):444–448. Used by permission of Mrs. E. J. Snell on behalf of the Biometrika Trustees. The table as printed here contains corrections given in W. J. Conover, *Practical Nonparametric Statistics*, © 1971, John Wiley & Sons, Inc.

TABLE H. WILCOXON SIGNED-RANK TEST

N	*Level of Significance for One-Tailed Test* .025	.01	.005
	Level of Significance for Two-Tailed Test .05	.02	.01
6	0	—	—
7	2	0	—
8	4	2	0
9	6	3	2
10	8	5	3
11	11	7	5
12	14	10	7
13	17	13	10
14	21	16	13
15	25	20	16
16	30	24	20
17	35	28	23
18	40	33	28
19	46	38	32
20	52	43	38
21	59	49	43
22	66	56	49
23	73	62	55
24	81	69	61
25	89	77	68

SOURCE: From F. Wilcoxon, *Some Rapid Approximate Statistical Procedures,* Table 2, p. 28, 1964. Used by permission of the American Cyanamid Company.

TABLE I. WILCOXON RANK SUM TEST

n_1, n_2	*Significance Level, Two-Tail* .05	.01	.001
2, 8	3, 19		
2, 9	3, 21		
2, 10	3, 23		
2, 11	4, 24		
2, 12	4, 26		
2, 13	4, 28		
2, 14	4, 30		
2, 15	4, 32		
2, 16	4, 34		
2, 17	5, 35		
2, 18	5, 37		
2, 19	5, 39	3, 41	
2, 20	5, 41	3, 43	
2, 21	6, 42	3, 45	
2, 22	6, 44	3, 47	
2, 23	6, 46	3, 49	
2, 24	6, 48	3, 51	
2, 25	6, 50	3, 53	
3, 5	6, 21		
3, 6	7, 23		
3, 7	7, 26		
3, 8	8, 28		
3, 9	8, 31	6, 33	
3, 10	9, 33	6, 36	
3, 11	9, 36	6, 39	
3, 12	10, 38	7, 41	
3, 13	10, 41	7, 44	
3, 14	11, 43	7, 47	
3, 15	11, 46	8, 49	
3, 16	12, 48	8, 52	
3, 17	12, 51	8, 55	
3, 18	13, 53	8, 58	
3, 19	13, 56	9, 60	
3, 20	14, 58	9, 63	
3, 21	14, 61	9, 66	6, 69
3, 22	15, 63	10, 68	6, 72
3, 23	15, 66	10, 71	6, 75
3, 24	16, 68	10, 74	6, 78
3, 25	16, 71	11, 76	6, 81
4, 4	10, 26		
4, 5	11, 29		
4, 6	12, 32	10, 34	
4, 7	13, 35	10, 38	
4, 8	14, 38	11, 41	
4, 9	15, 41	11, 45	
4, 10	15, 45	12, 48	
4, 11	16, 48	12, 52	
4, 12	17, 51	13, 55	
4, 13	18, 54	14, 58	10, 62
4, 14	19, 57	14, 62	10, 66
4, 15	20, 60	15, 65	10, 70
4, 16	21, 63	15, 69	11, 73
4, 17	21, 67	16, 72	11, 77
4, 18	22, 70	16, 76	11, 81
4, 19	23, 73	17, 79	12, 84
4, 20	24, 76	18, 82	12, 88
4, 21	25, 79	18, 86	12, 92
4, 22	26, 82	19, 89	13, 95
4, 23	27, 85	19, 93	13, 99
4, 24	28, 88	20, 96	13, 103
4, 25	28, 92	20, 100	14, 106
5, 5	17, 38	15, 40	
5, 6	18, 42	16, 44	
5, 7	20, 45	17, 48	
5, 8	21, 49	17, 53	
5, 9	22, 53	18, 57	15, 60
5, 10	23, 57	19, 61	15, 65

TABLE I. CONTINUED

n_1, n_2	Significance Level, Two-Tail .05	.01	.001	n_1, n_2	Significance Level, Two-Tail .05	.01	.001
5, 11	24, 61	20, 65	16, 69	7, 7	36, 69	32, 73	28, 77
5, 12	26, 64	21, 69	16, 74	7, 8	38, 74	34, 78	29, 83
5, 13	27, 68	22, 73	17, 78	7, 9	40, 79	35, 84	30, 89
5, 14	28, 72	22, 78	17, 83	7, 10	42, 84	37, 89	31, 95
5, 15	29, 76	23, 82	18, 87	7, 11	44, 89	38, 95	32, 101
5, 16	31, 79	24, 86	18, 92	7, 12	46, 94	40, 100	33, 107
5, 17	32, 83	25, 90	19, 96	7, 13	48, 99	41, 106	34, 113
5, 18	33, 87	26, 94	19, 101	7, 14	50, 104	43, 111	35, 119
5, 19	34, 91	27, 98	20, 105	7, 15	52, 109	44, 117	36, 125
5, 20	35, 95	28, 102	20, 110	7, 16	54, 114	46, 122	37, 131
5, 21	37, 98	29, 106	21, 114	7, 17	56, 119	47, 128	38, 137
5, 22	38, 102	29, 111	21, 119	7, 18	58, 124	49, 133	39, 143
5, 23	39, 106	30, 115	22, 123	7, 19	60, 129	50, 139	41, 148
5, 24	40, 110	31, 119	23, 127	7, 20	62, 134	52, 144	42, 154
5, 25	42, 113	32, 123	23, 132	7, 21	64, 139	53, 150	43, 160
				7, 22	66, 144	55, 155	44, 166
6, 6	26, 52	23, 55		7, 23	68, 149	57, 160	45, 172
6, 7	27, 57	24, 60					
6, 8	29, 61	25, 65	21, 69				
6, 9	31, 65	26, 70	22, 74				
6, 10	32, 70	27, 75	23, 79	8, 8	49, 87	43, 93	38, 98
6, 11	34, 74	28, 80	23, 85	8, 9	51, 93	45, 99	40, 104
6, 12	35, 79	30, 84	24, 90	8, 10	53, 99	47, 105	41, 111
6, 13	37, 83	31, 89	25, 95	8, 11	55, 105	49, 111	42, 118
6, 14	38, 88	32, 94	26, 100	8, 12	58, 110	51, 117	43, 125
6, 15	40, 92	33, 99	26, 106	8, 13	60, 116	53, 123	45, 131
6, 16	42, 96	34, 104	27, 111	8, 14	63, 121	54, 130	46, 138
6, 17	43, 101	36, 108	28, 116	8, 15	65, 127	56, 136	47, 145
6, 18	45, 105	37, 113	29, 121	8, 16	67, 133	58, 142	49, 151
6, 19	46, 110	38, 118	29, 127	8, 17	70, 138	60, 148	50, 158
6, 20	48, 114	39, 123	30, 132	8, 18	72, 144	62, 154	51, 165
6, 21	50, 118	40, 128	31, 137	8, 19	74, 150	64, 160	53, 171
6, 22	51, 123	42, 132	32, 142	8, 20	77, 155	66, 166	54, 178
6, 23	53, 127	43, 137	33, 147	8, 21	79, 161	68, 172	56, 184
6, 24	55, 131	44, 142	34, 152	8, 22	82, 166	70, 178	57, 191

TABLE I. CONTINUED

n_1, n_2	Significance Level, Two-Tail .05	.01	.001	n_1, n_2	Significance Level, Two-Tail .05	.01	.001
9, 9	63, 108	56, 115	50, 121	11, 13	103, 172	93, 182	83, 192
9, 10	65, 115	58, 122	52, 128	11, 14	106, 180	96, 190	85, 201
9, 11	68, 121	61, 128	53, 136	11, 15	110, 187	99, 198	87, 210
9, 12	71, 127	63, 135	55, 143	11, 16	114, 194	102, 206	90, 218
9, 13	73, 134	65, 142	56, 151	11, 17	117, 202	105, 214	92, 227
9, 14	76, 140	67, 149	58, 158	11, 18	121, 209	108, 222	94, 236
9, 15	79, 146	70, 155	60, 165	11, 19	124, 217	111, 230	97, 244
9, 16	82, 152	72, 162	61, 173				
9, 17	84, 159	74, 169	63, 180	12, 12	115, 185	106, 194	95, 205
9, 18	87, 165	76, 176	65, 187	12, 13	119, 193	109, 203	98, 214
9, 19	90, 171	78, 183	66, 195	12, 14	123, 201	112, 212	100, 224
9, 20	93, 177	81, 189	68, 202	12, 15	127, 209	115, 221	103, 233
9, 21	95, 184	83, 196	70, 209	12, 16	131, 217	119, 229	105, 243
				12, 17	135, 225	122, 238	108, 252
10, 10	78, 132	71, 139	63, 147	12, 18	139, 233	125, 247	111, 261
10, 11	81, 139	74, 146	65, 155				
10, 12	85, 145	76, 154	67, 163				
10, 13	88, 152	79, 161	69, 171	13, 13	137, 214	125, 226	114, 237
10, 14	91, 159	81, 169	71, 179	13, 14	141, 223	129, 235	116, 248
10, 15	94, 166	84, 176	73, 187	13, 15	145, 232	133, 244	119, 258
10, 16	97, 173	86, 184	75, 195	13, 16	150, 240	137, 253	122, 268
10, 17	100, 180	89, 191	77, 203	13, 17	154, 249	140, 263	125, 278
10, 18	103, 187	92, 198	79, 211				
10, 19	107, 193	94, 206	81, 219	14, 14	160, 246	147, 259	134, 272
10, 20	110, 200	97, 213	83, 227	14, 15	164, 256	151, 269	137, 283
				14, 16	169, 265	155, 279	140, 294
11, 11	96, 157	87, 166	78, 175				
11, 12	99, 165	90, 174	81, 183	15, 15	185, 280	171, 294	156, 309

SOURCE: From C. White, "The Use of Ranks in a Test of Significance for Comparing Two Treatments," *Biometrics 8*:37–39, 1951, as adapted by T. Colton, *Statistics in Medicine*, 1974, Little Brown and Co., Boston. Used by permission of the author and publisher.

TABLE J. THE SIGN TEST*

Level of Significance, P					*Level of Significance, P*				
N	*.20*	*.10*	*.05*	*.01*	*N*	*.20*	*.10*	*.05*	*.01*
4	0				28	10	9	8	6
5		0			29	10	9	8	7
6			0		30		10	9	7
7	1		0		31	11	10	9	7
8		1		0	32	11	10		8
9	2		1	0	33	12	11	10	8
10	2		1	0	34	12	11	10	9
11		2	1	0	35	13	12	11	9
12	3		2	1	36	13	12	11	9
13		3	2	1	37	14	13	12	10
14	4	3	2	1	38	14	13	12	10
15	4		3	2	39	15	13	12	11
16		4	3	2	40	15	14	13	11
17	5		4	2	41	15	14	13	11
18		5	4	3	42	16	15	14	12
19	6	5	4	3	43	16	15	14	12
20	6		5	3	44	17	16	15	13
21	7	6	5	4	45	17	16	15	13
22	7	6	5	4	46	18	16	15	13
23		7	6	4	47	18	17	16	14
24	8	7	6	5	48	19	17	16	14
25	8		7	5	49	19	18	17	15
26	9	8	7	6	50	20	18	17	15
27	9	8		6					

* The listed numbers under the *P* values are the *maximum* number of differences in the *minority* direction which may occur and still permit rejection of H_0 at that level of *P*. For $N > 50$, see Siegel.

SOURCE: From *Nonparametric and Shortcut Statistics,* by Merle W. Tate and Richard C. Clelland. Danville, Ill.: The Interstate Printers and Publishers, Inc., 1957, p. 140. Used by permission of the publisher.

TABLE K. RANDOM DIGITS

85967	73152	14511	85285	36009	95892	36962	67835	63314	50162
07483	51453	11649	86348	76431	81594	95848	36738	25014	15460
96283	01898	61414	83525	04231	13604	75339	11730	85423	60698
49174	12074	98551	37895	93547	24769	09404	76548	05393	96770
97366	39941	21225	93629	19574	71565	33413	56087	40875	13351
90474	41469	16812	81542	81652	45554	27931	93994	22375	00953
28599	64109	09497	76235	41383	31555	12639	00619	22909	29563
25254	16210	89717	65997	82667	74624	36348	44018	64732	93589
28785	02760	24359	99410	77319	73408	58993	61098	04393	48245
84725	86576	86944	93296	10081	82454	76810	52975	10324	15457
41059	66456	47679	66810	15941	84602	14493	65515	19251	41642
67434	41045	82830	47617	36932	46728	71183	36345	41404	81110
72766	68816	37643	19959	57550	49620	98480	25640	67257	18671
92079	46784	66125	94932	64451	29275	57669	66658	30818	58353
29187	40350	62533	73603	34075	16451	42885	03448	37390	96328
74220	17612	65522	80607	19184	64164	66962	82310	18163	63495
03786	02407	06098	92917	40434	60602	82175	04470	78754	90775
75085	55558	15520	27038	25471	76107	90832	10819	56797	33751
09161	33015	19155	11715	00551	24909	31894	37774	37953	78837
75707	48992	64998	87080	39333	00767	45637	12538	67439	94914
21333	48660	31288	00086	79889	75532	28704	62844	92337	99695
65626	50061	42539	14812	48895	11196	34335	60492	70650	51108
84380	07389	87891	76255	89604	41372	10837	66992	93183	56920
46479	32072	80083	63868	70930	89654	05359	47196	12452	38234
59847	97197	55147	76639	76971	55928	36441	95141	42333	67483
31416	11231	27904	57383	31852	69137	96667	14315	01007	31929
82066	83436	67914	21465	99605	83114	97885	74440	99622	87912
01850	42782	39202	18582	46214	99228	79541	78298	75404	63648
32315	89276	89582	87138	16165	15984	21466	63830	30475	74729
59388	42703	55198	80380	67067	97155	34160	85019	03527	78140

TABLE K. CONTINUED

58089	27632	50987	91373	07736	20436	96130	73483	85332	24384
61705	57285	30392	23660	75841	21931	04295	00875	09114	32101
18914	98982	60199	99275	41967	35208	30357	76772	92656	62318
11965	94089	34803	48941	69709	16784	44642	89761	66864	62803
85251	48111	80936	81781	93248	67877	16498	31924	51315	79921
66121	96986	84844	93873	46352	92183	51152	85878	30490	15974
53972	96642	24199	58080	35450	03482	66953	49521	63719	57614
14509	16594	78883	43222	23093	58645	60257	89250	63266	90858
37700	07688	65533	72126	23611	93993	01848	03910	38552	17472
85466	59392	72722	15473	73295	49759	56157	60477	83284	56367
52969	55863	42312	67842	05673	91878	82738	36563	79540	61935
42744	68315	17514	02878	97291	74851	42725	57894	81434	62041
26140	13336	67726	61876	29971	99294	96664	52817	90039	53211
95589	56319	14563	24071	06916	59555	18195	32280	79357	04224
39113	13217	59999	49952	83021	47709	53105	19295	88318	41626
41392	17622	18994	98283	07249	52289	24209	91139	30715	06604
54684	53645	79246	70183	87731	19185	08541	33519	07223	97413
89442	61001	36658	57444	95388	36682	38052	46719	09428	94012
36751	16778	54888	15357	68003	43564	90976	58904	40512	07725
98159	02564	21416	74944	53049	88749	02865	25772	89853	88714

SOURCE: From *A Million Random Digits with 100,000 Normal Deviates* by the Rand Corporation. Reprinted by permission.

TABLE L(a). STUDENT'S *t* DISTRIBUTION: NUMBER OF OBSERVATIONS FOR *t* TEST OF MEAN

	Level of t-Test																				
Single-Sided Test	$\alpha = 0.005$					$\alpha = 0.01$					$\alpha = 0.025$					$\alpha = 0.05$					
Double-Sided Test	$\alpha = 0.01$					$\alpha = 0.02$					$\alpha = 0.05$					$\alpha = 0.1$					
$\beta =$	0.01	0.05	0.1	0.2	0.5	0.01	0.05	0.1	0.2	0.5	0.01	0.05	0.1	0.2	0.5	0.01	0.05	0.1	0.2	0.5	
$\Delta = \frac{\mu - \mu_0}{\sigma}$ 0.05																					0.05
0.10																					0.10
0.15																				122	0.15
0.20										139					99					70	0.20
0.25					110					90				128	64			139	101	45	0.25
0.30				134	78				115	63			119	90	45		122	97	71	32	0.30
0.35			125	99	58			109	85	47		109	88	67	34		90	72	52	24	0.35
0.40		115	97	77	45		101	85	66	37	117	84	68	51	26	101	70	55	40	19	0.40
0.45		92	77	62	37	110	81	68	53	30	93	67	54	41	21	80	55	44	33	15	0.45
0.50	100	75	63	51	30	90	66	55	43	25	76	54	44	34	18	65	45	36	27	13	0.50
0.55	83	63	53	42	26	75	55	46	36	21	63	45	37	28	15	54	38	30	22	11	0.55
0.60	71	53	45	36	22	63	47	39	31	18	53	38	32	24	13	46	32	26	19	9	0.60
0.65	61	46	39	31	20	55	41	34	27	16	46	33	27	21	12	39	28	22	17	8	0.65
0.70	53	40	34	28	17	47	35	30	24	14	40	29	24	19	10	34	24	19	15	8	0.70
0.75	47	36	30	25	16	42	31	27	21	13	35	26	21	16	9	30	21	17	13	7	0.75
0.80	41	32	27	22	14	37	28	24	19	12	31	22	19	15	9	27	19	15	12	6	0.80
0.85	37	29	24	20	13	33	25	21	17	11	28	21	17	13	8	24	17	14	11	6	0.85
0.90	34	26	22	18	12	29	23	19	16	10	25	19	16	12	7	21	15	13	10	5	0.90

Value of	0.95	31	24	20	17	11	27	21	18	14	9	23	17	14	11	7	19	14	11	9	5	0.95
	1.00	28	22	19	16	10	25	19	16	13	9	21	16	13	10	6	18	13	11	8	5	1.00
$\Delta = \frac{\mu - \mu_0}{\sigma}$																						
	1.1	24	19	16	14	9	21	16	14	12	8	18	13	11	9	6	15	11	9	7		1.1
	1.2	21	16	14	12	8	18	14	12	10	7	15	12	10	8	5	13	10	8	6		1.2
	1.3	18	15	13	11	8	16	13	11	9	6	14	10	9	7		11	8	7	6		1.3
	1.4	16	13	12	10	7	14	11	10	9	6	12	9	8	7		10	8	7	5		1.4
	1.5	15	12	11	9	7	13	10	9	8	6	11	8	7	6		9	7	6			1.5
	1.6	13	11	10	8	6	12	10	9	7	5	10	8	7	6		8	6	6			1.6
	1.7	12	10	9	8	6	11	9	8	7		9	7	6	5		8	6	5			1.7
	1.8	12	10	9	8	6	10	8	7	7		8	7	6			7	6				1.8
	1.9	11	9	8	7	6	10	8	7	6		8	6	6			7	5				1.9
	2.0	10	8	8	7	5	9	7	7	6		7	6	5			6					2.0
	2.1	10	8	7	7		8	7	6	6		7	6				6					2.1
	2.2	9	8	7	6		8	7	6	5		7	6				6					2.2
	2.3	9	7	7	6		8	6	6			6	5				5					2.3
	2.4	8	7	7	6		7	6	6			6										2.4
	2.5	8	7	6	6		7	6	6			6										2.5
	3.0	7	6	6	5		6	5	5			5										3.0
	3.5	6	5	5			5															3.5
	4.0	6																				4.0

SOURCE: Reprinted with permission from *CRC Handbook of Tables for Probability and Statistics*, 2nd ed., 1968. Copyright The Chemical Rubber Co., CRC Press, Inc.

TABLE L(b). STUDENT'S t DISTRIBUTION: (CONTINUED): NUMBER OF OBSERVATIONS FOR t TEST OF DIFFERENCE BETWEEN TWO MEANS

	Level of t-Test																				
Single-Sided Test	$\alpha = 0.005$					$\alpha = 0.01$					$\alpha = 0.025$					$\alpha = 0.05$					
Double-Sided Test	$\alpha = 0.01$					$\alpha = 0.02$					$\alpha = 0.05$					$\alpha = 0.1$					
$\beta =$	0.01	0.05	0.1	0.2	0.5	0.01	0.05	0.1	0.2	0.5	0.01	0.05	0.1	0.2	0.5	0.01	0.05	0.1	0.2	0.5	
$\Delta = \frac{\mu_x - \mu_y}{\sigma}$ 0.05																					0.05
0.10																					0.10
0.15																					0.15
0.20																				137	0.20
0.25															124					88	0.25
0.30										123					87					61	0.30
0.35					110					90					64				102	45	0.35
0.40					85					70				100	50			108	78	35	0.40
0.45				118	68				101	55			105	79	39		108	86	62	28	0.45
0.50				96	55			106	82	45		106	86	64	32		88	70	51	23	0.50
0.55			101	79	46		106	88	68	38		87	71	53	27	112	73	58	42	19	0.55
0.60		101	85	67	39		90	74	58	32	104	74	60	45	23	89	61	49	36	16	0.60
0.65		87	73	57	34	104	77	64	49	27	88	63	51	39	20	76	52	42	30	14	0.65
0.70	100	75	63	50	29	90	66	55	43	24	76	55	44	34	17	66	45	36	26	12	0.70
0.75	88	66	55	44	26	79	58	48	38	21	67	48	39	29	15	57	40	32	23	11	0.75
0.80	77	58	49	39	23	70	51	43	33	19	59	42	34	26	14	50	35	28	21	10	0.80
0.85	69	51	43	35	21	62	46	38	30	17	52	37	31	23	12	45	31	25	18	9	0.85
0.90	62	46	39	31	19	55	41	34	27	15	47	34	27	21	11	40	28	22	16	8	0.90

Value of $\Delta = \frac{\mu_x - \mu_1}{\sigma}$																					
0.95	55	42	35	28	17	50	37	31	24	14	42	30	25	19	10	36	25	20	15	7	0.95
1.00	50	38	32	26	15	45	33	28	22	13	38	27	23	17	9	33	23	18	14	7	1.00
1.1	42	32	27	22	13	38	28	23	19	11	32	23	19	14	8	27	19	15	12	6	1.1
1.2	36	27	23	18	11	32	24	20	16	9	27	20	16	12	7	23	16	13	10	5	1.2
1.3	31	23	20	16	10	28	21	17	14	8	23	17	14	11	6	20	14	11	9	5	1.3
1.4	27	20	17	14	9	24	18	15	12	8	20	15	12	10	6	17	12	10	8	4	1.4
1.5	24	18	15	13	8	21	16	14	11	7	18	13	11	9	5	15	11	9	7	4	1.5
1.6	21	16	14	11	7	19	14	12	10	6	16	12	10	8	5	14	10	8	6	4	1.6
1.7	19	15	13	10	7	17	13	11	9	6	14	11	9	7	4	12	9	7	6	3	1.7
1.8	17	13	11	10	6	15	12	10	8	5	13	10	8	6	4	11	8	7	5		1.8
1.9	16	12	11	9	6	14	11	9	8	5	12	9	7	6	4	10	7	6	5		1.9
2.0	14	11	10	8	6	13	10	9	7	5	11	8	7	6	4	9	7	6	4		2.0
2.1	13	10	9	8	5	12	9	8	7	5	10	8	6	5	3	8	6	5	4		2.1
2.2	12	10	8	7	5	11	9	7	6	4	9	7	6	5		8	6	5	4		2.2
2.3	11	9	8	7	5	10	8	7	6	4	9	7	6	5		7	5	5	4		2.3
2.4	11	9	8	6	5	10	8	7	6	4	8	6	5	4		7	5	4	4		2.4
2.5	10	8	7	6	4	9	7	6	5	4	8	6	5	4		6	5	4	3		2.5
3.0	8	6	6	5	4	7	6	5	4	3	6	5	4	4		5	4	3			3.0
3.5	6	5	5	4	3	6	5	4	4		5	4	4	3		4	3				3.5
4.0	6	5	4	4		5	4	4	3		4	4	3			4					4.0

SOURCE: Reprinted with permission from *CRC Handbook of Tables for Probability and Statistics*, 2nd ed., 1968. Copyright The Chemical Rubber Co., CRC Press, Inc.

INDEX